Essentials in Robotics

Volume I

Essentials in Robotics
Volume I

Edited by **Rowland Wilson**

CLANRYE
INTERNATIONAL

New Jersey

Published by Clanrye International,
55 Van Reypen Street,
Jersey City, NJ 07306, USA
www.clanryeinternational.com

Essentials in Robotics: Volume I
Edited by Rowland Wilson

© 2015 Clanrye International

International Standard Book Number: 978-1-63240-229-5 (Hardback)

Printed in the United States of America.

Contents

Preface

Though there are hundreds of ways to define a robot. However, the Oxford English Dictionary defines a robot as "a machine capable of carrying out a complex series of actions automatically, especially one programmable by a computer". A robot can perform various functions at a time and can also save a lot of human effort and time. Due to the never ending human requirement to reduce efforts, various technical advances have been invented to accelerate the development of the science of robotics. In the present scenario, mobile robots have become the focus of recent researches. This particular area of robotics is a multidisciplinary field, which involves many engineering sectors such as electrical, electronic and mechanical engineering, computer, cognitive, and social sciences. The designing of robots is at the centre of few scientific approaches such as AI-artificial intelligence, Computation and Automata theory. Mobile robots have emerged at the center of present research developments due to their numerous capabilities such as manipulation, perception, communication, cognition. The characteristics of mobile robots that make them so important for the present, as well as the future are movement, sensory, and processing involved reasoning. The achievement of robotics today is such that from going to space to controlling manufacturing the smallest parts of an aircraft, there is nothing that can't be done by robots these days.

I especially wish to acknowledge the contributing authors, without whom a work of this magnitude would clearly not have been realizable. I thank all the authors for allocating much of their scarce time to this project. Not only do I appreciate their participation, but also their adherence as a group to the time parameters set for this publication. I believe, that the content of this book will serve as a valuable text to all those who work in the research and development of robots.

Editor

1

A Comparison between Position-Based and Image-Based Dynamic Visual Servoings in the Control of a Translating Parallel Manipulator

G. Palmieri,[1] M. Palpacelli,[2] M. Battistelli,[2] and M. Callegari[2]

[1] *Facoltà di Ingegneria, Università degli Studi e-Campus, 22060 Novedrate, Italy*
[2] *Dipartimento di Ingegneria Industriale e Scienze Matematiche, Università Politecnica delle Marche, 60121 Ancona, Italy*

Correspondence should be addressed to G. Palmieri, giacomo.palmieri@uniecampus.it

Academic Editor: Huosheng Hu

Two different visual servoing controls have been developed to govern a translating parallel manipulator with an eye-in-hand configuration, That is, a position-based and an image-based controller. The robot must be able to reach and grasp a target randomly positioned in the workspace; the control must be adaptive to compensate motions of the target in the 3D space. The trajectory planning strategy ensures the continuity of the velocity vector for both PBVS and IBVS controls, whereas a replanning event is needed. A comparison between the two approaches is given in terms of accuracy, fastness, and stability in relation to the robot peculiar characteristics.

1. Introduction

Visual servoing is the use of computer vision to control the motion of a robot; two basic approaches can be identified [1–4]: *position-based visual servo* (PBVS), in which vision data are used to reconstruct the 3D pose of the robot and a kinematic error is generated in the Cartesian space and mapped to actuators commands [5–7]; *image-based visual servo* (IBVS), in which the error is generated directly from image plane features [8–15]. Recently, a new family of hybrid or partitioned methods is growing, with the aim of combining advantages of PBVS and IBVS while trying to avoid their shortcomings [16, 17].

The principal advantage of using position-based control is the chance of defining tasks in a standard Cartesian frame. On the other hand, the control law strongly depends on the optical parameters of the vision system and can become widely sensitive to calibration errors. On the contrary, the image-based control is less sensitive to calibration errors; however, it is required the online calculation of the image Jacobian, that is, a quantity depending on the distance between the target and the camera which is difficult to evaluate. A control in the image plane results also to

be strongly nonlinear and coupled when mapped on the joint space of the robot and may cause problems when crossing points which are singular for the kinematics of the manipulator [1].

Visual servo systems can also be classified on the basis of their architecture in the following two categories [1]: the vision system provides an external input to the joint closed-loop control of the robot that stabilizes the mechanism (*dynamic look and move*); the vision system is directly used in the control loop to compute joints inputs, thus stabilizing autonomously the robot (*direct visual servo*). In general, most applications are of the dynamic look and move type; one of the reasons is the difference between vision systems and servo loops rates. The low frequency imposed by vision might cause problems on controller's stability, especially in cases where several DOFs are involved.

The aim of this work is to compare two different visual servoing controls, respectively, of the position-based and image-based type, implemented to govern a translating parallel manipulator with an eye-in-hand configuration. The robot must be able to reach and grasp a special target randomly positioned in the workspace; the control must be adaptive to compensate motions of the target in the 3D

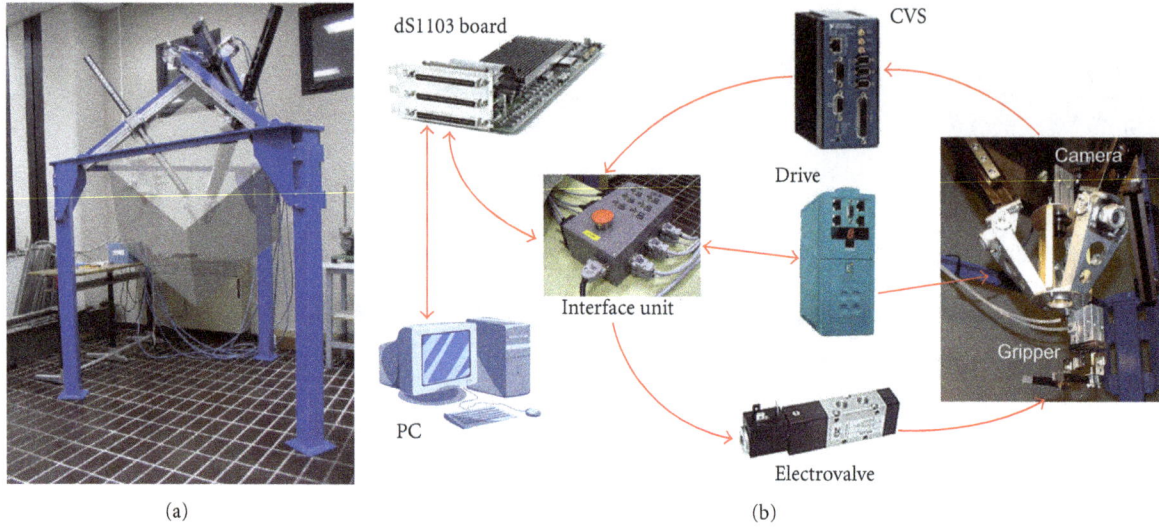

FIGURE 1: (a) ICaRo robot and workspace; (b) system architecture and connection diagram.

space. The use of Bézier curves [18, 19] in the trajectory planning algorithms ensures the continuity of the velocity vector, whereas a replanning event is needed. A dynamic look and move philosophy has been adopted to conjugate the low frame rate of the vision system (lower than 30 Hz) with the high control rate of the joint servo (about 1000 Hz).

2. Hardware Setup

The robot, which is called ICaRo, is a research prototype designed and realized at the Machine Mechanics Lab of Università Politecnica delle Marche [20]; the end-effector is in parallel actuated by 3 limbs whose kinematic structure only allows pure translations in the 3D space. The workspace is a cube of 0.6 m edge free of singular points (Figure 1(a)).

The eye-in-hand configuration has been chosen for the installation of the vision system. The optical axis of the camera is aligned with the end-effector vertical axis. The end-effector is also provided with a pneumatic gripper in order to grasp the target when the desired relative pose is reached (Figure 1(b)). The robot is managed by a central control unit, a DS1103 real-time board by dSPACE. The vision data, after image acquisition and preliminary processing made by the CVS real-time hardware of National Instruments, are sent via a serial interface to the central unit that runs the control algorithms.

3. PBVS Control

Using a position-based method, a 3D camera calibration is required in order to map the 2D data of the image features to the Cartesian space data. This is to say that intrinsic and extrinsic parameters of the camera must be evaluated. *Intrinsic parameters* depend exclusively on the optical characteristics, namely, lens and CCD sensor properties. The calibration of intrinsic parameters can be operated offline in the case that optical setup is fixed during the operative tasks of the robot. *Extrinsic parameters* indicate the relative pose of the camera reference system with respect to a generic world reference system. It is assumed that the world reference system is exactly the object frame, so that the extrinsic parameters give directly the pose of the camera with respect to the target. Obviously the extrinsic parameters are variable with robot or target motion, and an online estimation is needed in order to perform a dynamic look-and-move tracking task.

3.1. Estimation of Intrinsic Parameters. A standard pinhole camera model is here briefly described to introduce the notation used in this paper. A point in the 2D image space is denoted by $\mathbf{m} = [u \quad v]^T$, while a point in the 3D Cartesian space is denoted by $\mathbf{p} = [x \quad y \quad z]^T$. Introducing the homogeneous coordinates, it is $\tilde{\mathbf{m}} = [u \quad v \quad 1]^T$ and $\tilde{\mathbf{p}} = [x \quad y \quad z \quad 1]^T$. The image projection that relates a 3D point $\mathbf{p}$ to its corresponding point $\mathbf{m}$ in the image plane can be expressed as

$$s\tilde{\mathbf{m}} = \mathbf{A\Pi B\tilde{p}} \quad \mathbf{\Pi} = \begin{bmatrix} 1 & 0 & 0 & 0 \\ 0 & 1 & 0 & 0 \\ 0 & 0 & 1 & 0 \end{bmatrix}, \tag{1}$$

where s is an arbitrary scale factor and $\mathbf{B}$ is the 4×4 matrix of the extrinsic parameters that combines a rotation and a translation relating the world coordinate system to the camera coordinate system. $\mathbf{A}$ is the matrix of intrinsic parameters, defined as

$$\mathbf{A} = \begin{bmatrix} \alpha & \gamma & u_0 \\ 0 & \beta & v_0 \\ 0 & 0 & 1 \end{bmatrix}, \tag{2}$$

A Comparison between Position-Based and Image-Based Dynamic Visual Servoings in the Control of
a Translating Parallel Manipulator

3

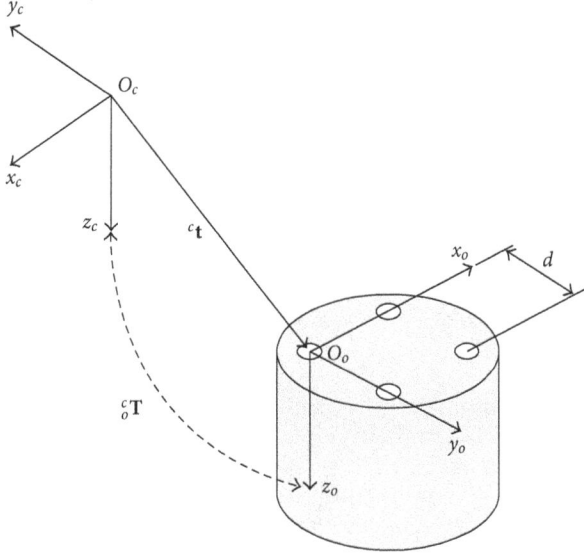

FIGURE 2: Object frame attached to the visual pattern and camera reference system.

where

(i) (u_0, v_0) are the coordinates of the principal point, that is, the projection of the optical axis on the CCD sensor;

(ii) α and β are the focal scale factors in the image u and v axes;

(iii) γ is the parameter describing the skewness of the two images axes, usually set to 0.

The pinhole model needs to be refined in order to correct image distortions introduced by the lens [21–25]. To this aim, normalized coordinates are conveniently introduced: given a point $\mathbf{p} = [x_c \quad y_c \quad z_c]^T$ in the camera reference system, the corresponding normalized pinhole projection is defined as $\mathbf{m}_n = [u_n \quad v_n]^T = [x_c/z_c \quad y_c/z_c]^T$. It can be demonstrated that, in the ideal case free of lens distortions, the intrinsic camera matrix $\mathbf{A}$ relates a point in the image plane with its normalized projection: $\tilde{\mathbf{m}}_n = \mathbf{A}^{-1}\tilde{\mathbf{m}}$. Actually, normalized coordinates obtained from previous equation are affected by image distortions, and the real case is distinguished from the ideal case by using the notation $\tilde{\mathbf{m}}_d = \mathbf{A}^{-1}\tilde{\mathbf{m}}$. Two components of distortion can be recognized: the radial lens distortion causes the actual image point to be displaced radially in the image plane; if centres of curvature of lens surfaces are not collinear, also a tangential distortion is introduced. The following expression is generally used to model both components of distortion:

$$\mathbf{m}_d = \mathbf{m}_n + \delta\mathbf{m}^{(r)} + \delta\mathbf{m}^{(t)}, \qquad (3)$$

where $\delta\mathbf{m}^{(r)}$ and $\delta\mathbf{m}^{(t)}$ are defined as high-order polynomial functions of the undistorted coordinates u_n, v_n [21]. Because of the high degree of the distortion model, there exists no general algebraic expression for the inverse problem of computing the normalized image projection vector $\mathbf{m}_n$ from the distorted coordinates of vector $\mathbf{m}_d$. If low-distortion lenses are used, a direct solution can be found making the approximation of computing $\delta\mathbf{m}^{(r)}$ and $\delta\mathbf{m}^{(t)}$ as functions of distorted coordinates. Inspiring to well-known distortion models [21–25], a rough correction of the image distortion can be quickly obtained by using the equation

$$\mathbf{m}_n = \frac{1}{a}\left(\mathbf{m}_d - \begin{bmatrix} b & c \end{bmatrix}^T\right), \qquad (4)$$

where

$$
\begin{aligned}
a &= 1 + k_1 r_d^2 + k_2 r_d^4 + k_3 r_d^6, \\
b &= 2p_1 u_d v_d + p_2\left(r_d^2 + 2u_d^2\right), \\
c &= p_1\left(r_d^2 + 2v_d^2\right) + 2p_2 u_d v_d, \\
r_d^2 &= u_d^2 + v_d^2,
\end{aligned}
\qquad (5)
$$

and k_1, k_2, k_3, p_1, p_2 are additional intrinsic parameters to be defined by calibration.

Several algorithms for the estimation of intrinsic parameters are available in literature [22–25]. A technique inspired to the Heikkilä algorithm has been adopted: after the settings of the optical system have been tuned for an optimal result of the vision, a number $n > 4$ of frames (15 in our application) of a grid of known dimensions are acquired; through an automatic extraction of the corners of the grids, the algorithm is able to estimate with an iterative procedure the intrinsic parameters of the camera including distortion parameters, according to (4) and (5).

3.2. Estimation of Extrinsic Parameters. The reconstruction of the relative pose between camera and object is possible through the estimation of extrinsic parameters of the camera with respect to a reference system that is coincident with the object frame $O_o - x_o y_o z_o$. Figure 2 shows the optical pattern realized on the top surface of the target object; it consists of four coplanar circles positioned at the corners of a square of known edge d. The object frame is attached to the pattern in the way shown in the figure.

The position of the origin O_o with respect to the camera frame is the vector ${}^c\mathbf{t} = {}^c(O_o - O_c)$, while the relative orientation between the two reference systems is the rotation ${}^c_o\mathbf{R}$. The homogeneous transformation ${}^c_o\mathbf{T}$ coincides with the extrinsic parameters matrix

$$\mathbf{B} = {}^c_o\mathbf{T} = \begin{bmatrix} {}^c_o\mathbf{R} & {}^c\mathbf{t} \\ \mathbf{0}^T & 1 \end{bmatrix}. \qquad (6)$$

The aim is to determine the extrinsic parameters, once the geometry of the optical pattern and the pixel coordinates of the centroids of the circles projected onto the image plane are known.

The problem of determining the pose, knowing the correspondence of n points in the world and camera frame reference systems, is typical in photogrammetry (*PnP* problem), and it is proved that for 4 coplanar points the solution is unique [26, 27]. While the calibration of intrinsic parameters is quite laborious, but does not need to be

Depth = 500 mm

(a)

Depth = 150 mm

■ x
■ y
■ z

(b)

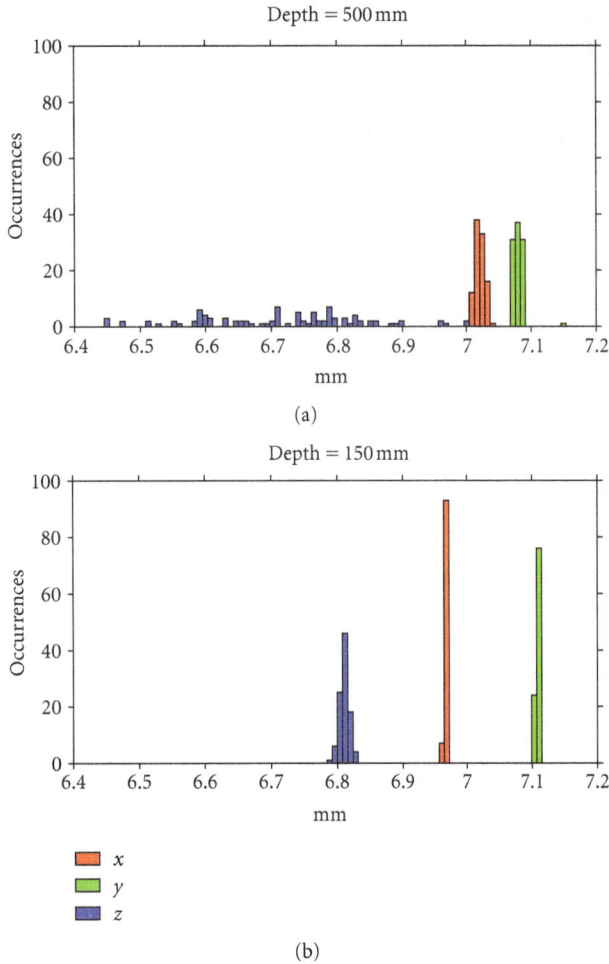

FIGURE 3: Distribution of displacement experimental data; (a) distance camera-target $\simeq 500$ mm; (b) distance camera-target $\simeq 150$ mm.

TABLE 1: Summary of displacement experimental data: mean values and standard deviations of measured data.

	Depth = 500 mm		
	x	y	z
Mean value [mm]	7.021	7.081	6.710
Standard deviation [mm]	0.007	0.009	0.134
	Depth = 150 mm		
	x	y	z
Mean value [mm]	6.967	7.108	6.810
Standard deviation [mm]	0.001	0.001	0.007

pose estimation increases approaching the target. The same conclusions can be drawn from Table 1, where experimental data are summarized in terms of mean values and standard deviations: passing from a distance of 500 mm to 150 mm, the standard deviation lowers nearly one order of magnitude in x and y and two orders in z.

3.3. Control Algorithm. A dynamic look and move control has been implemented following the position-based approach above described. The global architecture is shown in Figure 4, where **s** represents the vector of image features, **q** is the vector of joint angles, and $\mathbf{x}_m$, $\mathbf{x}_d$, and $\mathbf{x}_{\text{plan}}$ are, respectively, the measured, desired, and planned Cartesian coordinates vectors. Looking at the scheme of the control, two separate loops can be identified: the inner loop realizes a standard PD control in the operative space with gravity compensation (**g** term) by exploiting the information from encoders; the camera, working as an exteroceptive sensor, closes an outer loop where visual information is processed in order to plan and replan in real time the desired trajectory of the robot. It is important to remark that image features extraction is performed externally from the control loop by the real-time image processing hardware (CVS); only the pose estimation, that includes the solution of the *PnP* algorithm, is involved in the loop.

A control in the Cartesian space is realized imposing the following requirements to the planned motions:

(a) starting from an arbitrary status of the end-effector (position and velocity), the continuity of position and velocity must be ensured;

(b) the final point must be reached with a vertical tangent in order to facilitate the grasping of the target object.

To this purpose, a third-order Bézier curve has been adopted [18, 19]. This curve is defined by 4 points in the 3D space and has the fundamental property of being tangent to the extremities of the polyline defined by the four points (Figure 5(a)).

The parametric formulation of a cubic Bézier curve is

$$\mathbf{B}(s) = \mathbf{P}_0\left(1 - s^3\right) + 3\mathbf{P}_1 s(1 - s)^2 + \\ 3\mathbf{P}_2 s^2(1 - s) + \mathbf{P}_3 s^3 \quad s \in [0, 1]. \quad (7)$$

In the trajectory planner algorithm, $\mathbf{P}_0$ is the current position of the end-effector, while $\mathbf{P}_3$ is the target point

performed online, the solution of the *PnP* algorithm for 4 coplanar points is fast enough to be implemented online during the motion of the robot.

An experimental test is here presented in order to evaluate the accuracy of the implemented pose estimation method: while keeping fixed the camera, the optical target of Figure 2 is moved 7 mm along each axis using a precision micrometric stage. An image of the pattern is grabbed before the motion, and a series of 100 images are stored after the motion.

The relative displacements can be estimated by making the difference between the pose obtained from the first image and the pose obtained from each one of the further 100 images. The test has been repeated in two configurations, setting the initial z-distance between the camera and the object, respectively, at 500 mm and 150 mm.

Experimental data of the tests are plotted in Figure 3. It is clear that there is a larger dispersion of measurements at a bigger distance from the target; moreover, it can be noticed a larger sensitivity to noise along the z axis, parallel to the optical axis. It results also evident how the accuracy of the

A Comparison between Position-Based and Image-Based Dynamic Visual Servoings in the Control of
a Translating Parallel Manipulator

5

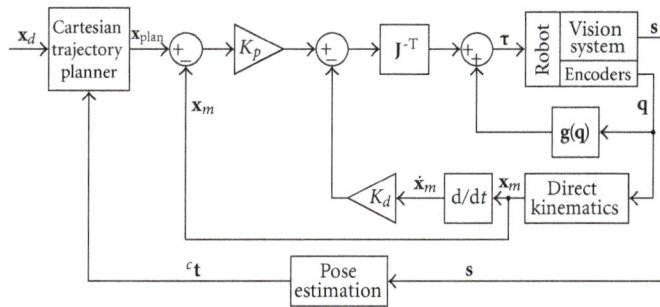

FIGURE 4: Dynamic look and move PBVS control.

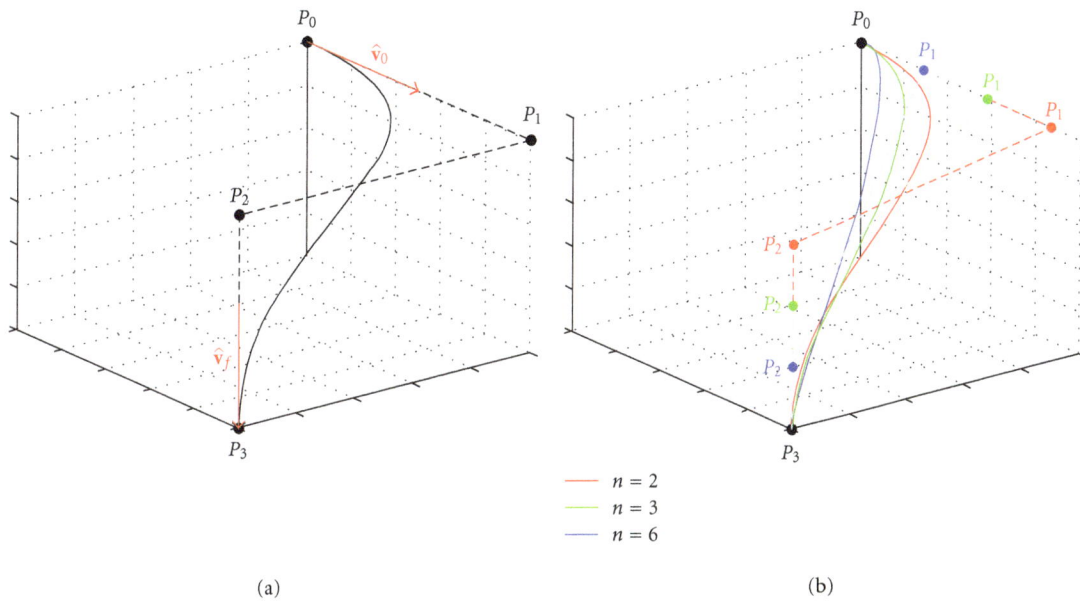

FIGURE 5: (a) Plot of a cubic 3D Beziér curve; (b) influence of the parameter n on the planned trajectory.

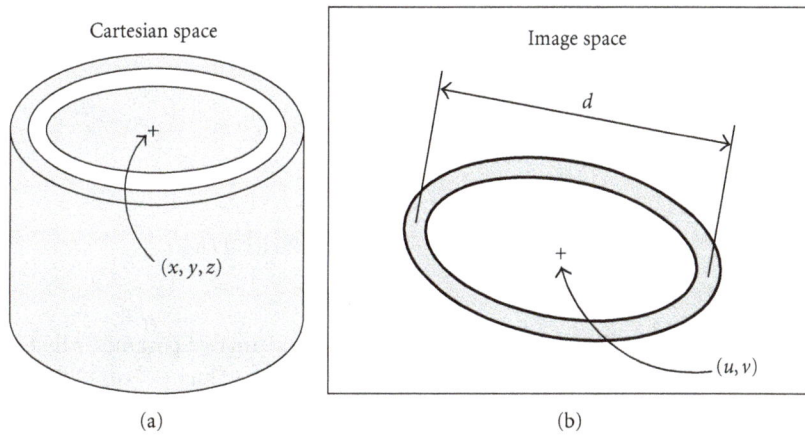

FIGURE 6: Visual pattern for the IBVS: coordinates in the Cartesian and image spaces.

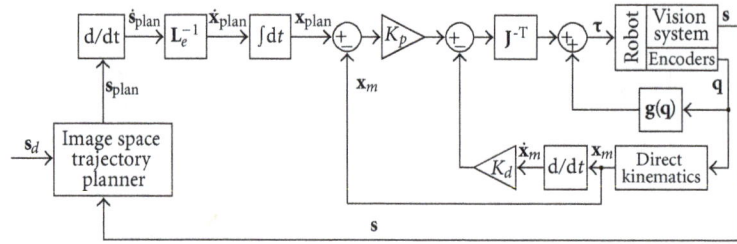

FIGURE 7: Dynamic look and move IBVS control.

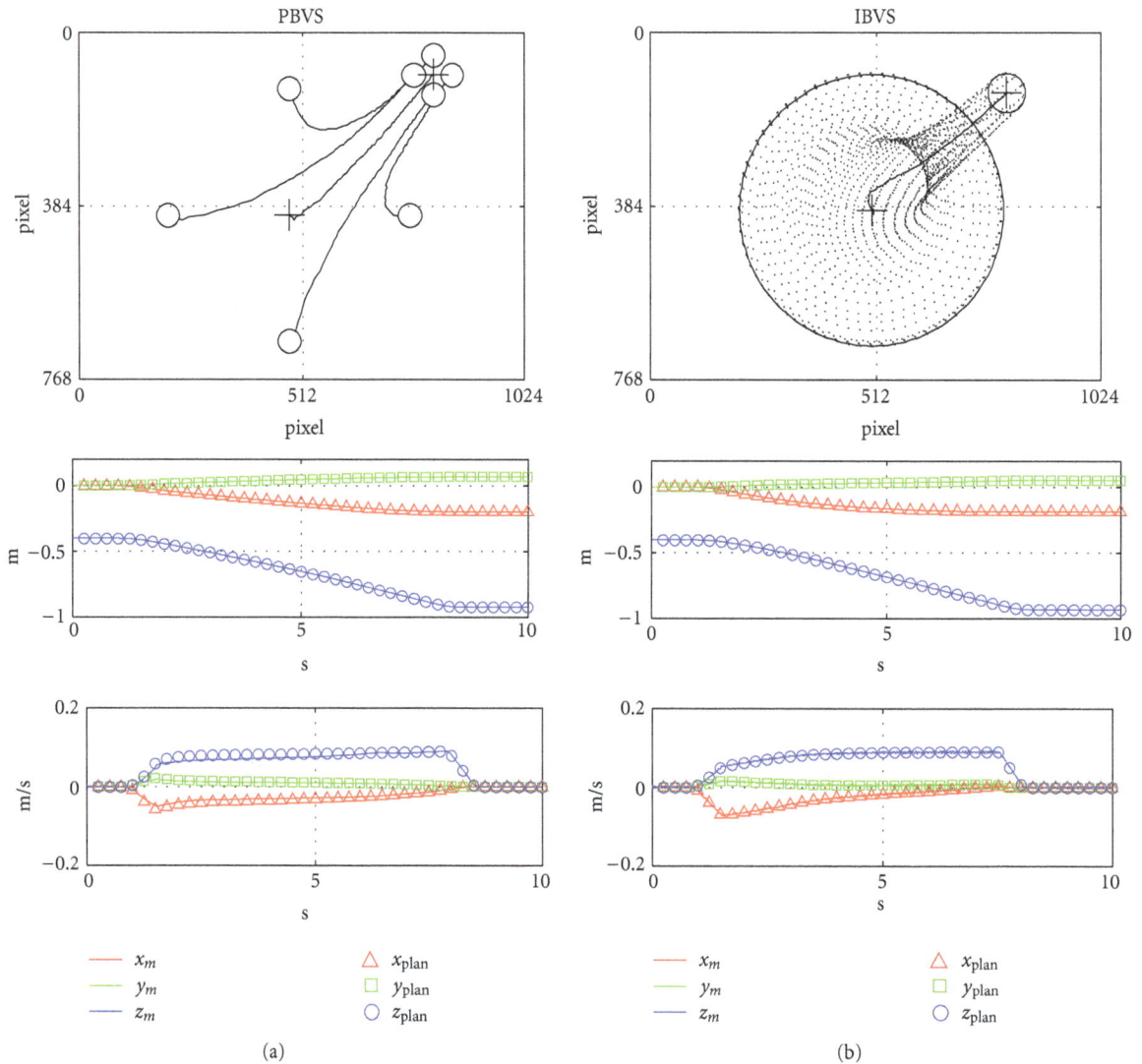

FIGURE 8: Slow tests ($v_{\max} = 10^{-1}$ m/s) with fixed target.

obtained from the vision system. Knowing the current velocity direction $\hat{\mathbf{v}}_0$ and the desired final velocity direction $\hat{\mathbf{v}}_f$, $\mathbf{P}_1$ and $\mathbf{P}_2$ are defined as

$$\mathbf{P}_1 = \mathbf{P}_0 + \frac{\|\mathbf{P}_3 - \mathbf{P}_0\|}{n}\hat{\mathbf{v}}_0,$$

$$\mathbf{P}_2 = \mathbf{P}_3 - \frac{\|\mathbf{P}_3 - \mathbf{P}_0\|}{n}\hat{\mathbf{v}}_f, \tag{8}$$

where n is a tunable parameter that influences the curvature of the trajectory (Figure 5(b)); all experimental data reported in the paper are obtained by setting $n = 3$.

Since the target $\mathbf{P}_3$ is continuously estimated during the motion of the robot, variations of its position due to motions of the object or to the reduction of the measurement errors (as shown in Figure 3) are compensated by a continuous replanning of the trajectory. A trapezoidal velocity profile

A Comparison between Position-Based and Image-Based Dynamic Visual Servoings in the Control of
a Translating Parallel Manipulator

7

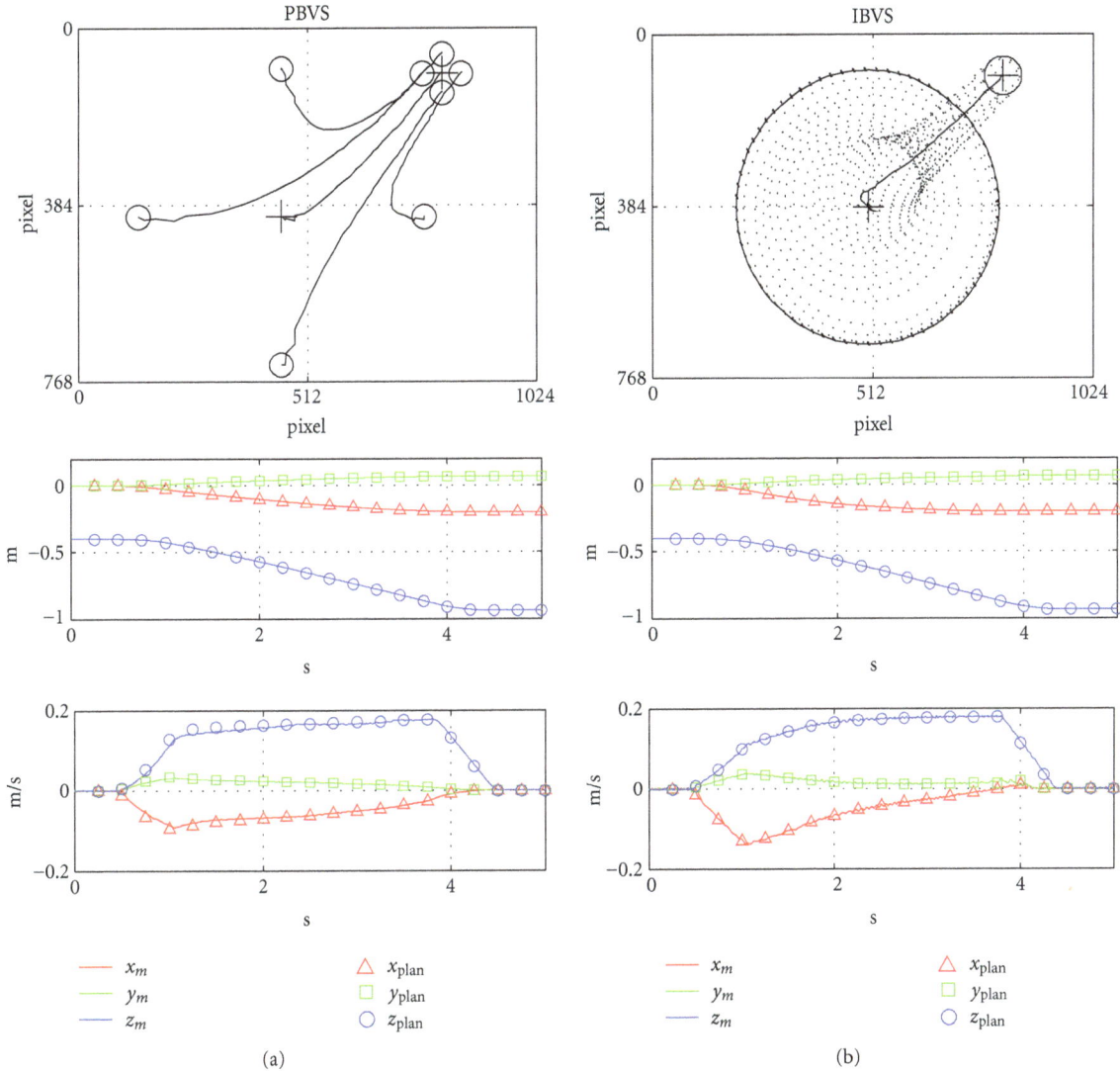

FIGURE 9: Fast tests ($v_{max} = 2 \cdot 10^{-1}$ m/s) with fixed target.

is used with a temporal planning algorithm that is able to modify the shape of the profile, while the length of the planned trajectory changes during the motion.

4. IBVS Control

In the image-based visual servo control, the error signal is defined directly in terms of image feature parameters. In the present application, the optical target is an annulus painted on the top surface of a cylinder (Figure 6).

The image features are the coordinates of the center and the major axis of the ellipse which best approximates the projection of the annulus on the image plane; they are collected in the image space vector $\mathbf{s} = \{u, v, d\}^T$ expressed in pixel. The relation between image space and Cartesian space is given by in the interaction matrix $\mathbf{L_e}$ [27]: being $\dot{\mathbf{x}} = \{\dot{x}, \dot{y}, \dot{z}\}^T$ the Cartesian velocity and $\dot{\mathbf{s}} = \{\dot{u}, \dot{v}, \dot{d}\}^T$ the image space velocity, it is

$$\dot{\mathbf{s}} = \mathbf{L_e}\dot{\mathbf{x}}. \tag{9}$$

Here, the interaction matrix is defined as

$$\mathbf{L_e} = \begin{bmatrix} f/z & 0 & -u/z \\ 0 & f/z & -v/z \\ 0 & 0 & -fD/z^2 \end{bmatrix}, \tag{10}$$

where f is the focal length (assumed equal in u and v directions) expressed in pixel and D is the metric value of the annulus diameter in the Cartesian space. It arises from (10) that the interaction matrix is a function of image features u, v and of the Cartesian variable z. The estimation of z is performed by a linear interpolation on the parameter d; the interpolation law is obtained in a preliminary calibration step. The inversion of (9) is performed at each time step of the control, so that

$$\dot{\mathbf{x}} = \mathbf{L_e}^{-1}\dot{\mathbf{s}}. \tag{11}$$

The trajectory planner operates in this case over the image space: in analogy with the PBVS control, it is defined

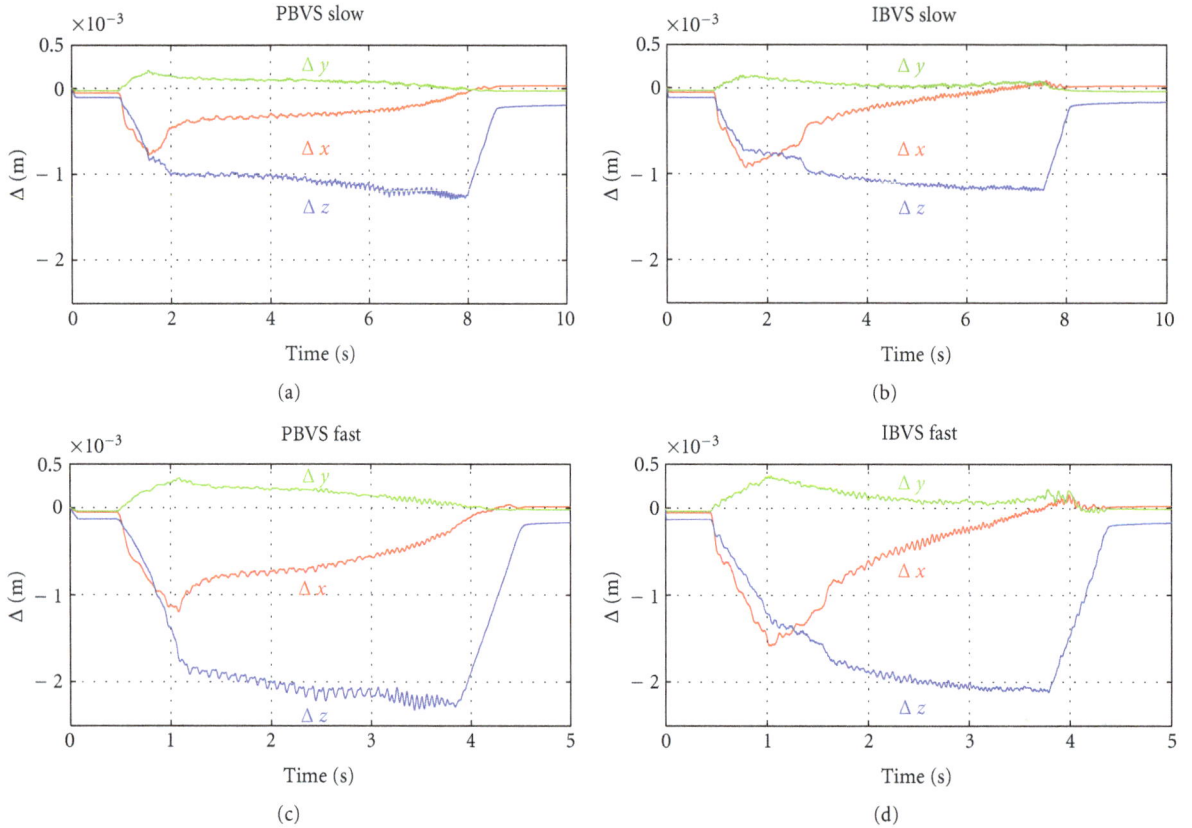

FIGURE 10: Errors between planned and measured Cartesian coordinates for PBVS and IBVS at slow and fast mode.

as a Bézier curve according to (2) through points defined by image coordinates $\{u, v, d\}$. Such kind of strategy allows for the continuity of the velocity in both image and Cartesian spaces, even in the case of motion of the target.

Particular attention is focused on the definition of the velocity profile: without going into details for the sake of brevity, it can be summarized that velocity profiles on the image space are properly planned in order to obtain an effective trapezoidal profile in the Cartesian space. This operation is required to prevent that any component of the Cartesian velocity may assume peaks that are too high or may approach asymptotically zero during the motion of the robot. As an example, we may look to the relation that results from the classic pinhole camera model

$$d = \frac{fD}{z}. \tag{12}$$

The derivative of the above expression is:

$$\dot{z} = -\frac{\dot{d}}{fD}z^2, \tag{13}$$

which could be find directly from the term $\mathbf{L_e}(3, 3)$.

Thus, if a constant $\dot{d}$ is imposed according to the trapezoidal planning, the resulting Cartesian vertical speed $\dot{z}$ will approach zero when z approaches zero (i.e., when the camera approaches the target). Therefore, if the vertical speed has to be controlled, it must be directly planned

in the Cartesian space, and analogously for the x and y components.

The scheme of the implemented IBVS, control is shown in Figure 7. As in PBVS it is possible to individuate the inner and outer loops. In the outer loop, the pose estimation block disappears, and the image features are directly the input of the trajectory planning block. Then the inverse of the interaction matrix is used to map the trajectory from image to Cartesian space according to (11).

5. Experimental Tests

A series of tests have been performed in order to compare the PB and IB approaches at different speeds and in cases of fixed or moving target. The starting position of the robot allows the camera to frame a wide area of the workspace; when the target is recognized by the vision system, the controller starts to govern the robot till the target is reached.

Figure 8 shows the results of the slow tests ($v_{\max} = 10^{-1}$ m/s) with a fixed target. The plots on the left side are referred to the PBVS, while IBVS results are on the right column; the position of the target is the same for both tests. The image plane trajectories of the features and the respective centroid are represented on the top of the figure; such features are the four coplanar circles for the PB and the annulus diameter (represented as a circle) for the IB. The Cartesian position and velocity are plotted, respectively, in

A Comparison between Position-Based and Image-Based Dynamic Visual Servoings in the Control of
a Translating Parallel Manipulator

9

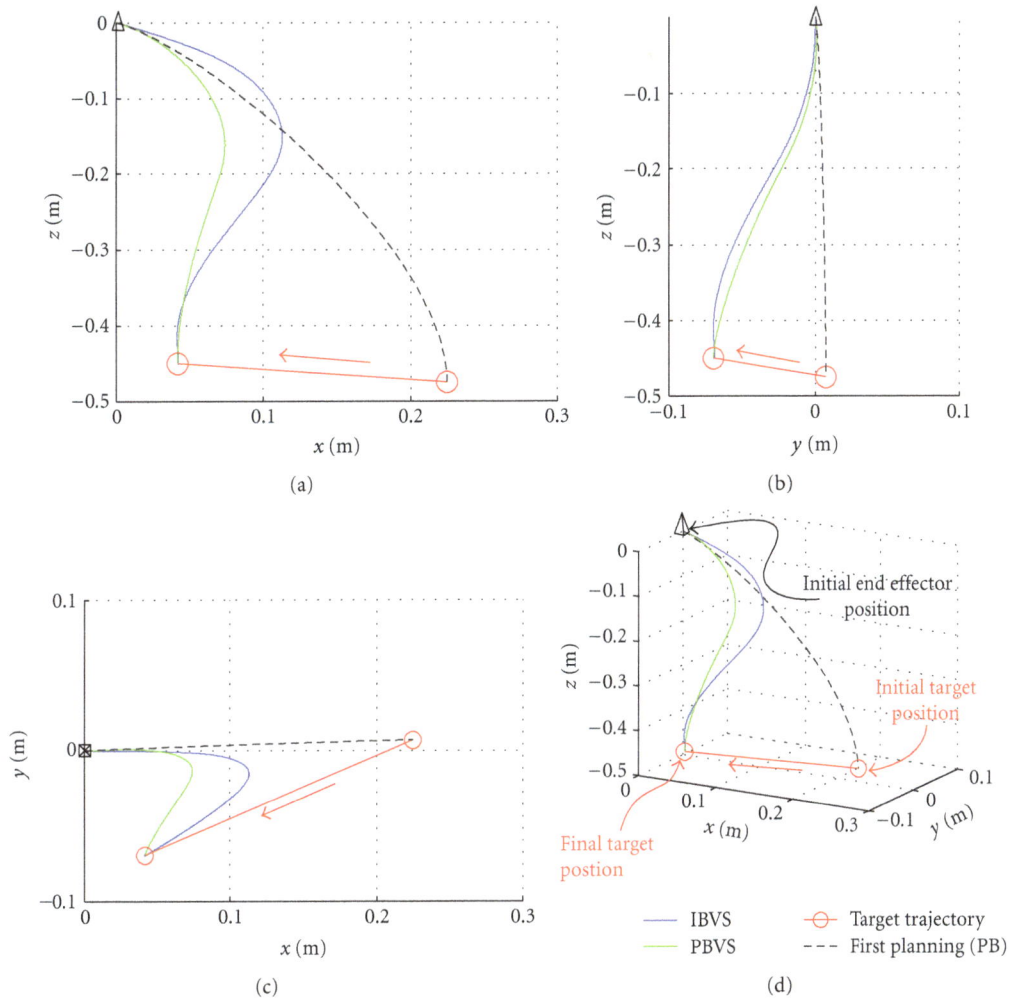

FIGURE 11: IBVS and PBVS tests with moving target.

the middle and at the bottom of the figure; lines refer to measured entities, while circular markers indicate planned values. Data show that the IB control has a slightly higher accuracy in centering the target. Further on, the IBVS presents higher x and y components of velocity in the initial part of the motion, which allows for quickly compensating the offset between camera and target in the horizontal plane.

Same considerations are even more evident in fast tests where the maximum velocity of the robot is twice that in the slow tests ($v_{max} = 2 \cdot 10^{-1}$ m/s). Results are plotted in Figure 9 in analogy with Figure 8. Against a strong decrease of the time required to reach the target, it is not noticed for the IB control an appreciable loss of accuracy. For the PBVS, on the contrary, a certain increment of the final error is evident in the image plane.

For the sake of completeness, the error between planned and measured Cartesian coordinates in slow and fast tests is plotted in Figure 10. The comparison between IB and PB controls at the same speed shows that Cartesian errors are substantially similar. On the other hand, the error increases of about a factor 2 passing from the slow to the fast mode.

A further experimental evaluation is here described to prove the ability of the control in compensating dynamically

a generic motion of the target: once the visual control has been started, the object is moved by means of a conveyor belt; the controller is able to replan continuously the trajectory ensuring the continuity of the velocity and quickly compensating the displacement imposed to the target. Tests are performed with $v_{max} = 2 \cdot 10^{-1}$ m/s, which implies a time of about 5 s to reach the target. The original (referred to the PB algorithm) and replanned trajectories are plotted in Figure 11 in orthographic and axonometric views. The higher sensitivity to offsets on the horizontal plane is once again noticeable for the IBVS, which translates in higher performances in path following tasks.

6. Conclusions

Two dynamic look and move visual servos, respectively of the position-based and image-based type have been implemented to govern a parallel translating robot. Both controls are able to detect a target in the 3D workspace and to dynamically compensate a generic motion of the target itself. A continuous replanning algorithm defines the trajectory of the end-effector ensuring the continuity of the velocity vector even if sudden displacements are imposed to the target;

trajectories are planned using cubic Bézier curves that give a smooth spatial path and allow to impose the final direction of the velocity in order to facilitate the grasping task; such curves are planned, respectively, in the Cartesian space for the PBVS and in the image space for the IBVS.

Even if accuracies of the two proposed controls seem to be substantially comparable, experimental results suggest a better behaviour of the IBVS. Furthermore, the IBVS shows a greater sensitivity to horizontal displacement proved by higher components of the velocity vector in the x and y directions in the first part of the trajectory.

Both methods require the knowledge of the geometric model of the optical target; specifically for the proposed controls, the edge of the square formed by the four circles in the PB case and the diameter of the annulus in the IB case must be known in metric units. Nevertheless, the PB approach requires also the 3D calibration of the camera in order to estimate the intrinsic parameters matrix, which is then required for the execution of the *PnP* algorithm; as known from literature, the sensitivity to calibration errors is probably the main drawback of the PBVS.

Further on, even if the solution to the *PnP* problem returns the homogeneous transformation matrix ${}^c_o\mathbf{T}$, only the vector ${}^c\mathbf{t}$ is a useful information for a pure translation robot. It means that the PB approach gives redundant information for our specific purpose, which translates in useless computational effort for the real-time hardware.

All these reasons make the IBVS preferable for the studied application: the method is simple and accurate and does not require a camera calibration. Moreover, the risk of passing through or nearby singularities, which is typical for IB controls, is avoided if, as in this case, the robot is free of singular points inside its workspace.

Finally, it is worth to remark as the above considerations are strictly related to the kinematic properties of the robot and may not be valid a priori for other architectures. In this sense, authors are intended to implement PB and IB controls on a parallel robot which confers to the end-effector a 3DOFs motion of pure rotation. Such robot (SpheIRo) has been studied and prototyped at Machine Mechanics Lab of Università Politecnica delle Marche [28] with the philosophy of complementary mobility with respect to ICaRo. Unlike the translation manipulator, SpheIRo presents singular points inside its workspace which may cause problems of stability for the IB control. Moreover, passing from a relative position to a relative orientation problem, the sensitivity of the two algorithms in detecting the orientation between frames has to be assessed.

References

[1] S. Hutchinson, G. D. Hager, and P. I. Corke, "A tutorial on visual servo control," *IEEE Transactions on Robotics and Automation*, vol. 12, no. 5, pp. 651–670, 1996.

[2] P. I. Corke and S. A. Hutchinson, "Real-time vision, tracking and control," in *Proceedings of the IEEE International Conference on Robotics and Automation (ICRA '00)*, pp. 622–629, April 2000.

[3] F. Chaumette and S. Hutchinson, "Visual servo control. I. Basic approaches," *IEEE Robotics and Automation Magazine*, vol. 13, no. 4, pp. 82–90, 2006.

[4] M. Staniak and C. Zieliński, "Structures of visual servos," *Robotics and Autonomous Systems*, 2010.

[5] P. Martinet, J. Gallice, and D. Khadraoui, "Vision based control law using 3D visual features," in *World Automation Congress, Robotics and Manufacturing Systems (WAC '96)*, vol. 3, pp. 497–502, 1996.

[6] W. J. Wilson, C. C. W. Hulls, and G. S. Bell, "Relative end-effector control using cartesian position based visual servoing," *IEEE Transactions on Robotics and Automation*, vol. 12, no. 5, pp. 684–696, 1996.

[7] N. R. Gans, A. P. Dani, and W. E. Dixon, "Visual servoing to an arbitrary pose with respect to an object given a single known length," in *Proceedings of the American Control Conference (ACC '08)*, pp. 1261–1267, June 2008.

[8] L. E. Weiss, A. C. Sanderson, and C. P. Neuman, "Dynamic visual servo control of robots: an adaptive image-based approach," in *Proceedings of the IEEE International Conference on Robotics and Automation*, pp. 662–668, 1985.

[9] J. T. Feddema and O. R. Mitchell, "Vision-guided servoing with feature-based trajectory generation," *IEEE Transactions on Robotics and Automation*, vol. 5, no. 5, pp. 691–700, 1989.

[10] K. Hashimoto, T. Kimoto, T. Ebine, and H. Kimura, "Manipulator control with image-based visual servo," in *Proceedings of the 1991 IEEE International Conference on Robotics and Automation*, pp. 2267–2271, April 1991.

[11] Z. Qi and J. E. McInroy, "Improved image based visual servoing with parallel robot," *Journal of Intelligent and Robotic Systems*, vol. 53, no. 4, pp. 359–379, 2008.

[12] O. Bourquardez, R. Mahony, N. Guenard, F. Chaumette, T. Hamel, and L. Eck, "Image-based visual servo control of the translation kinematics of a quadrotor aerial vehicle," *IEEE Transactions on Robotics*, vol. 25, no. 3, pp. 743–749, 2009.

[13] U. Khan, I. Jan, N. Iqbal, and J. Dai, "Uncalibrated eye-in-hand visual servoing: an LMI approach," *Industrial Robot*, vol. 38, no. 2, pp. 130–138, 2011.

[14] D. Fioravanti, B. Allotta, and A. Rindi, "Image based visual servoing for robot positioning tasks," *Meccanica*, vol. 43, no. 3, pp. 291–305, 2008.

[15] A. De Luca, G. Oriolo, and P. R. Giordano, "Image-based visual servoing schemes for nonholonomic mobile manipulators," *Robotica*, vol. 25, no. 2, pp. 131–145, 2007.

[16] F. Chaumette and S. Hutchinson, "Visual servo control. II. Advanced approaches [Tutorial]," *IEEE Robotics and Automation Magazine*, vol. 14, no. 1, pp. 109–118, 2007.

[17] N. R. Gans, S. A. Hutchinson, and P. I. Corke, "Performance tests for visual servo control systems with application to partitioned approaches to visual servo control," *International Journal of Robotics Research*, vol. 22, no. 10-11, pp. 955–981, 2003.

[18] T. W. Sederberg, "Computer aided geometric design," *BYU, Computer Aided Geometric Design Course Notes*, 2011.

[19] J. W. Choi, R. Curry, and G. Elkaim, "Path planning based on bézier curve for autonomous ground vehicles," in *Advances in Electrical and Electronics Engineering—IAENG Special Edition of the World Congress on Engineering and Computer Science (WCECS '08)*, pp. 158–166, October 2008.

[20] M. Callegari and M. C. Palpacelli, "Prototype design of a translating parallel robot," *Meccanica*, vol. 43, no. 2, pp. 133–151, 2008.

A Comparison between Position-Based and Image-Based Dynamic Visual Servoings in the Control of
a Translating Parallel Manipulator

11

[21] D. C. Brown, "Decentering distortion of lenses," *Photometric Engineering*, vol. 32, no. 3, pp. 444–462, 1966.

[22] J. Heikkila and O. Silven, "Four-step camera calibration procedure with implicit image correction," in *Proceedings of the IEEE Computer Society Conference on Computer Vision and Pattern Recognition (CVPR '97)*, pp. 1106–1112, June 1997.

[23] R. Y. Tsai, "A versatile camera calibration technique for high accuracy 3D machine vision metrology using off-the-shelf TV cameras and lenses," *IEEE Journal of Robotics and Automation*, vol. 3, no. 4, pp. 323–344, 1987.

[24] B. Bishop, S. Hutchinson, and M. Spong, "Camera modelling for visual servo control applications," *Mathematical and Computer Modelling*, vol. 24, no. 5-6, pp. 79–102, 1996.

[25] Z. Zhang, "A flexible new technique for camera calibration," *IEEE Transactions on Pattern Analysis and Machine Intelligence*, vol. 22, no. 11, pp. 1330–1334, 2000.

[26] O. Faugeras, *Three Dimensional Computer Vision: A Geometric Viewpoint*, MIT Press, Boston, Mass, USA, 1993.

[27] B. Siciliano and O. Khatib, *Springer Handbook of Robotics*, Springer, 2008.

[28] M. Callegari, "Design and prototyping of a spherical parallel machine based on 3-CPU kinematics," in *Parallel Manipulators: New Developments*, J.-H. Ryu, Ed., pp. 171–198, I-Tech, 2008.

Using the Functional Reach Test for Probing the Static Stability of Bipedal Standing in Humanoid Robots Based on the Passive Motion Paradigm

Jacopo Zenzeri,[1] **Dalia De Santis,**[1] **Vishwanathan Mohan,**[1] **Maura Casadio,**[2] **and Pietro Morasso**[1,2]

[1] *RBCS Department, Istituto Italiano di Tecnologia, Via Morego 30, 16163 Genoa, Italy*
[2] *DIBRIS Department, University of Genoa, Viale Causa, 13 16145 Genoa, Italy*

Correspondence should be addressed to Jacopo Zenzeri; jacopo.zenzeri@iit.it

Academic Editor: G. Muscato

The goal of this paper is to analyze the static stability of a computational architecture, based on the Passive Motion Paradigm, for coordinating the redundant degrees of freedom of a humanoid robot during whole-body reaching movements in bipedal standing. The analysis is based on a simulation study that implements the Functional Reach Test, originally developed for assessing the danger of falling in elderly people. The study is carried out in the YARP environment that allows realistic simulations with the iCub humanoid robot.

1. Introduction

In humans the ability to stand up on two legs is a necessary prerequisite for bipedal walking. Moreover, there is ample neurophysiological evidence that standing and walking are rather independent control mechanisms. Therefore, we suggest that also humanoid robots should be trained first to master the unstable standing posture in a generality of situations and then learn to walk.

We shall address this issue in relation with the humanoid robot iCub [1], which has the size of a three-year-old child (height is 105 cm and weight is 14.2 Kg) and has 53 degrees of freedom (DoF): 7 DoFs for each arm, 9 for each hand, 6 for the head, 3 for the trunk and spine, and 6 for each leg. iCub is still unable to stand or walk but only to crawl, as baby toddlers of the same age. Therefore, the goal of this paper is to carry out a preliminary study of the computational processes that may allow iCub to achieve the sensorimotor competence that is necessary for bipedal standing.

The study builds upon what has already been achieved in the bimanual coordination of iCub's movements [2, 3], using the Passive Motion Paradigm which is a biomimetic,

force-field based computational model based on the equilibrium point hypothesis. The model has been evaluated and validated both in a simulated environment and in real movements. However, the present study is limited to the simulation stage for "developmental constraints," because the sensorimotor system of iCub has not matured enough to achieve the features that are necessary for standing (postural control system) and walking (bipedal locomotion system) in a biomimetic, compliant way. Compliant motion is currently operating on the proximal joints of the upper and lower limbs. Biomimetic postural control requires a compliant ankle and this development will become available in the near future.

As a matter of fact, the whole body postural control system has two basic components: *static* (P1) and *dynamic* (P2). P1 requires that for each time instant the projection of the Center of Mass (CoM) on the ground remains inside the support base (the convex hull of the points of contact of the body with the ground). P1 is a problem of constrained coordination among the highly redundant DoFs of the human/ humanoid body, but satisfying such coordination constraints is not sufficient for maintaining balance unless the underlying dynamic controller operates with very high levels of stiffness of the joints, with particular reference to the ankle joints. This is not what happens in the human case, at least for healthy subjects who are characterized by a low level of stiffness, smaller than the rate of growth of the toppling torque due to gravity [4, 5], thus inducing persistent sway movements during standing. In contrast, high-stiffness upright posture characterizes pathological conditions such as Parkinson's disease [6]: this entails apparently enhanced stability (the size of the sway movements is smaller in PD patients than in controls) but also higher sensitivity to unexpected perturbations and thus higher danger of falling. In humans robust dynamic stabilization with low levels of stiffness is achieved by intermittent control mechanisms [7–9], which generate small, stabilizing control bursts on top of the body postures determined by whole-body synergies.

The focus of this paper is on P1 because the current implementation constraints of the robot do not allow a low-stiffness dynamic stabilization of the standing posture, but self-adjusting static stabilization is the necessary prerequisite for achieving a general purpose mastery of the standing posture. As already mentioned, we intend to address this problem by using a biomimetic, force-field based computational model, which takes inspiration from the Passive Motion Paradigm (PMP) [10], extended to include terminal attractor properties [11]. We already used this approach for the coordination of bimanual movements of the humanoid robot iCub and for modeling whole-body reaching (WBR) movements in humans [12]. Here we investigate the feasibility of applying this model to the coordination of WBR movements in iCub, with particular emphasis on a specific form of WBR, namely, the Functional Reach Test (FRT). FRT has been invented as a dynamic clinical measure of balance [13]: it measures the distance between the length of the arm and the maximal forward reach in the standing position, while maintaining a fixed base of support. FRT has been tested for both validity and reliability and is used in patients with diagnoses as different as stroke, Parkinson's disease, vestibular hypofunction, multiple sclerosis, and hip fractures. FRT has also been associated with an increased risk of fall and frailty in elderly people who are unable to reach out at least 15 cm during bipedal standing.

Since the current state of the iCub's competence for the standing posture has still some "pathological" aspects, as regards the danger of falling, the improvements coming from better design and better control could be appropriately evaluated with a test similar to FRT, used with humans.

2. FRT Network for iCub

The network architecture which has been applied for allowing iCub to carry out the Functional Reach Test is an extension of the architecture developed for modeling whole-body reaching movements in humans. The architecture is composed of four parts: (1) *task subnetwork*, (2) *focal subnetwork*, (3) *postural subnetwork*, and (4) *temporal coordination unit* (see Figure 1).

The three subnetworks are stable dynamical systems with terminal attractor characteristics, which are provided by the temporal coordination unit. This unit generates a

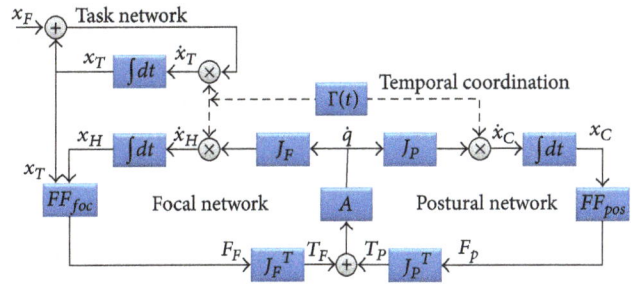

FIGURE 1: PMP network for the iCub humanoid robot that carries out the Functional Reach Test (FRT). x_F and x_T are the 3-dimensional position vectors of the final and moving targets, respectively; x_H is the position of the hand; x_C is the position of the CoM; q is the 5-dimensional joint rotation vector (ankle, knee, hip, shoulder, elbow). J_F is the 3×5 Jacobian matrix of the whole kinematic chain, used by the focal sub-network; J_P is the 3×3 Jacobian matrix of the partial kinematic chain (from ankle to hip) which is used by the postural subnetwork. FF_{foc}, FF_{pos} are the force field generators of the two subnetworks; F_F, F_P are the corresponding force fields; T_F, T_P are the related torque fields. A is the 5×5, diagonal admittance matrix. $\Gamma(t)$ is the temporal coordination function.

time-varying gain $\Gamma(t)$ which is transmitted to the three subnetworks and allows them to reach final equilibrium at the same time:

$$\Gamma(t) = \frac{\dot{\xi}}{(1 - \xi)},$$

$$\xi(t) = 6\left(\frac{t - t_0}{\tau}\right)^5 - 15\left(\frac{t - t_0}{\tau}\right)^4 + 10\left(\frac{t - t_0}{\tau}\right)^3.$$

(1)

Here t_0 is the initiation time and τ is the duration of the coordinated forward-reaching movement; $\xi(t)$ is a minimum jerk time base generator. The general rationale of PMP, as a synergy formation mechanism for highly redundant articulated systems, is to express the goal of an action as a force field applied to the end effector and task-specific constraints as additional force fields applied to task-related body parts. In our case, the task-specific constraint is to keep the projection of the center of mass on the ground inside the support base determined by the two feet and the associated force field is applied to the hip, according to a hip-balancing strategy. The task and focal subnetworks are responsible of the former force field and the postural subnetwork takes care of the latter.

2.1. Task Subnetwork. It generates a moving target $x_T(t)$ in 3D space which attracts both hands of iCub (represented by the time-varying vector $x_H(t)$) with a force field, induced by the focal sub-network. x_T evolves from the initial position of the hands, which are supposed to be jointed, to a final position x_F. In order to maximize the forward reach, in agreement with the nature of the FRT, the final position should be placed somehow beyond the reachable target area. In any case, a peculiar feature of the PMP approach to synergy formation is that it always implies well-formed transformations that do not collapse in the vicinity of singular configurations; in particular, if a chosen target is unreachable, given the values of robometric parameters and task-related constraints, the model guides the end effector to the final state which is closest to the target.

The investigated protocol corresponds to the most common form of FRT, that is, the bimanual test. We might also implement, in the same framework, a unimanual paradigm in which one hand is attracted by a forward-moving target and the other is either fixed to the body or is used as a further counterbalancing tool. The support base is a function of the position of the feet and in FRT they are supposed to be parallel and symmetric with respect to the body. In FRT simulations we positioned x_F just outside the reachable workspace, in the anterior-posterior direction. However the exact position is not critical, emphasizing the robustness of the computational model.

2.2. Focal Sub-Network. It generates an attractive force field F_{foc} of elastic type which is applied to both hands, implementing the focal part of the task whose goal is to allow the hands to reach or at least to approach as much as possible the target: $F_{foc} = K_{foc}(x_T - x_H)$. The force field F_{foc} is mapped into a torque field T_{foc}, from the task space to the joint space, by using the following transformation: $T_{foc} = J_F^T F_{foc}$, where J_F is the Jacobian matrix of the overall kinematic chain (from feet to hands).

The admittance matrix A transforms the torque field into a movement vector $\dot{q}$ of the overall kinematic chain (note that this is not a full admittance matrix but only its viscous component). A expresses the degree of participation of each individual DoF of the redundant kinematic chain of the whole body to the common synergy. Therefore, by modulating the relative values of this matrix, it is possible to implement different equivalent synergies, which may enhance the range of motion of a DoF with respect to the others. The movement vector $\dot{q}$ is then mapped from the joint space to the task space by the same Jacobian matrix, generating a prediction of the hand trajectory x_H and thus closing the loop.

The dynamics induced in the network by the temporal coordination unit allows the hand to reach the final position at the same time in which the final target is reached by the moving target, but there is no guarantee that, in the process, the CoM remains within the support base. Thus, if only driven by this mechanism, iCub would reach the final position but fall forward immediately after. The postural subnetwork is intended to prevent such unfortunate event.

2.3. Postural Subnetwork. It modifies the torque field T_{foc}, generated by the focal sub-network, by adding a "postural" component T_{pos} that takes into account the position of the CoM on the support base. The two torque fields are then superimposed, thus generating a total torque fields that combines the focal drive and the postural stabilization: $T_{tot} = T_{foc} + T_{pos}$. T_{pos} is computed by projecting from the task space to the joint space the postural field F_{pos} which is defined, in the vein of the so-called "hip strategy," as a force applied backward to the hip, in order to counteract the forward shift of the CoM induced by the focal field. This force was implemented in the model by a nonlinear function that diverges to high values when the CoM position x_C

approaches the forward limit x_{max} of the support base: $F_{pos} = -K_{pos} x_C/(x_{max} - x_C)$. The motion x_C of the CoM is derived from the motion of the whole kinematic chain by using a different Jacobian matrix J_P that takes into account only the ankle and knee joints. The activation of the postural field is meant to induce the following effects: (1) a smaller forward shift of the CoM; (2) a backward shift of the hip; (3) a forward tilt of the trunk associated with the lowering of the CoM. It is worth noticing that this complex control pattern is not explicitly programmed but is implicitly coded by the dynamics of the network. The postural field F_{pos} is then mapped from the task space to the joint space by the following transformation: $T_{pos} = J_P^T F_{pos}$. The focal and postural fields are superimposed, generating the combined motion through the common admittance matrix.

In summary, the integrated dynamics of the interacting subnetworks are characterized by the following equations, which achieve a balance between the forward pull, applied to the hand, and the backward pull, applied to the hip:

$$\dot{x}_T = \Gamma(t)(x_F - x_T),$$
$$\dot{x}_H = \Gamma(t) J_F A\, T_{tot},$$
$$\dot{x}_C = \Gamma(t) J_P A\, T_{tot},$$
$$T_{tot} = T_{foc} + T_{pos}, \tag{2}$$
$$T_{foc} = J_F^T K_{foc}(x_T - x_H),$$
$$T_{pos} = -J_P^T K_{pos} \frac{x_C}{x_{max} - x_C}.$$

The simulations of the iCub Functional Reach Test, which correspond to the integration of the equations above, used the iCub's simulator [14] implemented under the YARP open-source middleware [15].

3. Simulation Experiments with FRT Network

The computational architecture described in the previous section was tested by using the iCub simulator. The "robometric" parameters of iCub (length, mass) are summarized here:

(i) leg (21.3 cm, 0.95 Kg);

(ii) thigh (22.4 cm, 1.5 Kg);

(iii) trunk (12.7 cm, 4 Kg, including head);

(iv) humerus (15.2 cm, 1.15 Kg);

(v) forearm + hand (0.13.7 cm, 0.5 Kg).

As suggested by the FRT protocol, the following set of angles defines the initial posture of the test:

(i) ankle: 85°,

(ii) knee: 92°,

(iii) hip: 85°,

(iv) shoulder: 330°,

(v) elbow: 0°.

These angular values are absolute, referring to the horizontal line in the sagittal plane. With this posture, the initial position of the hand has a distance of 29.05 cm with respect to the vertical line passing through the ankle joint and the CoM is shifted 3.52 cm forward. The final position x_F of the target was set 50 cm forward, which is slightly beyond the maximum reachable forward distance, and the limit for the CoM displacement (x_{max}) was set equal to 13 cm, considering that the length of iCub's foot is 15 cm.

The control parameters of the FRT network and the values used in the simulations are as follows:

(i) gain of the focal field $K_{foc} = 700\,$N/m;

(ii) gain of the postural field $K_{pos} = 2\,$N;

(iii) admittance matrix of the whole kinematic chain A. In the reported experiments the matrix is 5×5 because the following five joints are involved: ankle, knee, hip, shoulder, elbow. For simplicity, we chose the matrix to be diagonal, also because this allows us to choose its values in a rational way. What is important, from the point of view of synergy formation, are the relative values of the matrix diagonal. For example, if they are all equal it means that all the joints have the same weight in the participation to the common action; if one is much smaller than the others, then the corresponding joint will change very little its angular value with respect to the other joints; if one admittance value is much greater than the others, then the corresponding joint will be the one that will move the most. The relative values of the matrix diagonal can be scaled up and down with little effect on the overall synergy because the nonlinear gating provided by the Γ function tends to normalize the overall gain. In the simulation example we set the five elements to the following values: A_1(ankle) = 0.02 rad/Nms; A_2(knee) = 0.01 rad/Nms; A_3(hip) = 0.3 rad/Nms; A_4(shoulder) = 0.1 rad/Nms; A_5(elbow) = 0.07 rad/Nms. The hip admittance has the highest value in order to facilitate the counterbalancing of the forward shift of the CoM, induced by the arm-reaching movement, with a suitable backward shift of the pelvis. For the same reason the admittance of the knee and elbow is relatively smaller. But we can change the pattern of admittance values in a large range as a function of specific task or physical constraints of the robot.

With these parameter values we could obtain the simulation results illustrated in Figures 2 and 3. In particular, Figure 2 shows the initial and final postures of the FRT. The end-effectors of iCub reach forward at a distance of 46.75 cm, which is shorter than the target distance (50 cm) because the backward pull of the postural force field allows the forward shift of the CoM (11.95 cm) to remain inside the planned limit of 13 cm. Thus static stability is preserved and the robot body is stretched forward as much as possible. The increase of the hand forward reach is 17.7 cm, with respect to the initial posture. Incidentally, this value is greater than the threshold of 15 cm which is considered clinically relevant, in the sense

that people who are unable to exceed such forward reach in the test have a significant risk of falling.

Figure 3 shows the time profile of the relevant variables. Figure 3(a) displays the intensities of the focal and postural force fields, respectively, together with the profile of the temporal coordination unit $\Gamma(t)$ (dashed), which provides terminal attractor characteristics to overall model. Figure 3(b) shows the rotations patterns of the five joints (ankle, knee, hip, shoulder, elbow) from the initial to the final posture. Please note that some angles evolve monotonously from initial to termination time whereas others do not. In particular, the elbow joint angle remains equal to 0 throughout the whole movement for two reasons: (1) it was set to 0 initially in agreement with the FRT protocol and (2) it remained 0 because both force fields were directed horizontally (the focal field forward and the postural field backward, resp.). Figure 3(c) plots the forward displacements of the hand and the CoM: these curves evolve monotonously, as should do, to the final shift values that must be compared to the final position of the target and the maximum admitted forward shift of the CoM, respectively. It turns out that the hand stops about 5 cm before the target, because the latter is outside the workspace of the robot; the CoM stops a few millimeters before the prefixed limit of stability. Finally, Figure 3(d) displays the speed profiles of the hand and the CoM, respectively: they appear to be bell shaped and synchronized, in agreement with the basic findings of the research in WBR in general [16–18], which shows indeed that the two parts of the WBR strategy are not independent but strictly coupled by a common action generation mechanism. Figures 3(a), 3(b), and 3(c) also display the time course of the Γ function, emphasizing its role in the ordered coordination and synchronization of so many different variables.

How robust is the proposed synergy formation model for testing the static stability of the standing posture in humanoid robots? The question can be analyzed from two points of view: (1) how effective is the model for inducing an optimal or quasi optimal forward reach? and (2) how capable is it in maintaining static stability, although with a narrow margin? In order to answer such questions we carried out a sensitivity analysis of the model with respect to the main parameters of the model: (1) the final position of the target x_F; (2) the gain factors of the two force fields, namely, K_{foc} and K_{pos}; (3) the relative values of the admittance matrix A.

Figure 4 shows how the modulation of x_F in a large range influences the variation of the final reach of the hand (x_H) and the forward shift of the CoM (x_C). For values of x_F which are inside the range of reachable positions, up to about 45 cm, there is a proportionality between x_F, on one side, and x_H or x_C, on the other. When x_F exceeds 55 cm there is a saturation for both x_H, at about 48 cm, and x_C, which approaches the threshold of 13 cm, without ever crossing it (the maximum value is 12.89 cm). Therefore, in the Functional Reach Test iCub cannot stretch the hand beyond 48 cm and such performance is weakly dependent on the position of the target, provided that it is greater than 55 cm. Moreover, with the nominal values of the parameters, used for the simulations illustrated in Figures 2 and 3, static stability is preserved for any final position of the target.

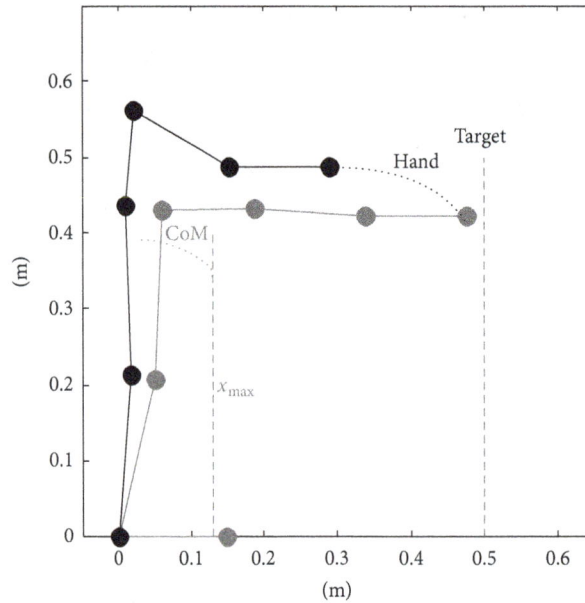

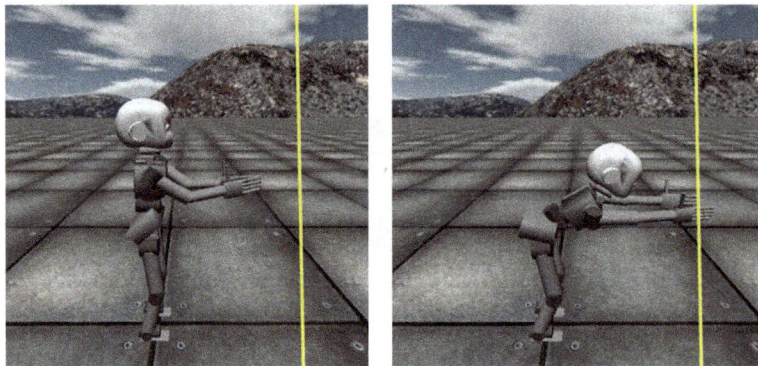

FIGURE 2: Initial and final poses of iCub in the Functional Reach Test.

Of the two gain parameters (K_{foc} and K_{pos}) the former one may have some influence on performance (how far iCub can reach forward, given a final position of the target) and the latter can affect, in principle, the static stability of the synergy. However, the influence is very mild. In particular, if we double the value of K_{foc} (from 700 to 1400 N/m), the forward reach is increased by less than 1 cm (from 46.75 cm to 47.29 cm); if it is halved to 350 N/m, the forward reach is decreased by about 6 cm. Therefore $K_{foc} = 700$ N/m seems to be a lower bound on the gain of the focal field. Stability is preserved in all cases in the sense that x_C remains always behind the threshold.

As regards the gain of the postural field, the nominal value $K_{pos} = 2$ N pushes the CoM quite close to the limit (11.95 cm vs. 13 cm) but preserves static stability. Without such field, that is, by setting $K_{pos} = 0$, the stability limit would be overcome by several centimeters. By reducing K_{pos} from the nominal value the CoM will be pushed closer and closer to the limit but it is necessary to go as low as $K_{pos} = 0.005$ N before losing static stability.

The admittance matrix A has an effect on the final posture of the body, when equilibrium is reached; however the influence on the task-related variables is very mild. For example, if all the elements of the diagonal are set all equal to 0.1 Nms/rad, the forward reach of the hand is changed from 46.75 cm to 46.80 cm and the forward shift of the CoM is slightly reduced from 11.95 cm to 11.93 cm. Similar results are obtained by changing the elements of the matrix by 50% or more in different combinations.

4. Discussion

The proposed synergy formation model is just an example of a large class of complex, whole-body tasks that can be captured by the Passive Motion Paradigm. In a recent paper [19], PMP was proposed as an alternative to optimal control for motor cognition in human and humanoid neuroscience. The next step will be to integrate in the formalism the dynamics derived from physical interaction between the body and the environment.

One of the most remarkable features of the model is that it is "self-adaptive" with respect to the articulation of the underlying body schema for the main following reasons: degrees of freedom can be added or deleted, for example, for incorporating in the body schema manipulated tools with internal dynamics or for taking into account the reduced mobility due to some kind of impairment; moreover, the synergy

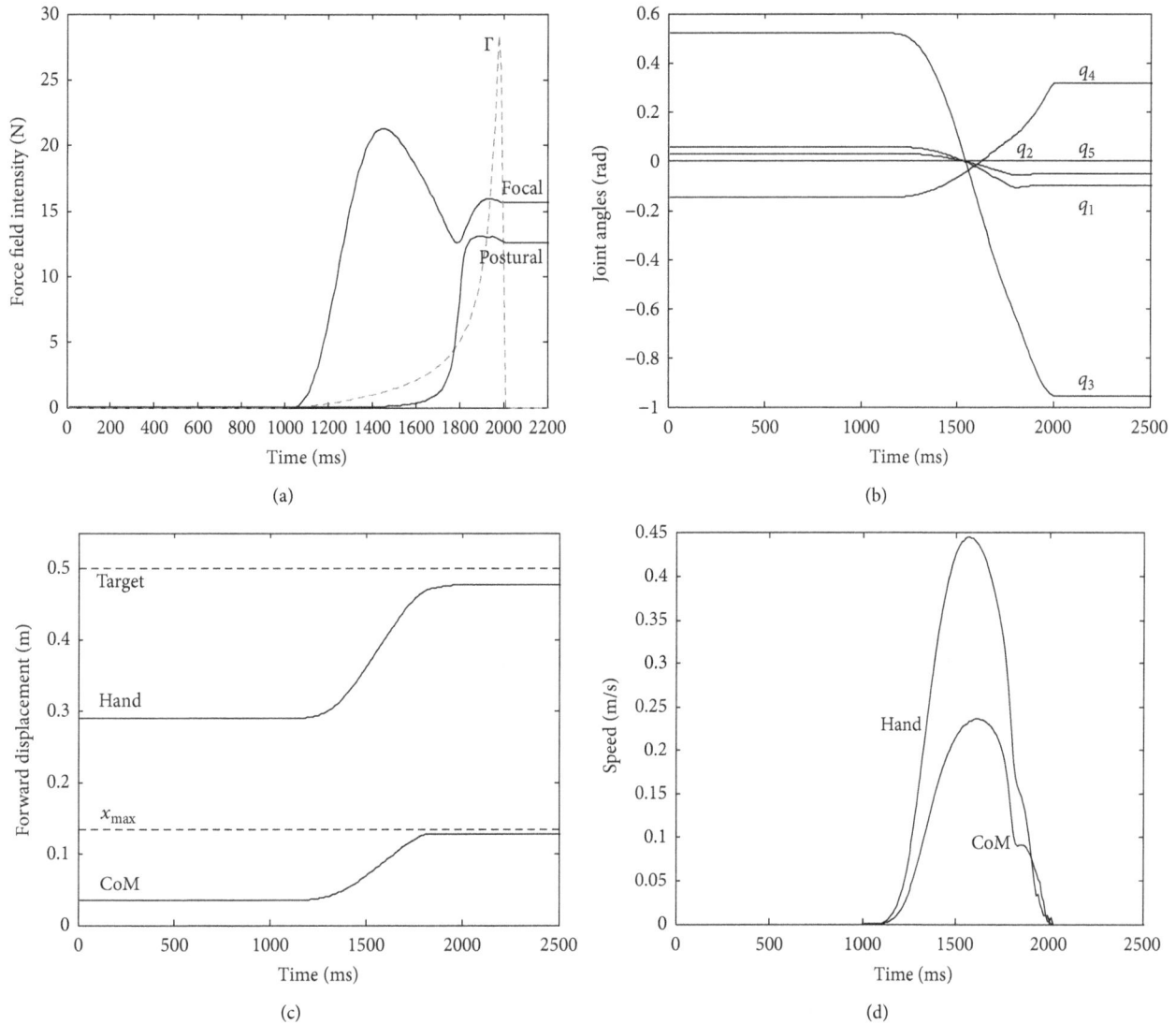

FIGURE 3: (a) Time course of the forces generated by the focal and the postural subnetworks, respectively, (continuous lines); time-base generator (dashed line). (b) Time course of the joint rotation angles, after subtracting the mean value: q_1(ankle) = 1.38 rad; q_2(knee) = 1.60 rad; q_3(hip) = 0.83 rad; q_4(shoulder) = 5.95 rad; q_5(elbow) = 0 rad. The angular values are absolute, relative to the horizontal line. (c) Forward shift of the hand (Functional Reach), related to the forward position of the target (50 cm with respect to the ankle) and forward shift of the CoM, related to the maximum stable position on the support base (x_{max} = 13 cm). (d) Velocity profiles of the hand and the CoM. Panel (a) also displays the time course of the Γ function (dashed line).

formation capabilities of the model remain intact, provided that the modifications of the Jacobian matrices are learned through an appropriate training: see the appendix for a possible, simple procedure of approximation and learning.

Another aspect of such computational robustness is that there is no need to compute the timing and the specific velocity profiles of all the joints, because they are implicit consequence of the internal model simulation process. In this sense, the curse of dimensionality that in most cases affects the efficiency of planning/control methods in highly redundant robots does not apply to the proposed model.

Clearly there is a link between WBR tasks like functional reaching and APAs (Anticipatory Postural Adjustments) which have been studied by many authors [20–22]. In fact,

any voluntary movement of the upper arms is intrinsically a source of disturbance to the posture and the stability of the whole body. It has been demonstrated that APAs are not reflexes but coordinative structures superimposed on the postural stabilization processes. This is also the case for WBR, in general, and for the FRT task, in particular. However, there is a difference: in WBR and FRT there is a strong coupling between the movements of the lower and upper extremities because they are both directly involved in both concurrent tasks, namely, the focal and postural task. In contrast, in APA experiments there are no targets to be reached and simple arm raises or trunk flexions/extensions are performed in order to perturb the standing posture: as a consequence, the postural system can operate independently of the arm system and this

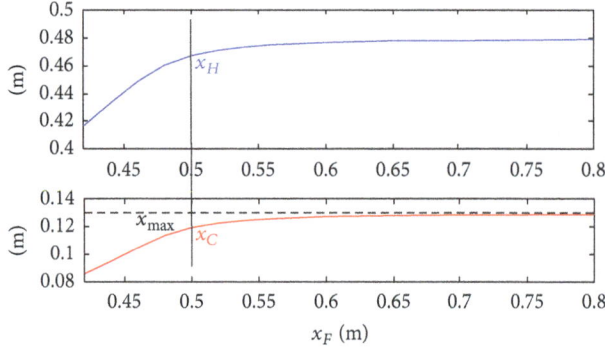

FIGURE 4: Top panel: final position of the hand in the forward direction (x_H) as a function of the final position of the target (x_F). Bottom panel: final position of the CoM (x_C) as a function of the final position of the target (x_F); stability requires that x_C is smaller than a critical threshold x_{max} (dashed line).

lowers the degree of correlation between the two kinds of movements.

Finally, let us consider the possible impact of biomedical robotics and biomechatronics technology on healthcare. There are many reasons for assuming that humanoid robotics can give a significant contribution to this theme in a very general sense. There is indeed a widely shared vision that a new generation of robotic technologies—*robot companions for citizens*—can help our society to come to terms with the special needs of an ageing population, in such a way to remain creative, productive, autonomous, and independent. For example, this is the vision of the FET Flagship Initiatives RoboCom, which envisages a new generation of soft, sentient machines that will help and assist humans in activities of daily living. We believe that the computational model presented in this paper goes in that direction because it allows a humanoid robot to acquire a degree of competence in focal/postural activities which is a prerequisite for safe, soft, friendly interaction between a robot companion and a needing human.

Appendix

Jacobian Matrix of the Whole-Body and Babbling Movements for Learning It

The whole body model has five degrees of freedom $q = [q_1, q_2, q_3, q_4, q_5]$ which identify the following joints: ankle, knee, hip, shoulder, elbow. A related vector $L = [L_1, L_2, L_3, L_4, L_5]$ stores the corresponding link lengths. The position of the end effector is identified by a three-dimensional vector $p = [p_1, p_2, p_3]$ where p_1 is the coordinate in the mediolateral direction, p_2 in the anteroposterior direction, and p_3 in the vertical direction. For the movements considered in this paper the kinematic function $p = f(q)$ can be written as follows:

$$p_1 = 0,$$

$$p_2 = L_1 c_1 + L_2 c_2 + L_3 c_3 + L_4 c_4 + L_5 c_5,$$

$$p_3 = L_1 s_1 + L_2 s_2 + L_3 s_3 + L_4 s_4 + L_5 s_5$$

$$\text{(A.1)}$$

with $c_1 = \cos q_1$, $s_1 = \sin q_1$, and so forth. From this we can immediately derive J_F:

$$J_F = \frac{\partial p}{\partial q}$$

$$= \begin{bmatrix} 0 & 0 & 0 & 0 & 0 \\ -L_1 s_1 & -L_2 s_2 & -L_3 s_3 & -L_4 s_4 & -L_5 s_5 \\ +L_1 s_1 & +L_2 s_2 & +L_3 s_3 & +L_4 s_4 & +L_5 s_5 \end{bmatrix}. \quad \text{(A.2)}$$

J_P is similar: it includes only the first three columns of J_F. If the precise robometric variable is not known, the kinematic function $p = f(q)$ and the corresponding Jacobian matrix can be approximated by means of a neural network, trained by means of babbling movements.

Given a generic kinematic chain, which maps the joint vector q into the position $p_{ee} = f(q)$ of the end-effector, the nonlinear function $p_{ee} = f(q)$ can be approximated by a neural network and the network parameter can be learned with the aid of a training set, namely, a large, representative set of input-output patterns: $\{p_{ee}(t_k), q(t_k); k = 1 \cdots n\}$, obtained experimentally via "babbling movements." For example, we can use a three-layered artificial neural network (ANN):

$$p = f(q)$$

$$\implies \begin{cases} h_j = \sum_i w_{ij} q_i, \\ z_j = g(h_j) \\ p_k = \sum_j w_{jk} z_j, \end{cases} \quad i = 1:5, j = 1:n, k = 1:3,$$

$$\text{(A.3)}$$

where n is the number of neurons of the hidden layer; h_j is an intermediate variable; z_j is the output of the hidden layer; $g(\cdot)$ is a sigmoid nonlinearity; w_{ij}, w_{jk} are the connection weights from input to hidden and from hidden to output layers, respectively; q_i and p_k are the inputs and outputs of the ANN. After training, by means of the standard back propagation method, we can extract the Jacobian matrix from the trained neural network in the following way:

$$J_{ki} = \frac{\partial p_k}{\partial q_i} = \sum_j \frac{\partial p_k}{\partial z_j} \frac{\partial z_j}{\partial h_j} \frac{\partial h_j}{\partial q_i} = \sum_j w_{jk} g'(h_j) w_{ij}. \quad \text{(A.4)}$$

Acknowledgments

The research leading to these results has received funding from the European Community's Seventh Framework Pro gramme (FP7/2007-2013) projects DARWIN (http://www .darwin-project.eu/, Grant no: FP7-270138), EFAA (http:// efaa.upf.edu/, Grant no: FP7-270490), and ROBOCOM (http://www.robotcompanions.eu/, Grant no: FP7-284951). This research is also supported by IIT (Istituto Italiano di Tecnologia, RBCS Dipartimento).

Using the Functional Reach Test for Probing the Static Stability of Bipedal Standing in Humanoid Robots
Based on the Passive Motion Paradigm

19

References

[1] G. Metta, L. Natale, F. Nori et al., "The iCub humanoid robot: an open-systems platform for research in cognitive development," *Neural Networks*, vol. 23, no. 8-9, pp. 1125–1134, 2010.

[2] V. Mohan, P. Morasso, G. Metta, and G. Sandini, "A biomimetic, force-field based computational model for motion planning and bimanual coordination in humanoid robots," *Autonomous Robots*, vol. 27, no. 3, pp. 291–307, 2009.

[3] V. Mohan, P. Morasso, J. Zenzeri, G. Metta, V. S. Chakravarthy, and G. Sandini, "Teaching a humanoid robot to draw 'Shapes'," *Auton Robots*, vol. 31, pp. 21–53, 2011.

[4] I. D. Loram and M. Lakie, "Direct measurement of human ankle stiffness during quiet standing: the intrinsic mechanical stiffness is insufficient for stability," *Journal of Physiology*, vol. 545, no. 3, pp. 1041–1053, 2002.

[5] M. Casadio, P. G. Morasso, and V. Sanguineti, "Direct measurement of ankle stiffness during quiet standing: implications for control modelling and clinical application," *Gait and Posture*, vol. 21, no. 4, pp. 410–424, 2005.

[6] M. G. Carpenter, J. H. J. Allum, F. Honegger, A. L. Adkin, and B. R. Bloem, "Postural abnormalities to multidirectional stance perturbations in Parkinson's disease," *Journal of Neurology, Neurosurgery and Psychiatry*, vol. 75, no. 9, pp. 1245–1254, 2004.

[7] I. D. Loram, C. N. Maganaris, and M. Lakie, "Human postural sway results from frequent, ballistic bias impulses by soleus and gastrocnemius," *Journal of Physiology*, vol. 564, no. 1, pp. 295–311, 2005.

[8] A. Bottaro, Y. Yasutake, T. Nomura, M. Casadio, and P. Morasso, "Bounded stability of the quiet standing posture: an intermittent control model," *Human Movement Science*, vol. 27, no. 3, pp. 473–495, 2008.

[9] Y. Asai, Y. Tasaka, K. Nomura, T. Nomura, M. Casadio, and P. Morasso, "A model of postural control in quiet standing: robust compensation of delay-induced instability using intermittent activation of feedback control," *PLoS ONE*, vol. 4, no. 7, Article ID e6169, 2009.

[10] F. A. M. Ivaldi, P. Morasso, and R. Zaccaria, "Kinematic networks—a distributed model for representing and regularizing motor redundancy," *Biological Cybernetics*, vol. 60, no. 1, pp. 1–16, 1988.

[11] M. Zak, "Terminal attractors for addressable memory in neural networks," *Physics Letters A*, vol. 133, no. 1-2, pp. 18–22, 1988.

[12] P. Morasso, M. Casadio, V. Mohan, and J. Zenzeri, "A neural mechanism of synergy formation for whole body reaching," *Biological Cybernetics*, vol. 102, no. 1, pp. 45–55, 2010.

[13] P. W. Duncan, D. K. Weiner, J. Chandler, and S. Studenski, "Functional reach: a new clinical measure of balance," *Journals of Gerontology*, vol. 45, no. 6, pp. M192–M197, 1990.

[14] V. Tikhanoff, A. Cangelosi, P. Fitzpatrick, G. Metta, L. Natale, and F. Nori, "An open-source simulator for cognitive robotics research," Cogprints 6238, 2008.

[15] G. Metta, P. Fitzpatrick, and L. Natale, "YARP: yet another robot platform," *International Journal of Advanced Robotic Systems*, vol. 3, no. 1, pp. 43–48, 2006.

[16] T. R. Kaminski, "The coupling between upper and lower extremity synergies during whole body reaching," *Gait and Posture*, vol. 26, no. 2, pp. 256–262, 2007.

[17] T. Pozzo, P. J. Stapley, and C. Papaxanthis, "Coordination between equilibrium and hand trajectories during whole body pointing movements," *Experimental Brain Research*, vol. 144, no. 3, pp. 343–350, 2002.

[18] P. J. Stapley, T. Pozzo, G. Cheron, and A. Grishin, "Does the coordination between posture and movement during human whole-body reaching ensure center of mass stabilization?" *Experimental Brain Research*, vol. 129, no. 1, pp. 134–146, 1999.

[19] V. Mohan and P. Morasso, "Passive motion paradigm: an alternative to optimal control," *Frontiers in Neurorobotics*, vol. 5, no. 4, pp. 1–28, 2011.

[20] A. S. Aruin, "The organization of anticipatory postural adjustments," *Journal of Automatic Control*, vol. 12, pp. 31–37, 2002.

[21] S. Bouisset and M. Zattara, "Biomechanical study of the programming of anticipatory postural adjustments associated with voluntary movement," *Journal of Biomechanics*, vol. 20, no. 8, pp. 735–742, 1987.

[22] W. A. Lee, T. S. Buchanan, and M. W. Rogers, "Effects of arm acceleration and behavioral conditions on the organization of postural adjustments during arm flexion," *Experimental Brain Research*, vol. 66, no. 2, pp. 257–270, 1987.

3

A Comparison between Two Force-Position Controllers with Gravity Compensation Simulated on a Humanoid Arm

Giovanni Gerardo Muscolo,[1] **Kenji Hashimoto,**[2] **Atsuo Takanishi,**[2,3] **and Paolo Dario**[4]

[1] *R&D Department, Creative Design Laboratory, Humanot s.r.l., via Modigliani 7-59100 Prato, Italy*
[2] *Department of Modern Mechanical Engineering, Waseda University, 17 Kikui-cho, Shinjuku-ku, Tokyo 162-0044, Japan*
[3] *Humanoid Robotics Institute, Waseda University, 2-2 Wakamatsu-cho, Shinjuku-ku, Tokyo 162-8480, Japan*
[4] *The BioRobotics Institute, Scuola Superiore Sant'Anna, Viale Rinaldo Piaggio 34, 56025 Pontedera, Italy*

Correspondence should be addressed to Giovanni Gerardo Muscolo; gmuscleg@hotmail.com

Academic Editor: Huosheng Hu

The authors propose a comparison between two force-position controllers with gravity compensation simulated on the DEXTER bioinspired robotic arm. The two controllers are both constituted by an internal proportional-derivative (PD) closed-loop for the position control. The force control of the two systems is composed of an external proportional (P) closed-loop for one system (P system) and an external proportional-integrative (PI) closed-loop for the other system (PI system). The simulation tests performed with the two systems on a planar representation of the DEXTER, an eight-DOF bioinspired arm, showed that by varying the stiffness of the environment, with a correct setting of parameters, both systems ensure the achievement of the desired force regime and with great precision the desired position. The two controllers do not have large differences in performance when interacting with a lower stiffness environment. In case of an environment with greater rigidity, the PI system is more stable. The subsequent implementation of these control systems on the DEXTER robotic bioinspired arm gives guidance on the design and control optimisation of the arms of the humanoid robot named SABIAN.

1. Introduction

The manipulation control [1] presents difficulties, especially in the variation of the compliance with the environment [2]. For the safety of persons that surround the manipulator, a variation of the stiffness of the humanoid robotic arms is necessary.

A real contact is a distributed phenomenon which involves the local elastic properties of both the manipulator and the environment. Many methodologies allow modifying the stiffness of the manipulator in relation to the task. The compliance inside the DC servo actuators is usually generated by mechanical systems such as linear or torsional springs. In these types of studies, a hardware modification is developed.

The actuators with variable stiffness are increasingly used in the field of humanoid robotics [3]; an example of this application on humanoid robot iCub is presented in [4]. A different kind of application of variable stiffness, which uses pneumatic or hydraulic systems as compliance element formed by the fluid, is presented in [5]. In [6], a control for regulation tasks of robot manipulators with flexible links is proposed.

In this paper, in order to modify the compliance of the manipulator, the authors modified only the software parameters.

Considering K_A as the environment stiffness matrix and by increasing or decreasing its value, it is possible to modify the compliance of all external part of the arm. Increasing the value of K_A allows the manipulator to reach a given position; thus, the arm encounters an obstacle that may not physically exist but in reality exerts a contact force which opposes the motion of the robotic arm and causes its arrest.

In this paper, the authors propose a comparison between two force-position control systems with gravity compensation simulated on a bioinspired manipulator of eight DOFs named DEXTER (Figure 1 and Tables 1, 2, 3, 4, and 5) [2, 6].

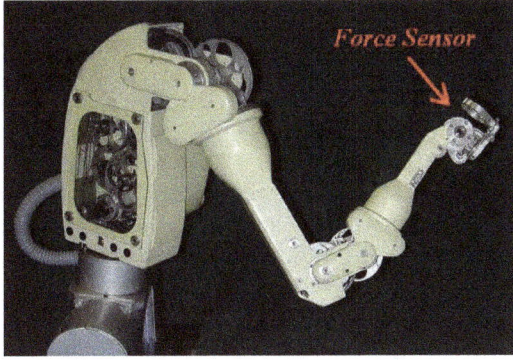

FIGURE 1: DEXTER bioinspired robotics arm.

TABLE 1: DEXTER characteristics.

Characteristics	DEXTER
Dimensions (mm) width-length-height	400-400-950
Weight (kg) (payload)	40 (2)
Workspace (mm-°)	1200-350°
Velocity (m/s)	0.2
DOF	Total: 8

The next steps will be the implementation of these systems into the DEXTER robotic arm with the ultimate aim of control and design of the two arms of the humanoid robot SABIAN (Figure 5) [7, 8].

The two proposed controllers are both constituted by an internal proportional-derivative (PD) closed-loop for the position control. The force controller of the two systems is composed of an external proportional (P) closed-loop for one system (P system) and an external proportional-integrative (PI) closed-loop for the other system (PI system).

This paper is composed of three parts. The first part describes the dimensional characteristics of the DEXTER and the SABIAN robots. In the second part, the two force-position controllers are proposed and discussed. In the third part, the analysis of simulational data and future applications based on these results are proposed.

2. Materials and Methods

2.1. DEXTER Bioinspired Robotics Arm. The DEXTER arm (Figure 1) has eight DOFs with eight rotational joints [2, 6]. The joints are implemented by many electric DC servo motors, located in the first link. The motors transmit torque through a cables and pulleys system. The motor-reducer for the coupling 0 is integrated in the mobile base, while the joint 1 is mounted on the link 0. The motors of the other joints are all in the link 1.

The advantage of this type of mechanical structure is to improve the dynamic performance of the robot because there is a reduction of the masses in motion. The disadvantage is related to the complexity of the mechanical transmission system.

TABLE 2: Motor data sheet.

Axis	Type of DC motor	Nominal voltage (V)	Torque constant (mNm/A)
0	PMI S9M4HI	24	84.10
1	PMI S9M4HI	24	84.10
2	3557 024 CR	24	42.90
3	3557 024 CR	24	42.90
4	3557 024 CR	24	42.90
5	3557 024 CR	24	42.90
6	3557 024 CS	24	41.00
7	3557 024 CS	24	41.00

TABLE 3: Drivers.

Axis	Type of driver	Continuous current (A)	Peak current (A)	$G_i = I_{peak}/\text{MaxInput}$ (A/V)
0	Elmo ISA 10/80	5	10	1
1	Elmo ISA 10/80	5	10	1
2	Elmo ISA 10/80	4	12	1.2
3	Elmo ISA 10/80	4	12	1.2
4	Elmo ISA 10/80	4	12	1.2
5	Elmo ISA 10/80	4	12	1.2
6	Elmo ISA 10/80	2	6	0.6
7	Elmo ISA 10/80	2	6	0.6

TABLE 4: Denavit-Hartenberg parameters.

Joint	a_i (mm)	d_i (mm)	α_i (rad)	θ_i (rad)
1	0	0	$\pi/2$	θ_1
2	144	450	$-\pi/2$	θ_2
3	0	0	$\pi/2$	θ_3
4	−100	350	$-\pi/2$	θ_4
5	0	0	$\pi/2$	θ_5
6	−24	250	$-\pi/2$	θ_6
7	0	0	$\pi/2$	θ_7
8	100	0	0	θ_8

TABLE 5: Center of mass of the links.

	r_x (mm)	r_y (mm)	r_z (mm)	m (kg)
Link 0	0	6.92	27.72	9.429
Link 1	−139.35	174.49	46.08	12.051
Link 2	0	−6.11	34.59	1.627
Link 3	90.72	133.77	−0.24	2.488
Link 4	0.01	−3.72	20.30	0.818
Link 5	−24.01	141.05	0.11	0.541
Link 6	−0.05	2.36	6.78	0.266
Link 7	10.35	1.81	33.26	0.095

In Tables 1, 2, 3, 4, and 5, the physical dimensions, the motor data sheet, the drivers, the parameters of Denavit-Hartenberg, and the coordinates of the centers of mass of the links are, respectively, represented.

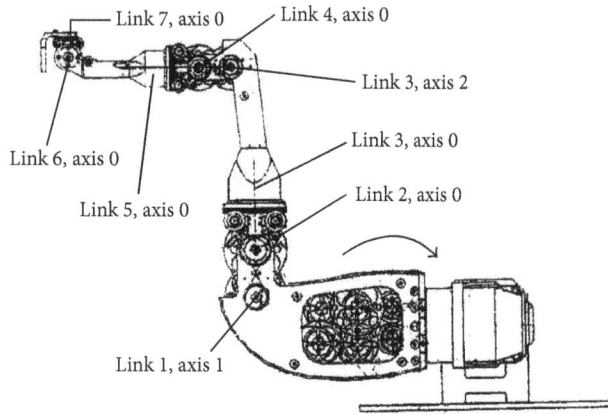

FIGURE 2: DEXTER robotic arm: axis, joints, and links position.

In Figures 2 and 3, the scheme of the manipulator is shown.

The DEXTER manipulator system includes the functional blocks of the Figure 4. The manipulation is constituted coordinating the joint movement of the arm with the movement of the hand mounted on the force sensor.

2.2. SABIAN Humanoid Robot.

SABIAN [7, 8] (Figure 5) has two 7-DOF legs, a 2-DOF waist, and a 2-DOF trunk. Each actuator system of the joint consists of a DC motor, a harmonic drive gear, a lug belt, and two pulleys. This double speed reduction mechanism allows a high reduction ratio and also a joint axis to be set apart from the motor axis. This mechanism provides designs for a human-like joint mechanism without a considerable projection. Figure 5(a) is a photograph of SABIAN. The height of the robot is 1480 mm and the weight is 40.7 kg without batteries. SABIAN is the Italian version of the WABIAN-2 robot [8] (see Figure 6). SABIAN does not have arms, and the head is a version of the ICUB head [4].

2.3. WABIAN-2's Arm: Characteristics.

The arm of WABIAN-2 has 7 DOFs, and Figure 7 shows the 3D-CAD model. The arms were designed based on a concept that the arms of the robot can hold the robot's weight while it leans on a walk-assist machine as shown in [9]. Since the robot can lean on a walk-assist machine, most of its weight will be distributed on both its forearms.

In general, a force/torque sensor is mounted on a robot's wrist in order to enable it to grasp, push, or pull something using a hand as an end-effector. But because one of the design concepts of WABIAN-2 is a robot that can lean against a walk-assist machine, a 6-axis force/torque sensor is mounted on each upper arm, which enables the robot to measure external forces acted on the forearm.

In order to realize a humanoid robot with great dexterity not only in locomotion as in WABIAN-2 but also in manipulation as in DEXTER, it is necessary to redesign the WABIAN-2's arm and implement it on the SABIAN robot.

2.4. WABIAN-2's Arm: Control Architecture.

A conventional position controller is used for WABIAN-2's arms. The six-dimensional hand trajectories are given, and each joint angle is calculated by solving the inverse kinematics, and the result is the ϑ_p value as input in Figure 8. The DC motor is driven by motor drivers (Model no.: TD12770-48W05) developed by TOKUSHU DENSO Co., Ltd, which enables speed control using an electrical governor without a tachogenerator with a 100 kHz PWM. The maximum output current range of TD12770-48W05 is greater than 15 A at 48 V. The current monitor port mounted on the motor drivers is utilized in energy consumption experiments. Figure 9 shows the photograph of the motor driver. The x_d is the fixed ideal value of the hand position.

3. Analysis of Control Schemes

3.1. DEXTER Bioinspired Robotics Arm: Control Architecture.

The matrix K_r of the reduction of the DEXTER robot motion is given by

$$K_r = K_r' A, \tag{1}$$

where A is the matrix of the speed reduction related to the mechanical transmission system; K_r' represents the diagonal matrix of the reduction coefficients. Consider

$$A = \begin{bmatrix} 1 & 0 & 0 & 0 & 0 & 0 & 0 & 0 \\ 0 & 1 & 0 & 0 & 0 & 0 & 0 & 0 \\ 0 & 0 & -4.375 & 0 & 0 & 0 & 0 & 0 \\ 0 & 0 & 1.162 & -3.375 & 0 & 0 & 0 & 0 \\ 0 & 0 & 0.583 & 0 & 3.424 & 0 & 0 & 0 \\ 0 & 0 & 0.483 & 0 & -1.005 & 2.882 & 0 & 0 \\ 0 & 0 & 0.894 & 0 & -0.638 & 0 & -1.106 & 0 \\ 0 & 0 & 0.894 & 0 & -0.638 & 0 & -1.106 & -1.557 \end{bmatrix},$$

$$K_r' = \begin{bmatrix} 282 & 0 & 0 & 0 & 0 & 0 & 0 & 0 \\ 0 & 141 & 0 & 0 & 0 & 0 & 0 & 0 \\ 0 & 0 & 66 & 0 & 0 & 0 & 0 & 0 \\ 0 & 0 & 0 & 66 & 0 & 0 & 0 & 0 \\ 0 & 0 & 0 & 0 & 43 & 0 & 0 & 0 \\ 0 & 0 & 0 & 0 & 0 & 43 & 0 & 0 \\ 0 & 0 & 0 & 0 & 0 & 0 & 66 & 0 \\ 0 & 0 & 0 & 0 & 0 & 0 & 0 & 43 \end{bmatrix}. \tag{2}$$

The reduction system is constituted by redactors in proximity of each joint. The generic element K_{rij} of the matrix K_r expresses the ratio of proportionality between the rotation of the joint j and the rotation of the rotor i. The relationship between the torque vector of the motors to the torque vector of the joints is given by

$$\tau = K_r^T \tau_m. \tag{3}$$

This relationship allows to implement the conversion block $\tau \rightarrow \tau_m$ and to directly move the motors. The size of K_r matrix is 7×7 for the elimination of the first row and first column relating to the joint 0 which is controlled by a PID controller. The standard PID controller used is expressed by the control law:

$$T_n = K_P E_n + K_D (E_n - E_{n-1}) + K_I S_n + K_U U_n + 64 K_Z Z_n + K_O. \tag{4}$$

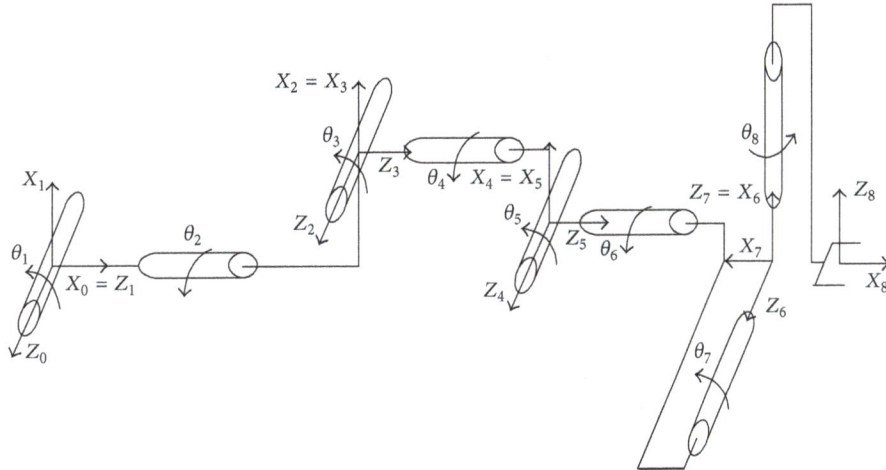

FIGURE 3: DEXTER DOF configuration.

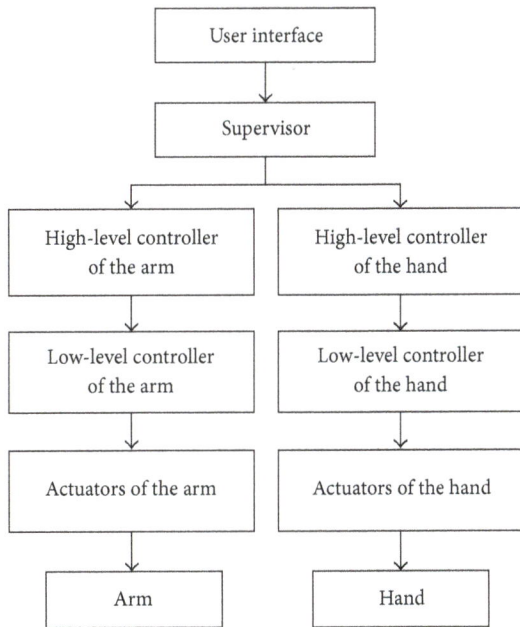

FIGURE 4: Functional architecture of the DEXTER manipulator system.

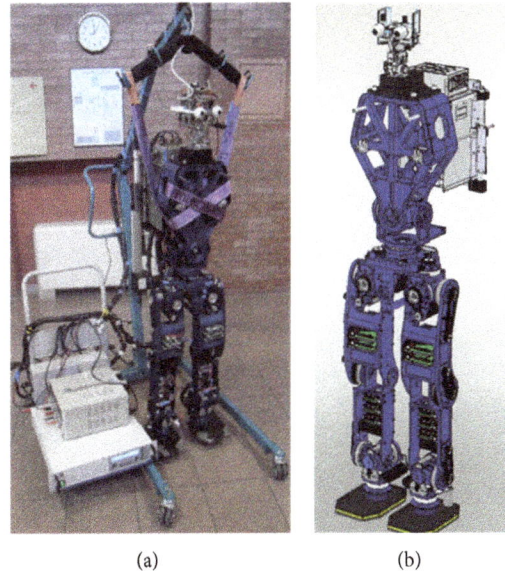

(a)   (b)

FIGURE 5: SABIAN humanoid robot.

T_n represents the analog voltage control of the generic motor output from the PID with a value ranging from −10 V to +10 V. K_P, K_D, and K_I are, respectively, the gains proportional, derivative, and integral. E_n represents the position error to the joints determined in each sampling instant. S_n is the integrated error, and in particular, conditions can be expressed as $S_n = S_{n-1} + E_n$; U_n and Z_n are the speed and the acceleration of the feed-forward terms; K_U and K_Z introduce the speed and the acceleration of the feed-forward terms. K_O is a static offset, used to compensate the small variations in the output voltage, due to the D/A converters or also to the offset of the amplifiers. After notifying the movement commands to the PID, the first of the 8 motors is implemented with a voltage value equal to T_n. Figures 10 and 11, respectively, show

the functional architecture of the subsystem of the arm and the functional scheme of the PID controller.

Figure 12 shows the PID controller of the joint 0. Figure 13 shows the solution used to bypass the PID control of the joint 0.

The control law used for the analysis of the position of the 1–7 joints is given by

$$\tau = K_P \tilde{q} - K_D \dot{q} + g(q), \tag{5}$$

where τ is the torque vector; K_P and K_D, are respectively, the proportional and derivative matrices; $g(q)$ is the gravity compensation, and q_d is the vector of the desired joint position. $\tilde{q} = q_d - q$ is the error. The relation (5) calculates the value of the torque τ but not of the speed d_q. Thus, it is necessary to bypass the PID control to be able to control directly the actuators (Figure 13).

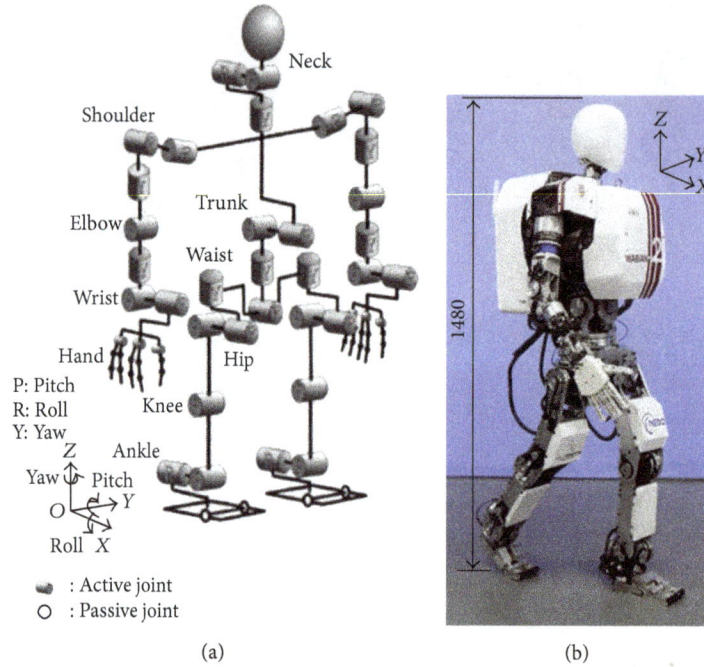

(a)   (b)

FIGURE 6: WABIAN-2 humanoid robot.

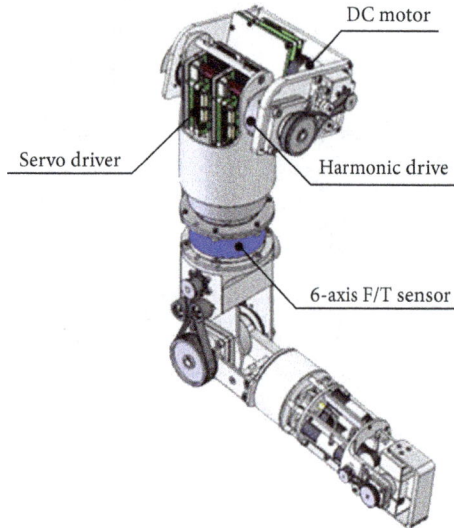

FIGURE 7: WABIAN-2's arm.

From (5), we obtain

$$\tau_m = \left(K_r^T\right)^{-1}\tau,$$

$$\tau_m = K_t i_m,$$

$$i_m = G_i v_m,$$ (6)

$$i_m = K_t^{-1}\tau_m,$$

$$v_m = G_i^{-1} i_m,$$

where K_t is the diagonal matrix of motor torque constant derived from data-sheet of the 7 motors (Table 2). G_i is the gain matrix of the power drivers. i_m and v_m are, respectively, the parameters of the currents and voltages in input to the power drivers. The technical specifications of the system implementation of the 8 joints are shown in Tables 2, 3, 4, and 5.

The software procedure achieves a conversion of the control voltage from volt (v_m) to increments (v_{m_inc}). The used relations are

$$\tilde{x} = x_d - x,$$ (7)

and if $\dot{x}_d = 0$,

$$\dot{\tilde{x}} = -J_A(q)\,\dot{q},$$ (8)

where $J_A(q)$ is the Jacobian matrix which allows to rewrite (6) in [1] as

$$\tau = J_A^T(q)\,K_P\tilde{x} - J_A^T(q)\,K_D J_A(q)\,\dot{q} + g(q).$$ (9)

The interaction force can be controlled in an indirect manner by acting on the x_d reference variable of the position controller of the manipulator; the interaction between manipulator and the environment is directly influenced by the characteristics of the compliance of the environment and of the manipulator. The measure of the force is corrupted by noise, for which the derivative control cannot be used.

In Figures 14 and 15, two examples of force control schemes, respectively, with position and velocity internal closed-loop as in [1] are shown. M_d is a generic diagonal positive matrix used for the impedance control.

Indicating with f_d the desired constant force, with C_F a diagonal matrix whose elements characterize the actions

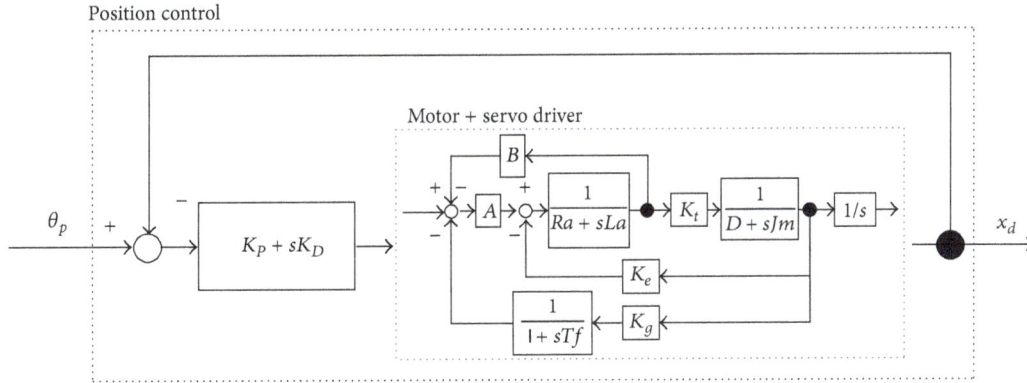

FIGURE 8: Position control scheme for WABIAN-2's arm.

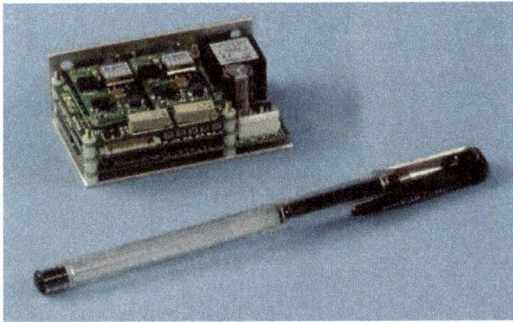

FIGURE 9: Servo driver used for WABIAN-2's arm (TD12770-48W05).

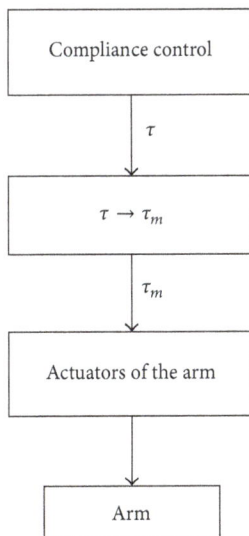

FIGURE 10: Functional architecture of the subsystem of the arm.

of control to exert along the directions of interest of the operative space, with x_e the equilibrium position of the not deformed environment, with K_A the stiffness matrix of the environment, and with x_F a reference that should then be

put in relation with an error of force, we have the following relationships, as in [1]:

$$x_F = C_F (f_d - f),$$
$$\tilde{f} = f_d - f. \quad (10)$$

The difference between the desired force and the force actually developed by the manipulator gives the error. Figure 14 shows that if C_F is proportional, f cannot be similar to f_d because x_e modifies the interaction force; if C_F also expresses an integral action on the force components, it is possible to obtain $f = f_d$ and at the same time to limit the influence of x_e on f. Therefore, an action proportional-integral (PI) can be chosen for C_F.

In Figure 16, an example is shown (as in [1]) of a force-position control. It is a modification of Figure 14 with x_d used in input to the position loop.

3.2. Force Controllers. Two force-control schemes were constructed as external closed-loop: a first system in which the force exerts a proportional-integral (PI System) controller (Figure 17) and a second system in which action developed by the controller is a proportional type (System P) (Figure 18).

3.2.1. PI System Force Controller. The block diagram of this system is shown in Figure 17. The action exerted by the force controller is proportional-integral (PI). For controlling the position of the joints, the law (5) has been used.

For the force control, the following law was used:

$$X_F = K_F \tilde{f} + K_I \int \tilde{f} \, dt. \quad (11)$$

The position transducer provides an electrical signal proportional to the angular displacement of link carried by the robot. The force sensor is used as the connecting member to the wrist between the last arm of DEXTER and the End-Effector. Through the use of the exteroceptive and proprioceptive sensors, it is possible to obtain the measured values of q and f. Using the control law of force (11), X_F is obtained, which allows to obtain in output the error in the

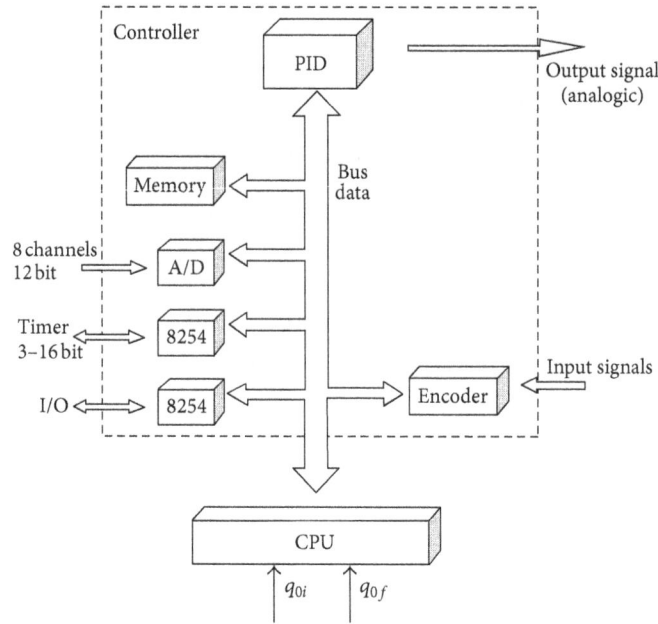

FIGURE 11: Functional scheme of the PID controller.

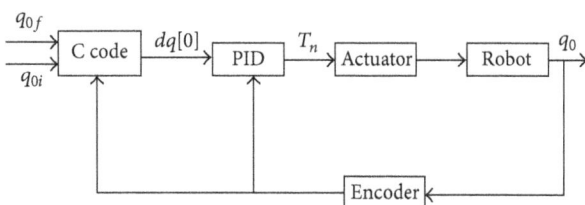

FIGURE 12: Joint 0.

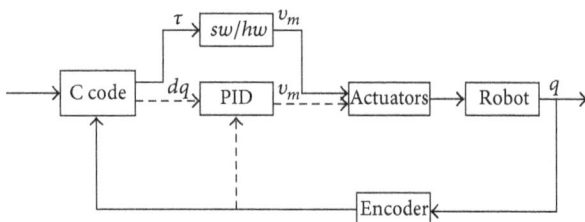

FIGURE 14: Classical force controller with internal position closed-loop [1].

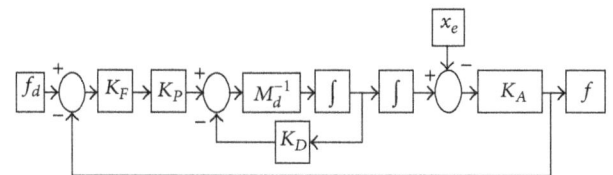

FIGURE 13: Joints 1–7.

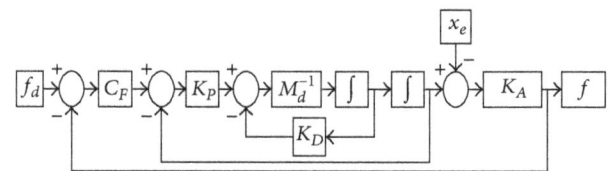

FIGURE 15: Classical force controller with internal velocity closed-loop [1].

operational space ΔX. The joints error $\tilde{q}$ is obtained by means of inverse kinematics, and the vector of torque to be supplied to the actuators is obtained by using the law for the position control.

3.2.2. P System Force Controller.
In this scheme (Figure 18), the force controller exerts a proportional (P) action. As for the PI system, the law (5) is used for the position control, while for the force control, the following has been used:

$$\dot{X}_F = K_I \tilde{f}. \tag{12}$$

In contrast to the scheme of Figure 17, this second system allows to derive the joints position error $\tilde{q}$ through

considerations on the speed and not on the positions in the operational space.

In the P system, a single block of inverse kinematics is present and this helps to reduce errors that can arise as a result of transformation from work space to the joints space.

4. Analysis of Experimental Data

The behaviour of the two control systems of Figures 17 and 18 was simulated using Matlab/Simulink software. In Figure 19(a), simulation of the PI System implemented in a planar representation of the DEXTER arm (3 DOF) and with and without an obstacle is shown.

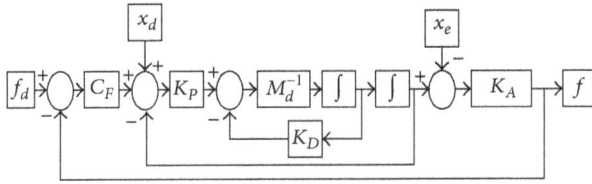

FIGURE 16: Classical force-position control scheme [1].

TABLE 6: Reference values of the PI System.

K_F	K_I	K_P	K_D	K_A
0.0005	0.08	10^6	6000	1000

Five gains (K_F, K_P, K_I, K_D, and K_A) were established as inputs to the simulation models. The simulation graphs are shown from Figure 20 to Figure 27. In particular, the force (N) value and the error position (m) value of the End-Effector of the DEXTER are presented. For each K gain, a maximum, a minimum or an intermediate value has been assigned in order to evaluate the behaviour of the system varying K gains. In Tables 6 and 7, the reference K values for the PI and P systems are presented. The reference K values are referred to the DEXTER robot during its movements.

4.1. PI System Force Controller. Figure 17 represents the block diagram of a force/position control system that uses a proportional-integrative (PI) and a proportional-derivative (PD) controller, respectively, to measure the force and the position of the End-Effector.

Considering Figure 20 and varying K_F, the measured (or real) force presents a peak value of 15 N, while the ideal (or desired) force has a value of 10 N. With the advancement of the simulation, after a time of 2000 ms, the measured force oscillates around the value of the ideal force.

Bringing the value of K_F ranging from 0.0005 to 0.005, a smaller difference between the measured and the ideal force is obtained with a consequent reduction of the amplitudes of oscillation after the 2000 ms. There are no substantial differences in the trends of the position and position error of the End-Effector varying K_F.

Figure 21 is made to vary only K_I maintaining the other values of Table 6. Decreasing this parameter from 0.08 (Figure 20(a)) to 0.04 (Figure 21(a)), the difference between the measured and ideal impulse force increases goes over the threshold of 18 N. Assigning to K_I a value equal to 0.12 (Figure 21(b)), the value of the difference between the measured and the ideal force is no more than 3.5 N.

Decreasing the value of K_P, from 10^6 (Figure 20(a)) to 250000 (Figure 22(a)), a higher peak pulse is created. Varying K_P value from 10^6 (Figure 20(a)) to 10^7 (Figure 22(b)), the maximum value of the measured force decreases.

Figure 23 shows the trends of the error position. Analysing the graphs, it is noted that the increase of K_P decreases the error. Differences in force, position, and position error were not noted in PI system varying K_D gain.

TABLE 7: Reference values of the P system.

K_P	K_I	K_D	K_A
10^6	0.05	6000	1000

4.2. P System Force Controller. Figure 18 represents the block diagram of a force/position control system that uses a proportional (P) and a proportional-derivative (PD) controller, respectively, to measure the force and the position of the End-Effector.

Using values of Table 7 and decreasing the value of K_I from 0.05 to 0.02, an increment of the force in Figure 24 is shown. The measured force value is equal to 26 N (Figure 24(a)). If K_I is equal to the 0.08 value (Figure 24(c)), the peak of the measured force in the P system decreases as it was decreased in the PI system.

Figure 25 shows the measured and the ideal force of the End-Effector of the P system varying K_P. If the K_P value decreases from 10^6 (Figure 24(b)) to 250000 (Figure 25(a)), the measured peak value will be equal to 20 N.

As for the PI System, in the P System, the position error decreases if K_P is equal to 10^7, and the simulation graphs are similar to the graphs of Figure 23.

A comparison between the simulation graphs of the two systems (PI and P) (Figures from 20 to 25) shows that in the P system the peak value of the measured force is bigger with respect to the other one.

4.3. Comparison between PI and P Systems: The Environment Interaction. In the last section, graphs were obtained from the simulation of the two systems by changing K_F, K_P, K_I, and K_D gains. In this section, the behaviour of the two controllers will be analysed by varying only the stiffness matrix value (K_A).

The authors considered K_A as the environment stiffness matrix, and by increasing or decreasing its value, it is possible to modify the compliance of all external part of the arm. Increasing the value of K_A allows the manipulator to reach a given position; thus, the arm encounters an obstacle that may not physically exist but in reality exerts a contact force which opposes the motion of the robotic arm and causes its arrest.

In the graphs of Figures 26 and 27, the force, the position, and the error position values, respectively, of the PI and P systems are presented.

Increasing the K_A value from 1000 (Figure 20(a)) to 10000 (Figure 26(a)), a high pulse of force was generated as shown in Figure 26. When the manipulator is working in a higher stiffness environment, the pulse of force that the End-Effector notes is naturally higher.

An increment of the real force from 18 N (Figure 24(b)) to 24 N (Figure 27(a)) is noted in Figure 27(a) assuming a K_A value equal to 5000.

The differences between the PI and P Systems are evident by a comparison of Figures 26(a) and 27(a). In particular, in Figure 26(a), the peak of the measured force is equal to 18 N, but no more oscillations were observed. In Figure 27(a), the peak of the measured force is 24 N, and the force value is variable before the 2000 ms. A comparison of these

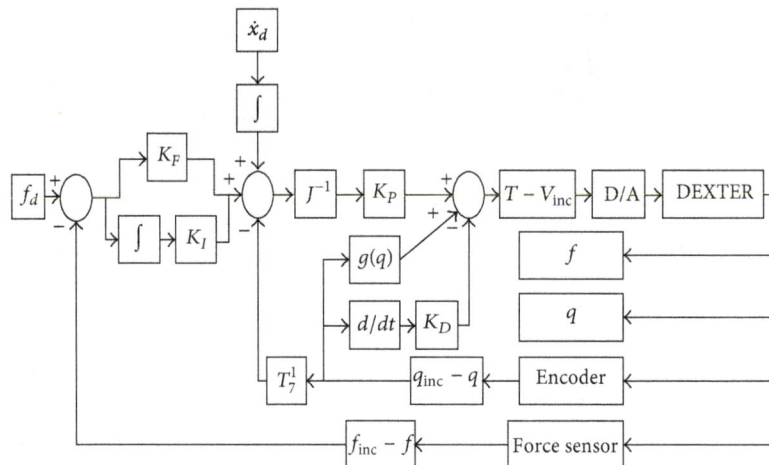

FIGURE 17: PI System.

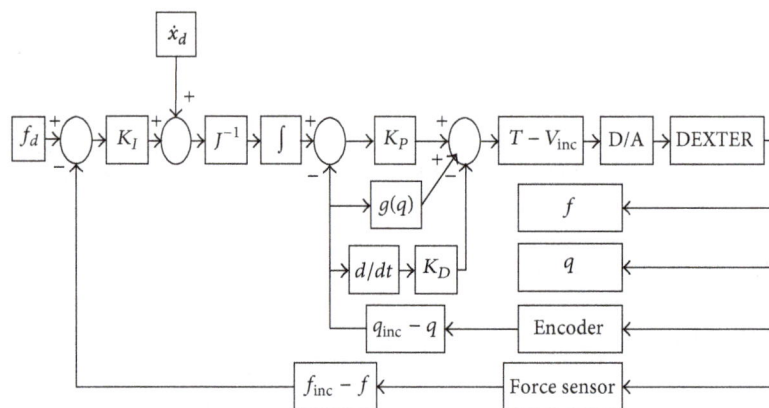

FIGURE 18: P System.

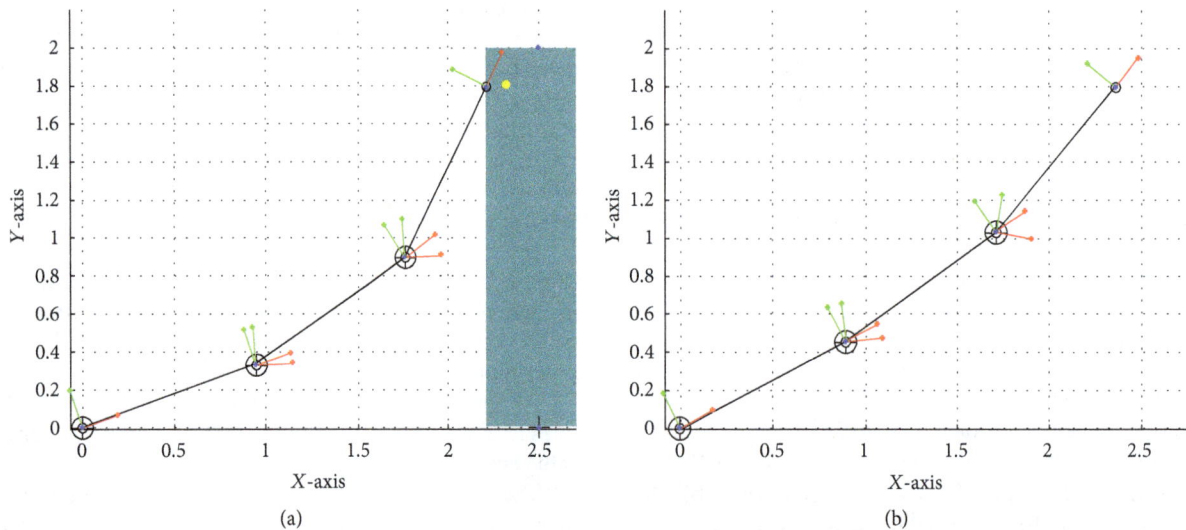

FIGURE 19: Simulation of the PI System in a planar representation of the DEXTER arm (3DOF) with (a) and without (b) an obstacle: $x = 1.86$, $y = 1.5$ is the initial position of the End-Effector; $x = 2.35$, $y = 1.8$ is the final position of the End-Effector; $x = 2.2$ is the position of the obstacle.

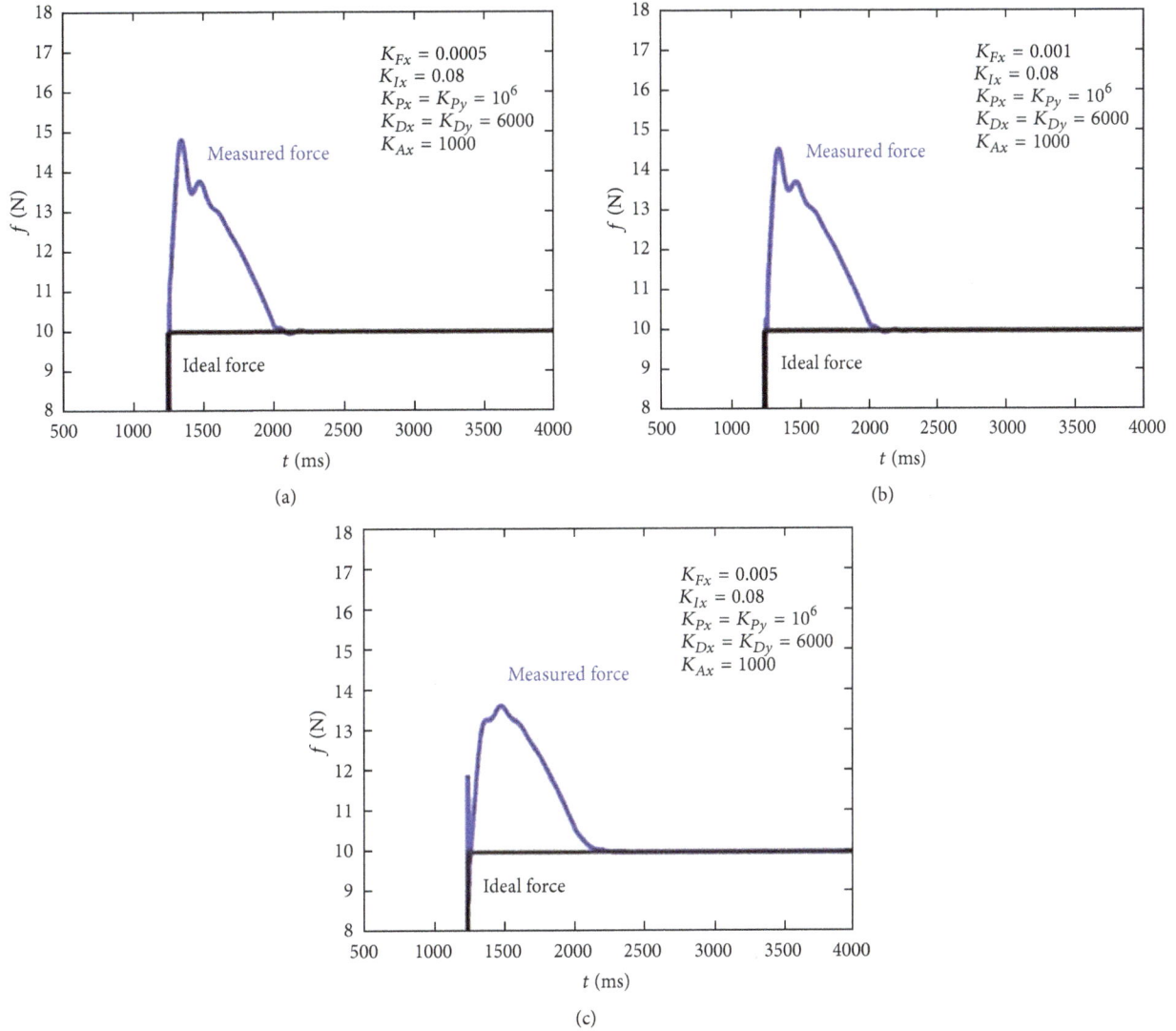

FIGURE 20: End-Effector force value (N) varying the K_F gain in the PI system: (a) $K_F = 0.0005$; (b) $K_F = 0.001$; (c) $K_F = 0.005$.

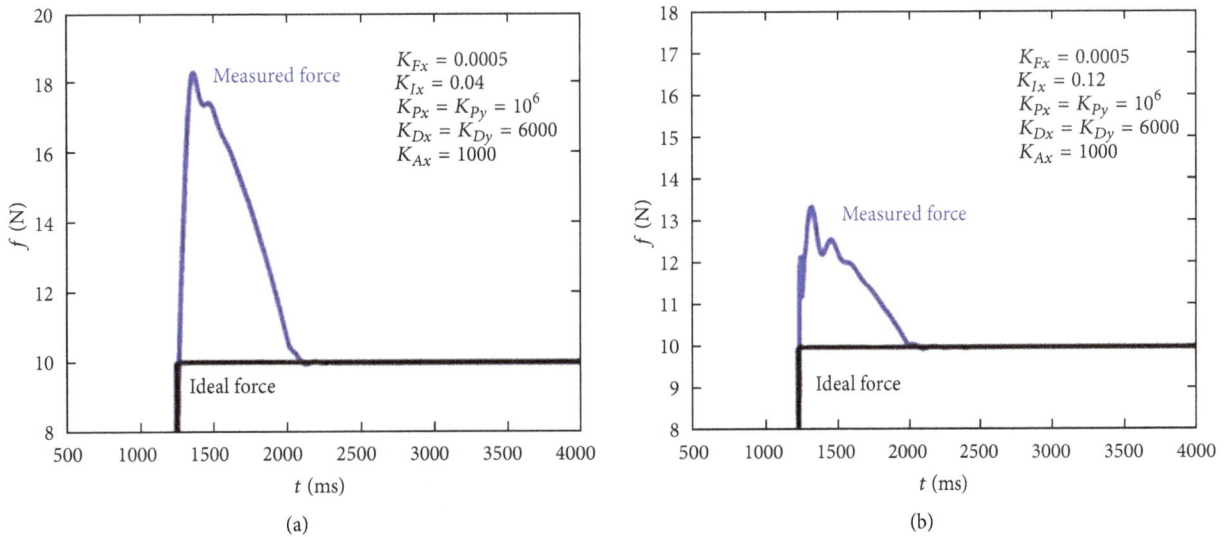

FIGURE 21: End-Effector force value (N) varying the K_I gain in the PI system: (a) $K_I = 0.04$; (b) $K_I = 0.12$.

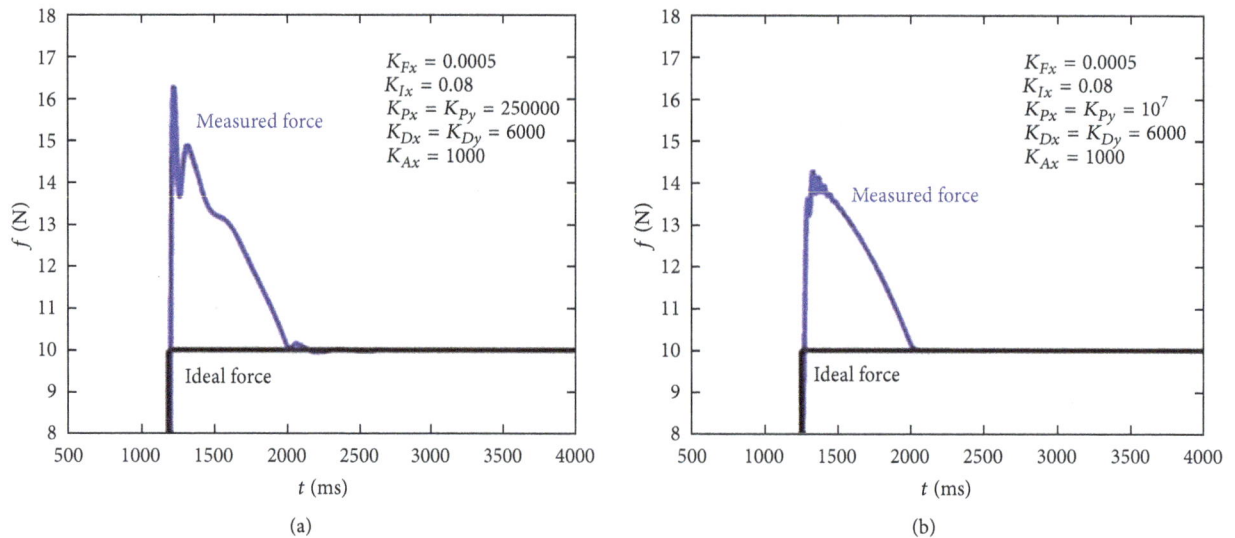

FIGURE 22: End-Effector force value (N) varying the K_P gain in the PI system: (a) $K_P = 250000$; (b) $K_P = 10^7$.

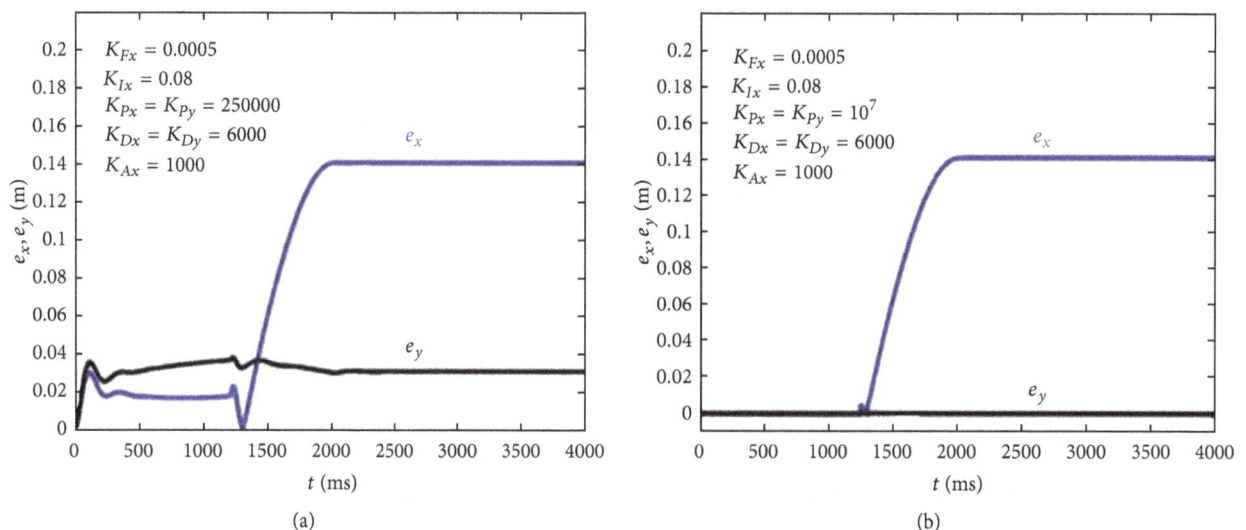

FIGURE 23: End-Effector error position value (m) varying the K_P gain in the PI system: (a) $K_P = 250000$; (b) $K_P = 10^7$.

graphs indicates that the P system is more influenced by the environment (varying K_A) with respect to the PI system. By means of the analysis of Figures 26(b), 26(c), 27(b), and 27(c), it was noted that the increase or decrease of the K_A value influences only the trends of the force and not the position of the End-Effector of the manipulator.

5. Conclusions and Future Developments

In this paper, the authors propose a comparison between two force-position control systems with gravity compensation, designed for an eight-DOF bioinspired robotic arm, named DEXTER. The two position controllers are both proportional-derivative (PD); the two force controllers are one proportional (P system) and one proportional-integrative (PI system).

The simulation tests performed with the two systems on a planar representation of the DEXTER (3-DOF arm in the plane) show that by varying the stiffness of the environment, with a correct setting of parameters, both systems ensure the achievement of the desired force regime and with great precision the desired position. The two controllers do not have large differences in performance when interacting with a lower stiffness environment. In case of an environment with greater rigidity, the PI system is more stable.

The next step of this work will be the implementation of these systems into the DEXTER robotics arm with the ultimate aim of control and design of the two arms of the humanoid robot SABIAN.

In this paper, the authors explained the differences between the proposed method and other studies. The proposed approach is oriented to execute a software modification of the compliance in humanoid robotics in opposition to

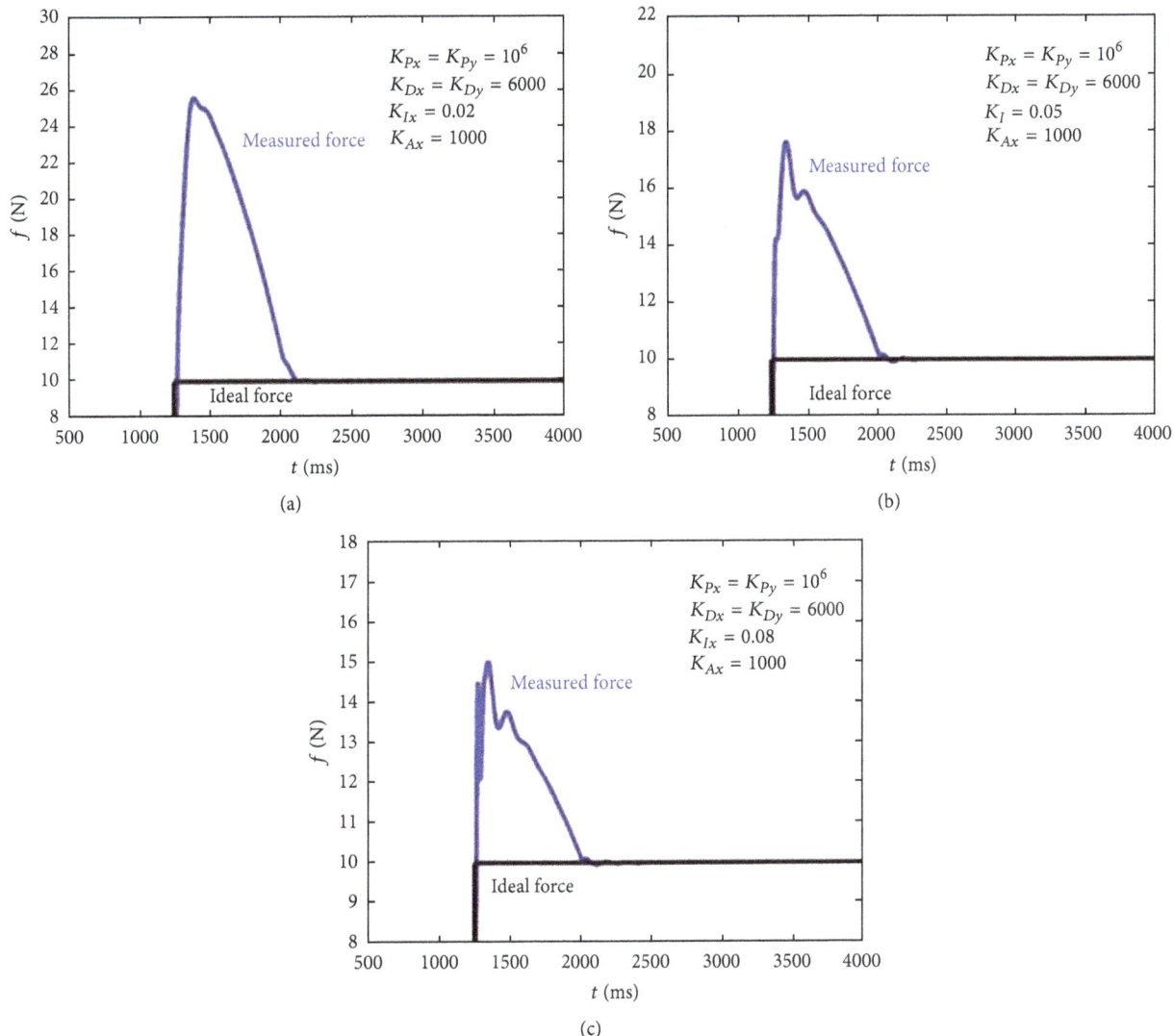

FIGURE 24: End-Effector force value (N) varying the K_I gain in the P system: (a) $K_I = 0.02$; (b) $K_I = 0.05$; (c) $K_I = 0.08$.

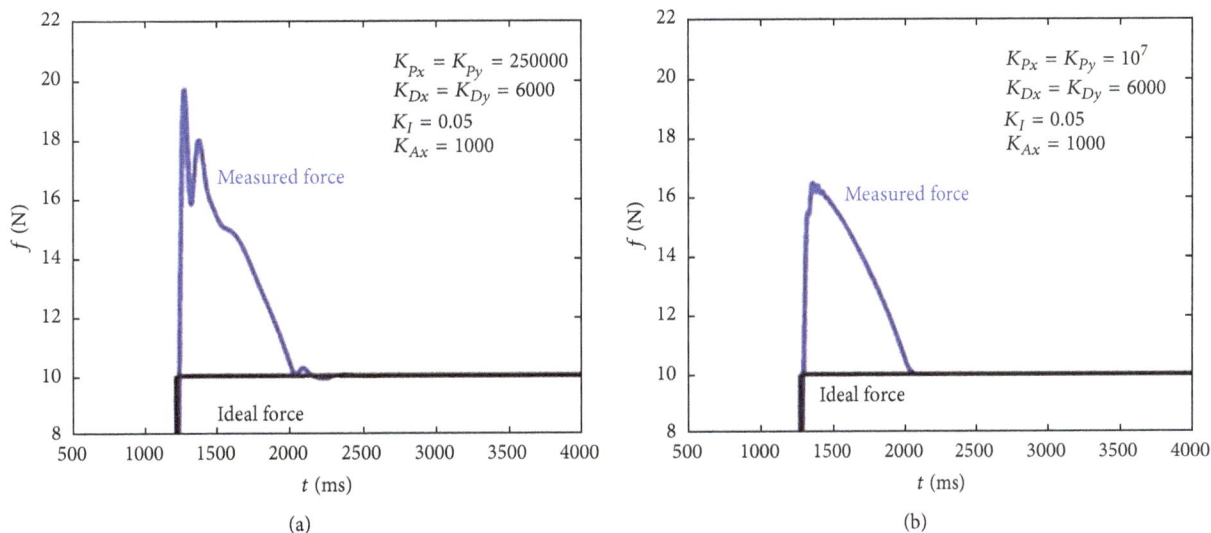

FIGURE 25: End-Effector force value (N) varying the K_P gain in the P system: (a) $K_P = 250000$; (b) $K_P = 10^7$.

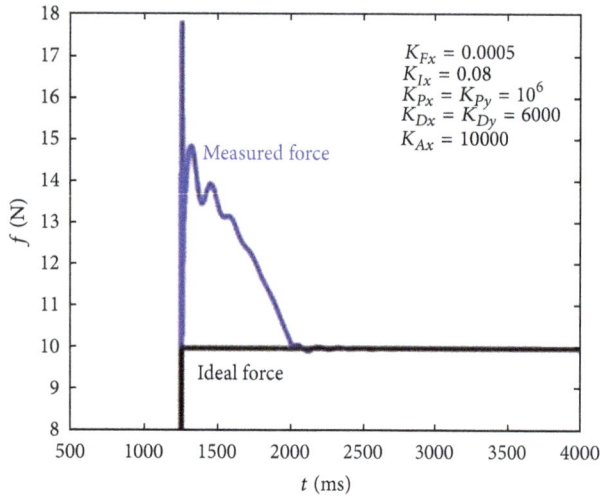

(a)

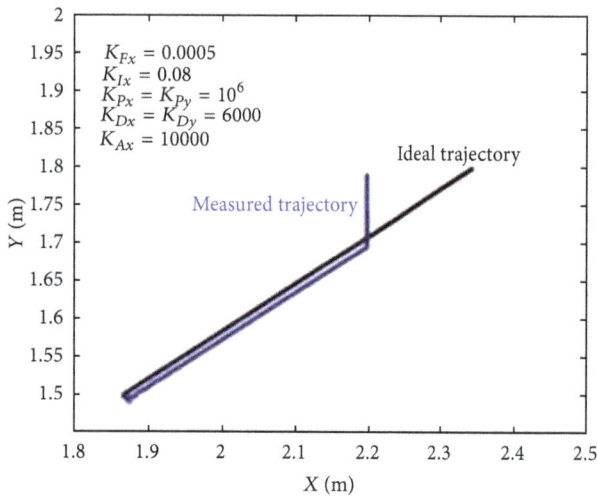

(b)

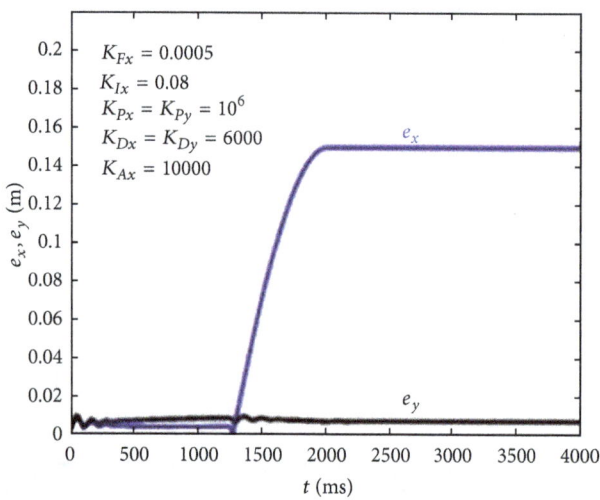

(c)

FIGURE 26: Force (N), position (m), and error position (m) of PI system assuming that $K_A = 10000$.

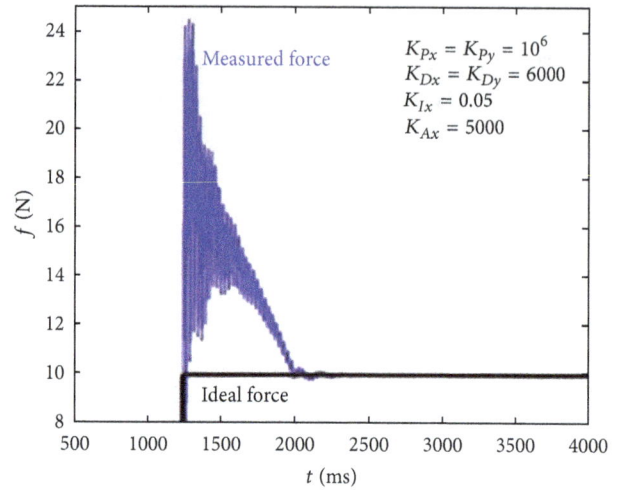

(a)

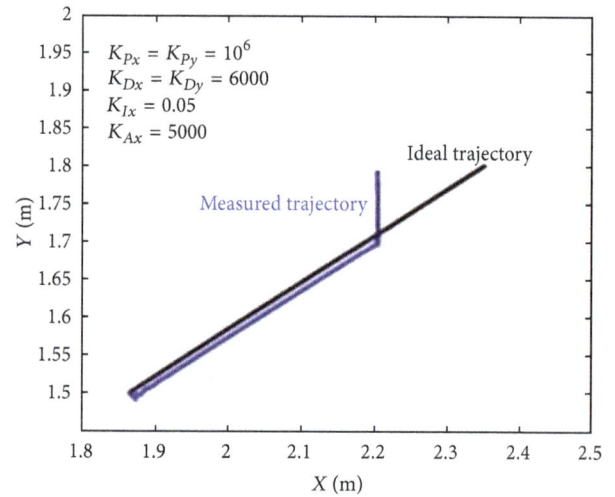

(b)

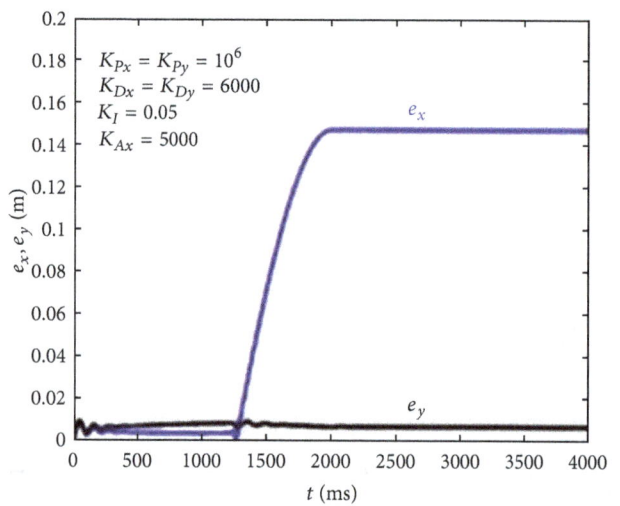

(c)

FIGURE 27: Force (N), position (m), and error position (m) of P system assuming that $K_A = 5000$.

the hardware modification of the compliance developed by other studies. On the other hand, a comparison between two force-position controllers could be used by the science and technology community.

Acknowledgments

This research was conducted at the RobotAn Lab, in the BioRobotics Institute of the Scuola Superiore Sant'Anna, and was partly supported by the Italian Ministry of Foreign Affairs, supporting the RobotAn and the RoboCasa joint labs of the Scuola Superiore Sant'Anna and Waseda University, as well as by the European Commission in the ICT STREP RoboSoM Project (Contract no. 248366).

References

[1] L. Sciavicco and B. Siciliano, *Modelling and Control of Robot Manipulators*, Springer, London, UK, 2nd edition, 2000.

[2] L. Zollo, B. Siciliano, E. Guglielmelli, and P. Dario, "A bio-inspired approach for regulating visco-elastic properties of a robot arm," in *Proceedings of the IEEE International Conference on Robotics and Automation*, pp. 3576–3581, Taipei, Taiwan, September 2003.

[3] G. Tonietti, R. Schiavi, and A. Bicchi, "Design and control of a variable stiffness actuator for safe and fast physical human/robot interaction," in *Proceedings of the IEEE International Conference on Robotics and Automation*, pp. 526–531, Barcelona, Spain, April 2005.

[4] N. G. Tsagarakis, M. Laffranchi, B. Vanderborght, and D. G. Caldwell, "A compact soft actuator unit for small scale human friendly robots," in *IEEE International Conference on Robotics and Automation. Kobe International Conference Center*, Kobe, Japan, May 2009.

[5] D. G. Caldwell, G. A. Medrano-Cerda, and M. Goodwin, "Control of pneumatic muscle actuators," *IEEE Control Systems Magazine*, vol. 15, no. 1, pp. 40–48, 1995.

[6] L. Zollo, B. Siciliano, A. De Luca, and E. Guglielmelli, "PD control with online gravity compensation for robots with flexible links," in *Proceedings of the European Control Conference*, Kos, Greece, July 2007.

[7] G. G. Muscolo, C. T. Recchiuto, K. Hashimoto, C. Laschi, P. Dario, and A. Takanishi, "A method for the calculation of the effective Center of Mass of humanoid robots," in *Proceedings of the 11th IEEE-RAS International Conference on Humanoid Robots (Humanoids '11)*, pp. 371–376, Bled, Slovenia, October 2011.

[8] G. G. Muscolo, C. T. Recchiuto, K. Hashimoto, P. Dario, and A. Takanishi, "Towards an improvement of the SABIAN humanoid robot: from design to optimisation," *Journal of Mechanical Engineering and Automation, Scientific & Academic Publishing*, vol. 2, no. 4, pp. 80–84, 2012.

[9] Y. Ogura, H. Aikawa, K. Shimomura et al., "Development of a humanoid robot capable of leaning on a walk-assist machine," in *Proceedings of the 1st IEEE/RAS-EMBS International Conference on Biomedical Robotics and Biomechatronics (BioRob '06)*, pp. 835–840, Pisa, Italy, February 2006.

4

3D Assembly Group Analysis for Cognitive Automation

Christian Brecher,[1] **Thomas Breitbach,**[1] **Simon Müller,**[1] **Marcel Ph. Mayer,**[2]
Barbara Odenthal,[2] **Christopher M. Schlick,**[2] **and Werner Herfs**[1]

[1] *Laboratory for Machine Tools and Production Engineering (WZL), RWTH Aachen University, 52074 Aachen, Germany*
[2] *Institute for Industrial Engineering and Ergonomics, RWTH Aachen University, 52062 Aachen, Germany*

Correspondence should be addressed to Thomas Breitbach, th.breitbach@wzl.rwth-aachen.de

Academic Editor: Ivo Bukovsky

A concept that allows the cognitive automation of robotic assembly processes is introduced. An assembly cell comprised of two robots was designed to verify the concept. For the purpose of validation a customer-defined part group consisting of Hubelino bricks is assembled. One of the key aspects for this process is the verification of the assembly group. Hence a software component was designed that utilizes the Microsoft Kinect to perceive both depth and color data in the assembly area. This information is used to determine the current state of the assembly group and is compared to a CAD model for validation purposes. In order to efficiently resolve erroneous situations, the results are interactively accessible to a human expert. The implications for an industrial application are demonstrated by transferring the developed concepts to an assembly scenario for switch-cabinet systems.

1. Introduction

One of the effects of globalization in public view is the reduction of production in high-wage countries especially due to job relocation abroad to low-wage countries, for example, towards Eastern Europe or Asia [1–3]. Based on this, a competition between manufacturing companies in high-wage and low-wage countries typically occurs within two dimensions: value-orientation and planning-orientation. Possible disadvantages of production in low-wage countries concerning process times, factor consumption and process mastering are compensated by low productive factor costs.

In contrast, companies in high-wage countries try to utilize the relatively expensive productivity factors by maximizing the output (economies of scale). Another way to compensate the arising unit cost disadvantages is customization or fast adaptation to market needs (economies of scope), even though the escape into sophisticated niche markets does not seem to be a promising way for the future anymore.

Within the dimension planning-orientation companies in high-wage countries try to optimize processes with sophisticated, investment-intensive planning approaches, and production systems while value-orientation offers the benefit of shop floor-oriented production with little planning effort. Since processes and production systems do not exceed the limits of an optimal operating range, additional competitive disadvantages for high-wage countries emerge.

In order to achieve a sustainable competitive advantage for manufacturing companies in high-wage countries with their highly skilled workers, it is therefore not promising to further increase the planning orientation of the manufacturing systems and simultaneously improve the economies of scale. The primary goal should be to wholly resolve the so-called polylemma of production, which is analyzed in detail by Klocke [4]. Economies of scale and economies of scope must be maximized at the same time, while additionally the share of added-value activities must be further maximized without neglecting the planning quality. Therefore, according to the "law of diminishing returns" a naive increase in automation will likely not lead to a significant increase in productivity but can also have adverse effects. According to Kinkel et al. [5], the amount of process errors is in average significantly reduced by automation, but the severity of potential consequences of a single error increases disproportionately. These "Ironies of Automation" [6] which were identified by Lisanne Bainbridge as early as 1987 can be considered as a vicious circle [7], where a function that was allocated to a human operator due to poor human reliability is automated. This automation results in higher function complexity, finally increasing the demands on the human operator for planning, teaching, and monitoring, and hence leading to a more error-prone system.

In order to break the cited vicious circle, one essential step is the application of cognitive control mechanisms by means of simulation of human cognition within the technical system. Such cognitive production cells can generally be understood as a further development of autonomous production cells. Admittedly, autonomous production cells only possess limited abilities in self-optimization and self-adaption to changing production tasks. These abilities are the fundamental approach of cognitive production cells and are currently one challenge in research and development [8]. Based on these functions, the concept of cognitive automation was introduced by Onken and Schulte in 2010 [7]. However, their original concept was strongly influenced by the research field of unmanned vehicles. The corresponding concept of the "cognitive plant" by Zaeh et al. [9] transfers the cognitive approach onto production systems. This concept successfully integrates cognitive mechanism into manufacturing systems, but the superior subject of using cognitive modules including the "human factor" as an operator and surveillant, however, still remains unexplored. Based on artificial cognition, technical systems shall not only be able to (semi)autonomously perform process planning, adapt to changing manufacturing environments or objectives, and be able to learn from experience, but also to simulate goal-directed human behavior and therefore significantly increase the conformity with operator expectations. Within this focus, a highly debated issue is the software architecture of a cognitive system. For this purpose, various architectures were proposed as a basic framework for the simulation of cognitive functions [10, 11]. Herein, a popular approach is the three-layer model with a cognitive, an associative, and a reactive layer of regulation [12, 13]. Comparative structures can be found within the Collaborative Research Centre 614 "Self-optimizing concepts and structures in mechanical engineering" (CRC 614) [14] as well as within the "cognitive controller" at the Technical University of Munich [15, 16]. Further broad researches within the field of cognitive technical systems can be found in Onken and Schulte [7] and, with special focus on the production environment, in Ding et al. [17]. Herein, the implementation of cognitive abilities within security systems for plant control focused. In this context, especially the safety of human-machine interaction and safety at work are taken into account. Additional concepts and methods can be found within the automotive sector as well as within space and aeronautic research [18, 19].

Within the Cluster of Excellence "Integrative Production Technology for High-Wage Countries" at RWTH Aachen University, a Cognitive Control Unit (CCU) and its ergonomic user-centered human machine interface are developed for a robotized production unit [20–23] which partially transfers non-value-adding planning and implementation tasks of the skilled worker to the technical cognitive system. The validation and interaction with the human expert

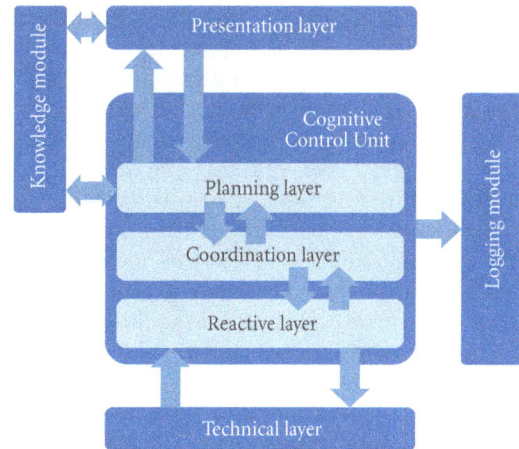

FIGURE 1: Concept of the cognitive architecture.

requires a technical recognition system which is able to measure the current state of the environment and provide feedback upon that information. This system needs to be integrated into the software architecture of the CCU.

2. Cognitive Control Unit and Evaluation Scenario

Due to the fact that the CCU is a system, which needs to be comprised of several different software components, the design required a modular structure. It allows for an integration and enhancement through distributed software modules.

2.1. Cognitive Architecture. The software architecture of the CCU is separated into five layers and shown in Figure 1. The presentation layer incorporates the human machine interface and an interface for editing the knowledge base. The planning layer is the deliberative layer in which the actual decision for the next action in the assembly process is made. The services that the coordination layer provides can be invoked by the planning layer to start action execution. The reactive layer is responsible for a low response time reaction of the whole system, for example, in order to efficiently respond to emergency situations. The knowledge module contains the necessary domain knowledge of the system in terms of production rules.

2.2. Human Operator. In regard to the role that humans play in standard automated production, the main task involves managing and monitoring the manufacturing system. In the advent of malfunction, they must be able to take over manual control and return the system to a safe, productive state. This concept, termed "supervisory control" by Sheridan [24], involves five typical, separate subtasks that exist in a cascading relationship to one another: plan, teach, monitor, intervene, and learn.

After receiving an (assembly) order, the human operator's first task usually involves planning the assembly process.

To do so, he or she must first understand the functions in the relevant machine and the physical actions involved to be able to construct a mental model of the process. Using this basic understanding, the operator then develops a concrete plan that contains all specific subtargets and tasks necessary. "Teaching" involves translating these targets and tasks into machine-readable format—for example, NC or RC programs—which allow for a (partially) automated process. The resulting automation must be monitored to ensure that it runs properly and generates products of the desired quality. The expectations for the process are drawn from the mental model the operator created at the start. In cases where reality significantly deviates from this model or where there are anomalies, the human operator can intervene for example, by modifying the NC or RC program or by manually optimizing the process parameters. Ultimately, every intervention involves the human operator continually adapting his/her mental model, while existing process information, characteristic values, and trend analyses help the operator to better understand the process and develop a more detailed mental model.

With a cognitively automated system, the tasks change gradually, but in a conceptually relevant way. In this system, the human operator defines the assembly tasks based on the status of the subproduct or end product, carries out adaptations or sets priorities as needed, compiles rough process plans, and sets initial and boundary conditions. The information-related stress on the human operator is considerably reduced in the areas of detailed planning and teaching, since they are handled by the cognitive system. But shifting this load from the human to the machine can result in the human operator forming an insufficient mental model of the state variables and state transition functions in the assembly process. In order to ensure the conformity with the operator's expectations during the supervision of the assembly process [25], the first step is the use of motion descriptors to plan and execute the assembly process, since motions are familiar to the human operator from manually performed assembly tasks [26]. Therefore, the Methods-Time Measurement (MTM) system as a library of fundamental movements was chosen [25, 27]. Even though the sequence of fundamental movements (e.g., reach, grasp, move, position, release) is explainable a posteriori, the sequence of parts positioned after another is not predictable a priori due to a lack of elaboration knowledge [22]. Odenthal et al. [28] and Mayer et al. [29] identified human assembly strategies that were formulated as production rules. When the reasoning component is enriched with these human heuristics, a significant increase of the robot's predictability when assembling the products can be achieved.

Further, if an error occurs which the system cannot identify or solve, the human operator must receive all information relevant to the situation in an easily understandable form so that he/she can intervene correctly and enable system recovery.

2.3. Assembly Scenario. To test and develop a CCU in a near-reality production environment in a variety of different assembly operations, a robotic assembly cell was set up [21].

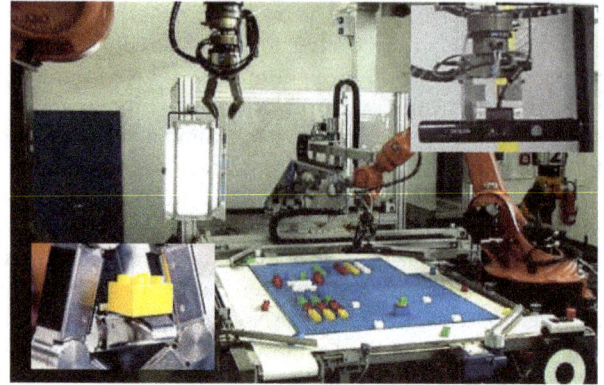

FIGURE 2: Assembly cell.

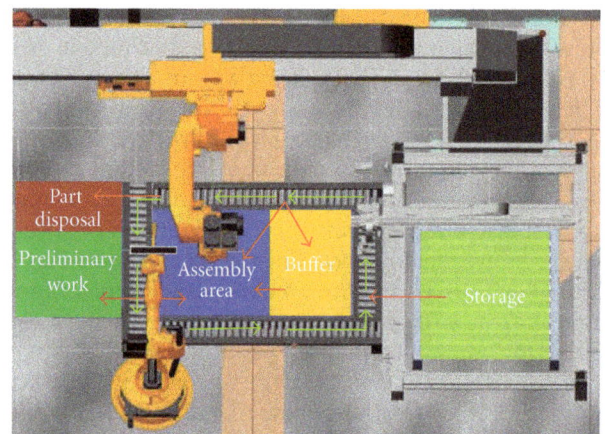

FIGURE 3: Assembly and storage areas in the assembly cell.

The layout of this cell is shown in Figure 2. The scenario was selected to address major aspects of an industrial application ("relevance") and at the same time to easily illustrate the potential of a cognitive control system ("transparency") [30].

2.3.1. General Setup of the Assembly Cell. The main function of the demonstrator cell is the assembly of predefined objects. Part of the cell is made up of a circulating conveyor system comprising six individually controllable linear belt sections. Several photoelectric sensors are arranged along the conveyor route for detection of components. Furthermore, two switches allow components to be diverted onto and from the conveyor route. Two robots are provided, with one robot travelling on a linear axis and carrying a tool (a flexible multifinger gripper) and a color camera. Several areas were provided alongside the conveyor for demand-driven storage of components and as a defined location for the assembly (see Figure 3). One area is provided for possible preliminary work by a human operator. This is currently separated from the working area by an optical safety barrier. The workstation has a multimodal human-machine interface that displays process information ergonomically, allowing it to provide information on the system state as well as help for solving problems, if necessary. To simultaneously achieve a high level of transparency, variability, and scalability in an (approximate) abstraction of an industrial assembly process, building an assembly of Hubelino bricks was selected as the assembly task. These are in size and shape very similar to LEGO Duplo bricks. To take into account the criterion of flexibility for changing boundary conditions, the bricks are delivered at random. In terms of automation components, the system consists of two robot controllers, a motion controller, and a higher-ranking sequencer.

The initial state provides for a random delivery of required and nonrequired components on a pallet. A FESTO handling system successively places the components onto the conveyor. The automatic-control task now consists of coordinating and executing the material flow, using all the technical components in a way such that only the assembled product is in the assembly area at the end.

2.3.2. Actions and Sequences. The assembly scenario is as follows. An engineer has designed a mechanical assembly of medium complexity by composing it, for example, with a CAD system containing any number of subcomponents. The human operator assigns the desired assembly goal to the cognitive system via the presentation layer (see Figure 1). The desired goal is transferred to the planning layer where the SOAR-based reasoning component derives the next action based on the actual environmental state (current state on the conveyor, the assembly area, and the buffer) and the desired goal. The environmental state is based on the measured vector from the sensors in the technical application system (TAS). In the coordination layer the raw sensor data is aggregated to an environmental state. The next best action derived in the planning layer is sent back to the coordination layer, where the abstract description of that action is translated into a sequence of actor commands which are sent to the TAS. There, the sequence of commands is executed and the changed environmental state is measured again by the sensors.

2.3.3. Motivation for Assembly Group Analysis. The last step, an image-based recognition process of the assembly object's state in the assembly area, is focused on this contribution and is described in detail later on. If the current state differs from the target state, the human operator is informed so that he/she can detect and correct occurred errors.

Generally four types of errors are possible, when positioning a brick in the assembly group:

(1) It might occur during assembly that a brick is not placed in the assembly group at all. This is the case, for example, when the gripper loses the brick during the transportation from conveyor to assembly area.

(2) A generated assembly sequence might not be correct and a brick is placed at a false location. In practice, this error has never occurred, but its existence needs to be considered.

(3) The brick is placed at the correct position, but not fitted properly. This error case refers to possible tolerances for both the brick and the position of the

robot. For example, in situations where a brick has to be positioned between two other bricks, the accuracy of the robot's position might not be sufficient. For the most part this leads to a part lying on top of the assembly board rather than being assembled onto that board.

(4) At last it is also possible that a brick has correctly been placed and fitted at the right location—but it was the wrong brick, for example, a blue brick instead of a green one. These errors are mostly related to the image processing system, which detects the moving part on the conveyor, failing at high conveyor speeds.

It can be stated that in practice most of the errors that occur during the course of the assembly are related to the failure of some sort of component. In order to efficiently interact with a human expert, the type of error needs to be identified. Since error type 1 should be reflected in the data as a whole set of missing data points and for error type 3 only as a small shift at a specific location, it becomes apparent that the reliability of this classification varies for the different error types.

The human operator must be supported with more information in case of an assembly error during the operation, for example, if the image-based control of the assembly step leads to a deviation between the current and the target state. Under this assumption, a first prototype of a supporting system was developed dealing with the task of error identification in an assembly object (incorrect construction of the assembly object). More precisely in this case, a prototype of an augmented vision system (AVS) was developed and implemented with the focus on the presentation of the assembly information. The aim of using this system was to place the human operator or skilled worker, respectively, in a position to detect the construction errors in a fast and adequate way. Therefore, a laboratory test was carried out in order to investigate different display types and different modes of visualizable assembly information from an ergonomic point of view [23, 31, 32]. Within a second step, the AVS was extended to assist the human operator in the disassembly of the erroneous object in order to correct the detected error in cooperation with the robot [33].

While a detailed overview of the cognitive control can be found in Kempf [30], there was still the need for an automated verification of the assembly group. In the following, the technical recognition of an assembly group within the assembly area of the cognitively automated assembly cell is described in detail.

3. Recognition Process

In this scenario, individual assembly groups are established which consist of an arbitrary combination of different basic elements. For demonstration, the actual assembly groups consist of Hubelino parts which differ in color (yellow, orange, red, light green, dark green, and blue) and length. The length of each part varies from 32 mm to 192 mm in steps of 32 mm. Each element has a width of 32 mm and a height of 16 mm.

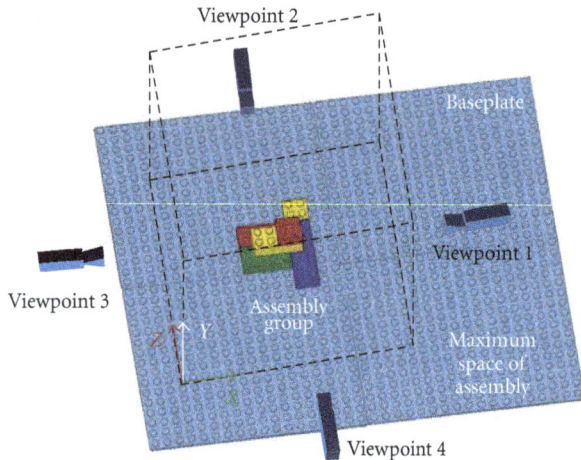

FIGURE 4: General design of the recognition process.

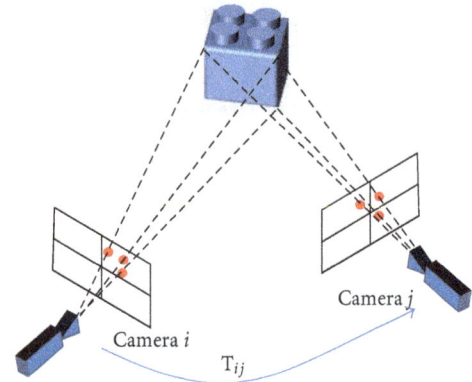

(a) Triangulating color images

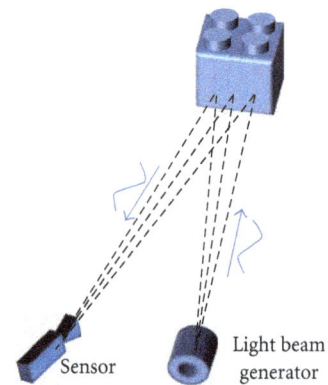

(b) Time-of-flight method

FIGURE 5: Principle of both contactless 3D measurement techniques.

Since the Hubelino parts are plugged onto the baseplate, respectively, onto each other, their positions are defined by the round shapes on top of each Hubelino part (see Figure 4). Each Hubelino part is laminated with a glossy surface that has a high light-reflecting coefficient.

3.1. Technical Recognition Systems. Within the recognition process, the assembly has to be checked if it is constructed correctly. Therefore, the assembly group is analyzed from four different viewpoints using a contactless 3D measurement system. Figure 4 shows the general design of the scenario. The dashed cuboid defines the maximum space of assembly.

At each position the Hubelino brick, that is present in the real assembly group, potentially differs in color as well as in size from the corresponding part defined in the virtual model. Thus information about these two properties needs to be measured for all of the possible assembly positions.

3.1.1. Contactless Measurement Methods. In the past few years, contactless 3D measurement systems have become an important tool for quality control. The two most used contactless measurement methods [34] are (i) triangulating the image data with several color images, (ii) time-of-flight method.

Within the first method, a scene is inspected from several viewpoints and identical "landmarks" inside each color images are identified (e.g., edges or corners). The pixel position of those landmarks combined with the camera's intrinsic calibration (relation between a pixel position and the directions relative to the camera center) yields in the vectors pointing from the camera center to the direction of each landmark. Determining this vector in at least two camera images and considering the relative position between these cameras, the special position of this landmark can be triangulated (compare Figure 5(a)). For this measurement method, at least two RGB cameras are necessary as well as an accurate spatial transformation T_{ij} between the cameras. The disadvantages of this technology are the huge computing

power necessary to calculate several high-resolution color images as well as finding concurrent and unique landmarks in each image. Consequently, this measurement method fails in image regions with a structured surface. Details concerning this technology are described in [34].

The second widespread contactless measurement method is the time-of-flight technique (see Figure 5(b)). Here, the time that a frequency-modulated beam of light needs to reach an object and to get back to the sensor is measured indirectly by comparing the phase of the emitted light with the phase of the received signal. For that reason, this method requires a device emitting the light beam as well as a light sensitive sensor detecting the reflected light beam. Accordingly, this method depends on the light-reflecting coefficient of the measured surface. Hence, objects with a low reflectivity coefficient (i.e., windows) are not detected as well as objects with a very high reflectivity coefficient (i.e., bright surfaces).

Widely used types of application for this technology are 2D laserscanners. Within 2D-laserscanners, a single light beam is diverted by a mirror. Concerning the angular position of the mirror and the time of flight of the single light beam, the depth values of a line can be measured. By panning the 2D-laserscanner, a complete 3D depth image of the environment can be computed. Recently, time of flight sensors are available measuring the depth values not only of

TABLE 1: Sensor requirements.

Requested attribute for assembly group analysis	Available for the 3D time-of-flight method	Available for triangulating color images
Depth image and color image available	Only depth information available	Both supported
High accuracy of the depth image	About 1 cm	Depends on resolution and calibration accuracy; recently ca. 1 mm
Depth image independent of the surface	Depends on light-reflecting coefficient	Fails in region with structured surfaces
Mapping between color and depth image	Not available	Available for each camera
High resolution of the depth image	Up to 320×240 Pixels	Equals the number of identified landmarks
Dependency on ambient light	Independent	Fails if scene is under/overexposed

a single point or line, but also of a complete matrix. Hereby, a 3D depth image is measured directly without requiring mechanical parts for panning the sensor [35, 36].

Concerning the recognition process, both described measurement methods have disadvantages. For the recognition both the color and the exact spatial dimensions of the assembly are considered. Therefore a color image as well as an accurate 3D depth image and the mapping between the two are required. In Table 1, the requirements for analyzing an assembly are faced with the characteristics of both contactless measurement methods.

According to Table 1 the main disadvantages of the 3D time-of-flight method are that it does not meet the requirements in terms of accuracy for depth data and that it does not provide a color image. However, compared to the triangulation method, it provides a higher measurement frequency and is independent of ambient light.

Still, both measurement methods are not completely independent of the surface of the measured objects. On the one hand, the triangulation method cannot detect the depth values of structured surfaces, while on the other hand the time-of-flight method relies on the reflectivity of the surface.

However, the actual measurement task requires the detection of the color and the exact spatial dimensions of small devices with a completely structured and glossy surface. Hence a combination of both measurement methods is required.

3.1.2. The Kinect as a Recognition Device.

Within the field of 3D measuring, the Kinect sensor, developed by Microsoft and PrimeSense, was presented to the public. Initially developed for game consoles, this device combines a 3D depth sensor with a resolution of 640×480 pixels and a standard color camera with a resolution up to 1280×1024 pixels. Additionally, a microphone array, a position sensor which measures the vector of gravitation, and an electrical motor for tilting the unit are integrated. Table 2 gives a short overview of the relevant specifications [37]. Combining the measured data of those devices, the Kinect sensor provides multiple innovative opportunities for research and development within the field of environment recognition as well as in the field of man-machine interaction [38].

The Kinect's 3D depth sensor combines aspects of both contactless 3D measurement methods described above, whereby the scene is continuously illuminated with infrared structured light [39]. Structured light measures a 3D scene

TABLE 2: Specifications of the Kinect.

Device	Specifications
Color sensor	32 bit RGB 1280×1024 at 15 fps
Depth sensor	16 bit Depth 640×480 at 30 fps
Depth sensor range	650 mm–3500 mm
Spatial Resolution (at 2 m distance)	x/y resolution: 2 mm, z resolution: 1 cm
Field of view	Horizontal 57 degree, vertical 43 degree
Tilt range	27 degree
Microphones	2 microphones, each 16 bit audio at 16 kHz

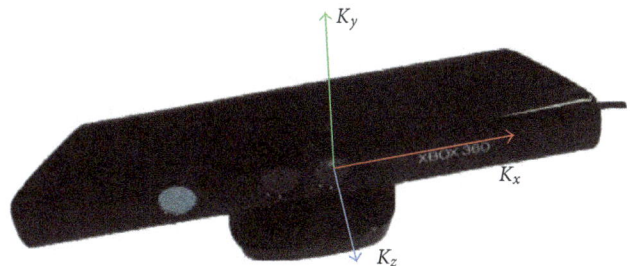

FIGURE 6: Coordinate system defined by OpenNI.

by projecting a known pattern of light onto the environment and recording it with a standard camera. The way this pattern deforms when striking surfaces allows calculating the depth information of the objects in the environment. Song [40] provides a detailed description of the computations. The Kinect projects a matrix of single IR dots onto the scene and provides a depth image with an accuracy of 1 cm at 2 m distance.

In order to accomplish the assembly group analysis, only depth and color information of the Kinect are merged. Communication and data exchange with the Kinect sensor are realized with the official driver from PrimeSense and its modifications for OpenNI. OpenNI uses a right-hand coordinate system for processing the depth data (see Figure 6).

As the Kinect depth sensor illuminates the room with an "IR Light Coding" and evaluates the reflected image,

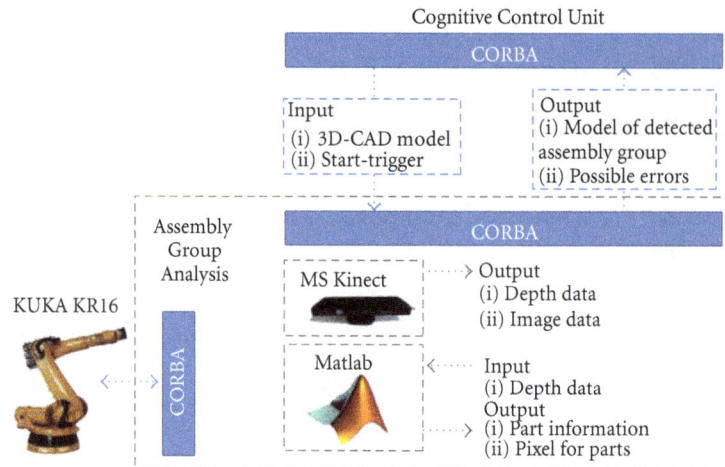

FIGURE 7: Software structure of the AGA.

the efficiency depends on the reflectivity coefficient of the measured surface. Additionally, the correct operation of the Kinect depth sensor relies on the incidence angle between the emitted light beam and the measured object. If the incidence angle is undersized, the light beam cannot be reflected back to the sensor with a sufficient intensity to be detected. The other way round, an incidence angle, which is nearly 90 degrees on an object with a high reflectivity coefficient, reflects the light beam with a very high intensity. This combination results in a cross-talk between adjacent pixels, whereby a blind spot within the Kinect's depth data occurs. As the blind spot occurs only under the described conditions, these need to be taken into consideration when the Kinect is used for analyzing an object. However, since the Kinect allows for a seamless integration of both depth and color data, it presents an appropriate device for the analysis of an assembly group.

3.2. Assembly Group Analysis (AGA). In order to create a perceived model, the assembly group is observed from multiple perspectives. For the generation of this model two approaches will be presented. In the first approach only general assumptions regarding the structure of a Hubelino model will be used. This approach is able to create a perceived model for any Hubelino model without previous knowledge about the bricks that should have been assembled, but is only valid in Hubelino scenarios where the general assumptions can be applied. A second approach takes advantage of the given customer-defined CAD data to derive recognizable patterns. On the one hand this approach is not able to construct a perceived model, if no CAD data is present. On the other hand it eliminates the general assumptions and thus is applicable not only in Hubelino scenarios, but also in any scenario, provided the condition that CAD data exists. Both approaches include the perception of both depth and color data and perform error detection. The superior goal of verifying a real assembly group of Hubelino bricks is achieved by comparing the perceived model with the virtual model from the customer-defined CAD file. The AGA is implemented as

a separate software component within the framework of the Cognitive Control Unit.

3.2.1. Software Structure of the AGA. In order to communicate with multiple components without having to implement a new interface for every new component, an abstract interface was used, which is able to transparently connect multiple devices. This abstract interface uses the CORBA Middleware and thus provides a real-time connection between devices and the CCU. This leads to an architecture, where the commands that the CCU issues are translated into CORBA calls that are directed to a specific device, in this case the AGA component. Hence the AGA component needs to provide interfaces to the CCU, which allow for a comprehensive communication. Since Matlab is a common tool for data visualization and the computation of complex algorithms, it was used to perform the computationally intensive tasks. A complete overview of the resulting architecture and their connections is provided in Figure 7.

In order to be able to verify the correct composition of an assembly group, the AGA component needs to know what the correct model looks like. Thus the CCU needs to submit a CAD file of the assembly group's expected state to the analysis component. This communication takes place for every verification, since it is possible for a model to change over time. The CAD model is given as an LDRAW file, which is an appropriate file format for the assembly of Hubelino parts [41]. After the analysis is completed, the AGA provides feedback to the presentation layer of the cognitive control framework (see Figure 1). This feedback contains the detected parts within the assembly group as well as possible errors. This allows for the CCU to present the results in the presentation layer, hence enabling a human operator to take action based on the received results.

In order to calculate a 3D model with a single sensor, the sensor needs to change its position relative to the object to examine all sides. Consequently, the Kinect sensor is attached to the flange of the KUKA KR 16 which can explore the assembly from an arbitrary position. Once the CCU requests a new

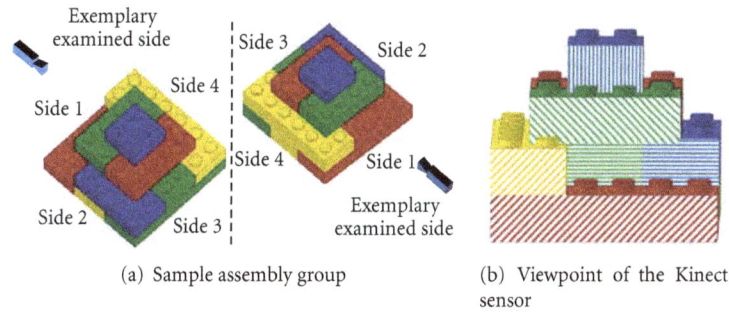

(a) Sample assembly group   (b) Viewpoint of the Kinect sensor

FIGURE 8: Assembly group.

analysis, the AGA component creates a connection to the Kinect as well as to the KUKA KR16 via CORBA. Within the analysis process, the assembly is examined from four different positions (compare Figure 4). Each viewpoint is aligned perpendicular to the particular side of the assembly with the intention of avoiding perspective masking of single Hubelino parts. For each viewpoint, the assembly group is analyzed and a discrete model for each side is calculated. Afterwards, these four discrete models are combined into one complete model of the assembly group.

3.2.2. Data Acquisition. The following description of the recognition process is based on the assembly group as shown in Figure 8. For demonstration purposes, side 1 is examined.

Figure 8(a) displays the exemplarily used assembly from two different points of view. Figure 8(b) represents the same assembly but presented from the viewpoint of the Kinect sensor. Concerning the Kinect's viewpoint, the Hubelino parts constitute planes parallel to the Kinect's x-y-layer. Those planes are marked by different striped patterns.

For the data acquisition, the positioning of the viewpoints is essential, thus the sensor characteristic needs to be taken into account. As described above, the Kinect sensor fails if the emitted light beam is reflected perpendicularly by an object with a high reflectivity coefficient. Within the given task, all objects to be recognized have a high reflectivity coefficient. Therefore, each viewpoint needs to fulfill the following conditions:

(1) The distance to the measured object has to be greater than 650 mm.

(2) The light beams should not be reflected perpendicularly.

(3) The particular side of the assembly group needs to be examined perpendicularly to avoid perspective masking of single Hubelino bricks.

Consequently, each viewpoint is at least 700 mm away from the assembly group and the assembly group itself is about 200 mm beneath the center of the depth image. Figure 4 shows their position and chronology. This choice of viewpoints has another important effect: in those viewpoints, the baseplate does not reflect the emitted light beams, according to the undersized angle of incidence. Hence the

baseplate will not be recognized by the Kinect, whereby the effort of interpreting the measured depth values is reduced.

After defining the viewpoints, the depth values need to be acquired. Thus, the maximum dimensions of the assembly group are limited by a virtual cube and only depth values within this cube are used for identifying the assembly (see Figure 4 dashed cuboid).

3.2.3. Prefiltering. Before starting to fit Hubelino parts, the depth data needs to be filtered. The first step is to identify the assembly group within the scatter plot. Within the defined cube, some parts of the baseplate or some object not belonging to the assembly, respectively, may be perceived by the Kinect sensor. Those objects disturb the recognition of the assembly group and need to be filtered.

Therefore, the data points are classified by their distance to each other. If the distance between two points is larger than a specified value, both points get a different classification. Otherwise they will be sorted into the same class. The distance for separating different scatter plots is 100 mm.

After passing this filter step, the data points are separated into their particular scatter plots. Based on the notion that the assembly group is the largest object within the defined cube, only the class with the most members is used for fitting the Hubelino parts. The result is shown in Figure 9. From the incoming measurement data (Figure 9(a)) only the red marked points are used for fitting the Hubelino parts (Figure 9(b)).

As described, the Hubelino parts can only be positioned in discrete places and the viewpoint of the sensor is perpendicular to each side of the assembly group. Hence the measured depth values are positioned on planes parallel to the sensors x-y-plane (compare Figures 8(b) and 9(b)). Since all sensors are disturbed by noise, the depth values are not measured exactly and the planes of the assembly group need to be reconstructed.

3.2.4. Identifying Planes. For this task the RANSAC algorithm is used and optimized for finding planes. The RANSAC algorithm is an iterative method to estimate parameters of a mathematical model from a set of sensor data [42, 43]. Within the scope of reconstructing planes out of disturbed depth data, the mathematical model of a plane is described by three 3D points.

(a) All depth data within the cuboid

(b) Filtered depth data

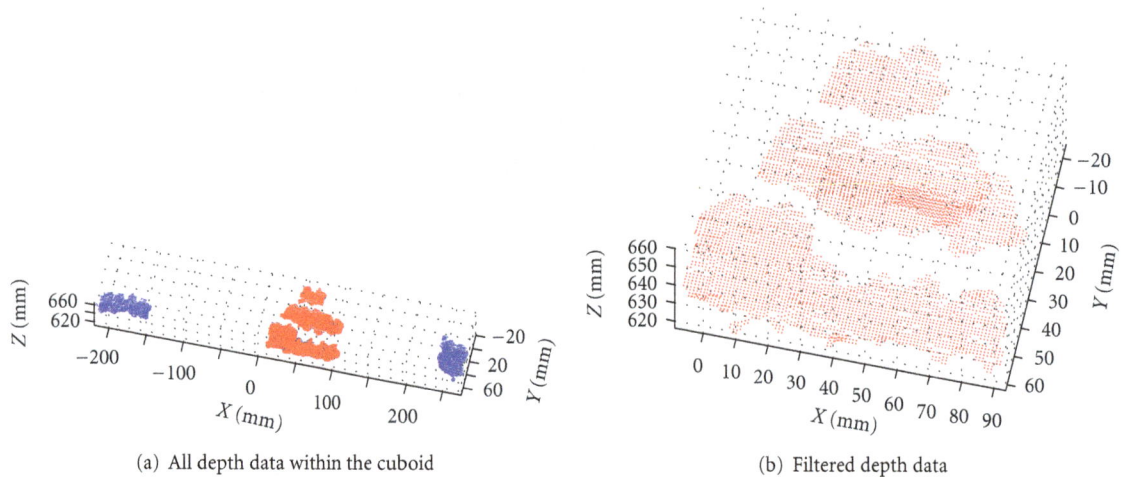

FIGURE 9: Prefiltering measured data.

The algorithm finds the plane as follows:

(1) Define a plane by randomly taking three points from the depth data.

(2) If this plane is parallel to the sensor's x-y-plane then continue, otherwise choose different points.

(3) Calculate the distance between each 3D point within the depth data and the randomly generated plane.

(4) Determine the score S of the plane as follows:

$$S = \sum_{i=0}^{i=n} p(\text{distance}) \quad \text{with } p = \begin{cases} 1 - \dfrac{\text{distance}}{\text{maximum distance}}, & \text{if distance} \leq \text{maximum distance}, \\ 0, & \text{if distance} > \text{maximum distance}. \end{cases} \qquad (1)$$

(5) Repeat steps 1 to 4 for a predefined number of iterations.

(6) The plane with the highest score is the best estimation of a plane within the depth data. If the highest plane score is beneath 80% of the theoretical minimum plane score, no plane is found.

The value assigned to the maximum distance results from the discrete positions of the Hubelino parts. As those positions are defined by the round shapes on top of each part, it is obvious that the minimum distance between two planes within the depth image is 16 mm (compare Figure 8). In order to allocate each depth value to a plane, the maximum distance between a measured point and the corresponding plane is 8 mm.

The decision, if a found plane is valid, depends on the minimum theoretical plane score. This parameter is equal to the theoretical number of measurements on the surface of the smallest possible plane (see Figure 10).

The theoretical number of depth data within a certain area yields from the sensor characteristics. As the Kinect depth sensor has a resolution of 640×480 pixels and the lens

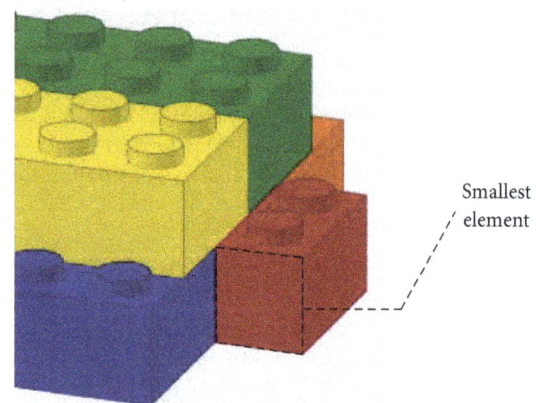

FIGURE 10: Definition of the smallest element.

has an opening angle of 57 degree horizontally, 43 degree vertically, respectively, the spatial resolution decreases with increasing distance. This interrelationship is given in Figure 11.

As Figure 11 shows, the theoretical minimum plane score depends on the distance of the plane and thus needs to be determined individually for each plane. Considering the fact

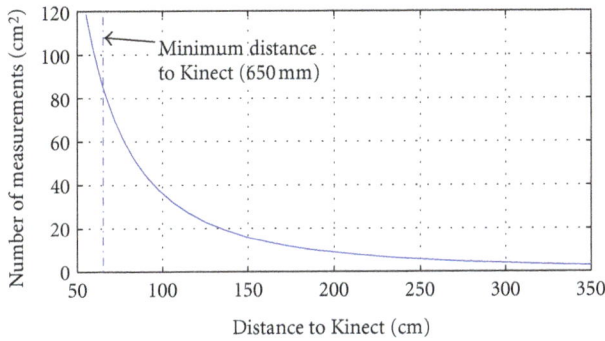

FIGURE 11: Interrelationship of spatial resolution and distance.

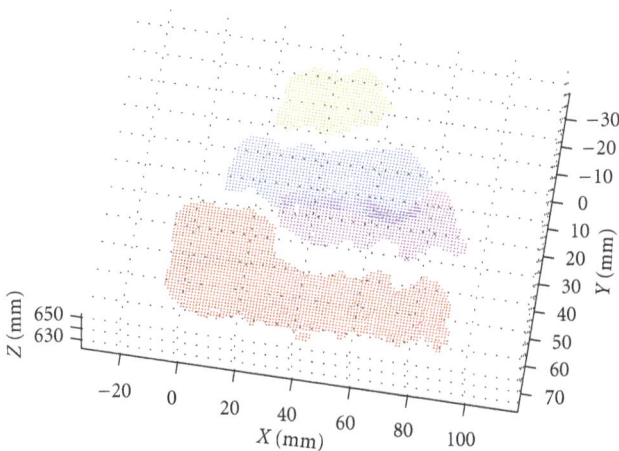

FIGURE 13: Fitting elements into plane number 4.

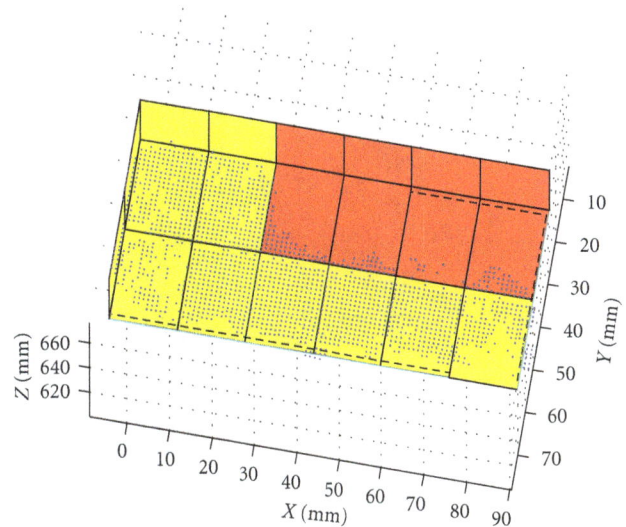

FIGURE 12: Found planes.

is approximated towards the nearest multiple of the smallest element's size (black dashed line). Afterwards, the resulting rectangle is filled with smaller rectangles that have the shape of the smallest elements (red and yellow cuboids). This procedure is repeated for each found plane.

Since not all of those fitted elements really exist, the next step is to decide whether an element belongs to the assembly group or not. This decision is based on the quotient of the actual number of measured 3D depth points on the element and the theoretical maximum number of data points on the element. This maximum is given by Figure 11. Since the edge line of the found planes is irregular, the quotient is generally not equal to one. Thus a threshold of 90% is defined. Concerning this threshold, the decision whether a fitted element belongs to the assembly group can be evaluated by the following steps:

(1) Determine the number of measured depth data within the element.

(2) Based on the distance of the element calculate how many data points should be on the element.

(3) If the result of step 1 is at least 90% of the result of step 2, then the element belongs to the assembly group.

This procedure is repeated for each element on each found plane and results in a model for the examined side. Within Figure 13, the red marked cuboids do not belong to the assembly group.

that the Kinect sensor is disturbed by noise, the found plane is valid, if its score is at least 80% of the theoretical score.

As the assembly group generally consists of more than one plane, the RANSAC algorithm has to be executed several times. Therefore the depth values belonging to the already found planes are deleted, and within the remaining depth data another plane is searched. This procedure is repeated until no further plane can be found. The result is shown in Figure 12.

3.2.5. Fitting of Virtual Hubelino Bricks. After finding the planes within the depth data, the single Hubelino parts can be fitted. Since the Hubelino parts can only be placed at discrete positions they can occlude each other. Therefore, the smallest visible element is half of one side of a 32 mm × 32 mm Hubelino brick (see Figure 10). Hence, only rectangles with a height of 32 mm and a width of 16 mm are positioned.

The fitting process itself is based on the assumption that size and shape of the found planes can only consist of multiples of the smallest elements (compare Figure 13). Hence for each plane the surrounding rectangle is calculated (light blue line). Regarding the sensor disturbance, the dimensions of this rectangle are not necessarily a multiple of the smallest element. Consequently, the size of the rectangle

3.2.6. Error Correction for a Single Side. After the Hubelino parts are positioned according to the established planes, the resulting model is checked for errors. This step proved to be necessary, after several initial evaluation tests failed due to frayed shapes of the identified planes (compare Figure 13).

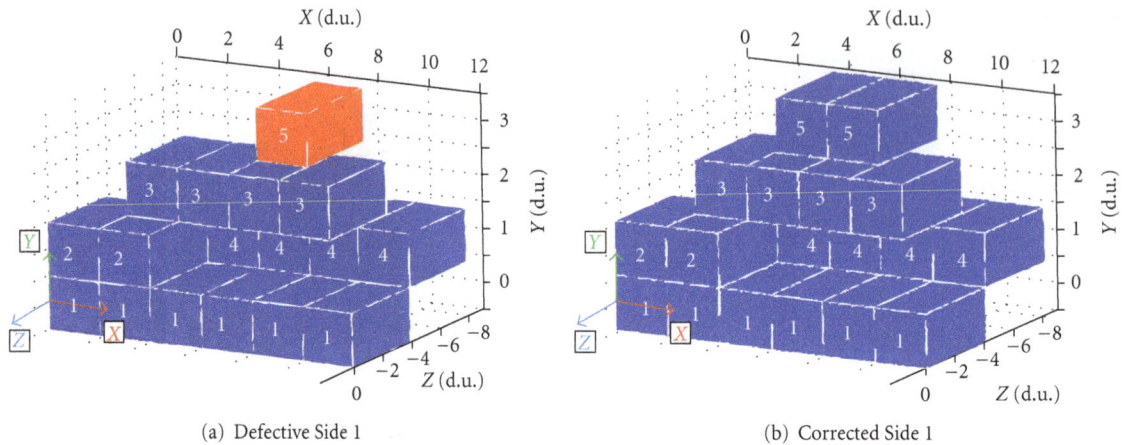

(a) Defective Side 1

(b) Corrected Side 1

FIGURE 14: Correction of side 1.

Within the correction process, the found model has to meet the following requirements:

(1) A single smallest element can only appear if the rest of the Hubelino part is masked (compare Figure 10).

(2) Each line of the model must consist of an even number of smallest elements.

(3) All found elements have to be placed at a discrete position.

In order to check these conditions, first the coordinate system is changed. Since the presented calculations contain the measured data, they are based on the coordinate system defined by OpenNI. This measured data is represented by the fitted elements which are placed in discrete positions. Hence, the unit of the new coordinate system is given in dimension units (d.u.). Thereby, the size of two dimension units equals the width of the smallest element. The new coordinate system is placed on the lower left element of the scanned side (see Figure 14). Within this coordinate system the values of the x- and y-axis are positive and those for the z-axis are negative. In Matlab the Hubelino model is contained in an array according to discrete positions of smallest element's multiples (see Figure 10).

The error correction process is illustrated in Figure 14. First, the found element are grouped (numbers on the elements in Figure 14) based on their position within the model. Therefore, all elements within the same coordinates in y and z are grouped together. For each group with a noneven number of members (red marked element) the fitting of the element inside the plane is rechecked. Within this check, the unused elements of a plane are analyzed. Of special interest are the elements at the left and right of the defective group, respectively. Those elements meet only one conditions listed in Table 3 and result in the corresponding error correction.

Each increase as well as decrease refers to the width of an element. For each correction process only the dimension of the surrounding rectangle changes. The position of its center remains the same. The advantage of this error correction process is that it can be calculated straightforward without iteratively searching the best solution.

TABLE 3: Possible errors and corresponding corrections.

Error	Correction
Both elements were not chosen	Increase the width of the surrounding rectangle
Both elements were chosen	Decrease the width of the surrounding rectangle
One element was chosen	Decrease the width of the surrounding rectangle

Figure 15 demonstrates this error correction process for the defectively fitted group presented in Figure 14(a). For this figure it is obvious that the error correction finds a better mapping between the surrounding rectangle (light blue line) and the measured data.

Since the calculated model is based on disturbed sensor data, it is generally not the exact model of the actual assembly group. Hence, the last step is to reconstruct the complete 3D model of the assembly group and to check if all four sides match.

3.2.7. Reconstruction of the Complete Model. In order to reconstruct the complete model of the assembly group out of the models for each side, first the transformations between adjacent sides have to be calculated. Considering the anticlockwise chronology of the scan positions (see Figure 4), the values of the right-hand rotation matrix to rotate side B into the coordinate system of side A are given as follows:

(i) Rotation around the x-axis: 0°.

(ii) Rotation around the y-axis: 90°.

(iii) Rotation around the z-axis: 0°.

The translation vector between both sides is calculated as follows:

(a) Starting from the assumption that model A is correct, calculate those parts within model A that are expected to be part of side B.

(b) Within side B, calculate those elements, which side A would expect to be part of side B.

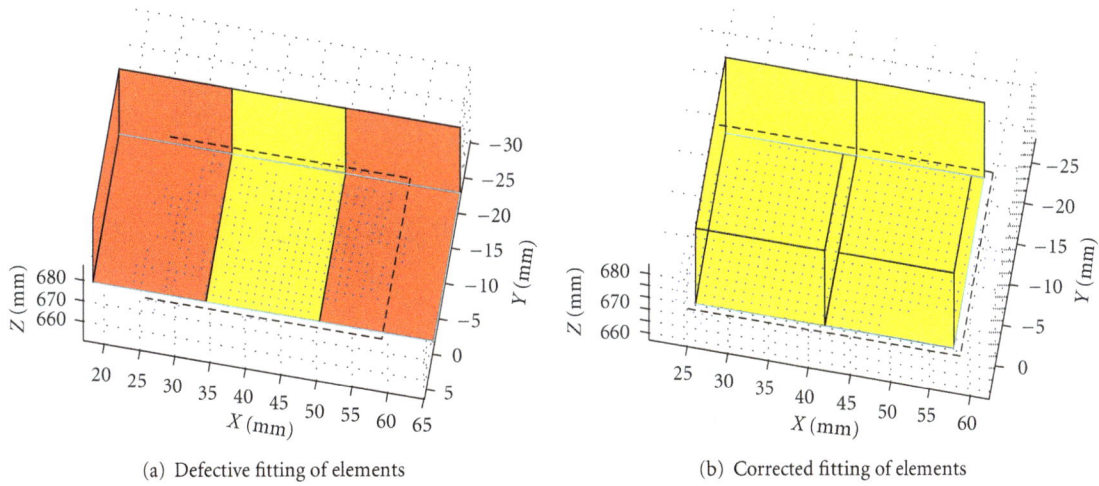

(a) Defective fitting of elements

(b) Corrected fitting of elements

FIGURE 15: Failure correction for plane 1 within side 1.

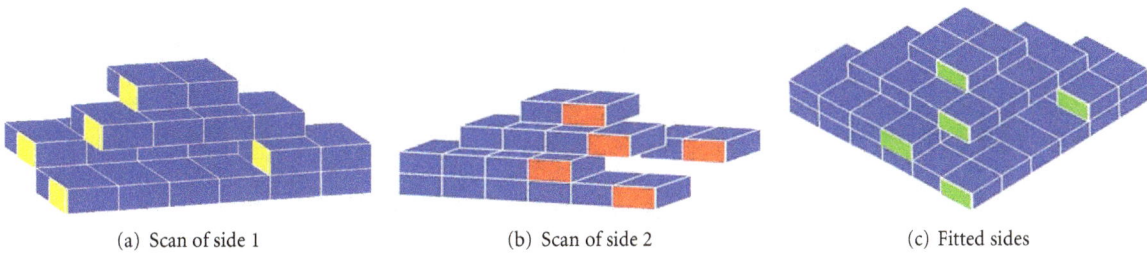

(a) Scan of side 1

(b) Scan of side 2

(c) Fitted sides

FIGURE 16: Fitting of side 1 and 2.

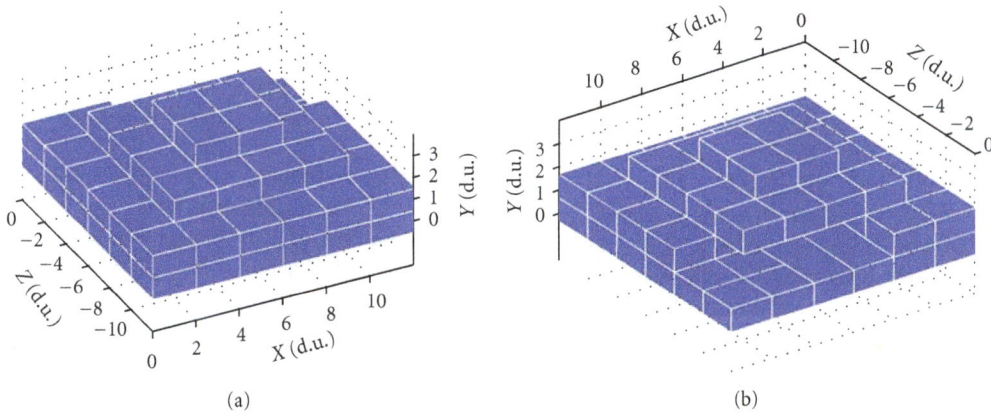

(a)

(b)

FIGURE 17: Reconstructed assembly group.

(c) Rotate side B into the coordinate system of side A.

(d) Find the translation vector mapping most elements of step 1 to those of step 2.

If not all elements of step (b) are mapped to those of step (a), at least one of the involved sides contains errors. Figure 16 shows this procedure for side 1 and 2 (compare Figure 8). The calculated elements of steps (a) and (b) are marked in yellow and red, respectively. Afterwards those elements are mapped onto each other by the found transformation (green elements).

Those transformations are calculated between all adjacent sides. If none of the sides contains errors, the 3D model of the complete assembly group can be calculated. The 3D model of the analyzed assembly group is shown in Figure 17.

3.2.8. Color Detection. In order to recreate an exact virtual model of the assembly group, information about the color of the detected parts needs to be acquired. In case of a scenario, that involves a color camera as the sole source of information, usually very complex algorithms have to be applied in order to efficiently detect regions of certain colors. In the given

TABLE 4: RGB values for the main colors used in the scenario.

	R	G	B
Red	196	40	27
Green	40	127	70
Blue	13	105	171
Yellow	245	205	47

case, however, this task is extremely simplified, since it is already known which pixels belong to a certain part. This information can be used to identify the color for a detected part. As an example, it can be assumed that an identified pixel region $x_m \cdots x_n$ contains j pixels for one side of a detected part. For the assumption that all of those pixels belong to the same part and should thus roughly contain the same color information, the basic approach of calculating the mean RGB value should suffice. Thus the resulting RGB value for the detected part can easily be calculated as:

$$cd.R = \frac{\sum_{k=m}^{n} x_k.R}{j} \qquad cd.G = \frac{\sum_{k=m}^{n} x_k.G}{j}$$

$$cd.B = \frac{\sum_{k=m}^{n} x_k.B}{j}. \tag{2}$$

Since the RGB values for most Hubelino parts are publicly available, see for example, [44], these can be used to identify the corresponding part color. The RGB values for the parts used in this example are presented in Table 4.

Given the fact that the different colors are saved in an array z, the resulting part color can be calculated:

$$\|(z_i.R - cd.R)\| + \|(z_i.G - cd.G)\|$$
$$+ \|(z_i.B - cd.B)\| = \min. \tag{3}$$

However, since only four different part colors are possible, the significant distance between the different RGB values partly contributes to the efficiency of this approach. Since the RGB values are determined by the manufactured parts and thus do not relate to the specific conditions (i.e., lighting, reflection), even better results can be achieved if these factors are taken into consideration. Hence, sample values for the different colors have to be recorded manually. Since lighting is a significant factor, the best results can be achieved by recording different color values for each of the four viewpoints.

In order to achieve even better results, the calculation of the RGB values for a detected part can be improved. In the previous approach, all of the points that were considered as belonging to a specific part were used for the determination of that part's color. Inasmuch as all of those color values are roughly the same, this approach is sufficient. However, the approach can be improved if the mean RGB value for a part is calculated using only the ten most central points for a specific brick. On the one hand this approach eliminates the problem that the part's color tends to change at the part's border. On the other hand this might lead to false detections in case that local reflection takes place near a part's center.

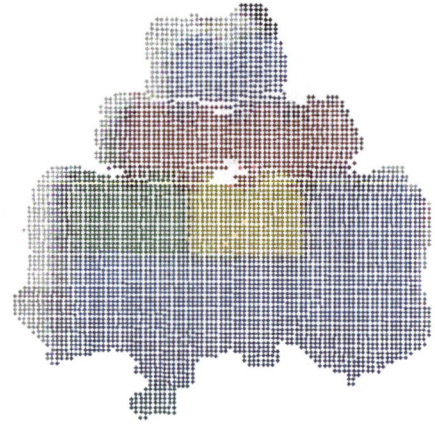

(a) Mapping without calibration

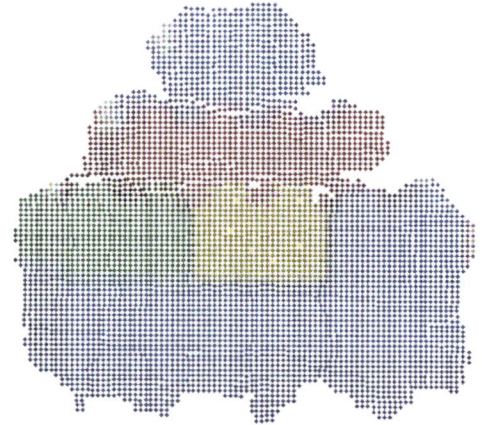

(b) Final mapping

FIGURE 18: Mapped sensor data.

Given the situation that a limited solution space of colors is present, the problem of reflection can be reduced by comparing each of a part's pixels with the possible colors. A color is assigned to a pixel if

$$\|(z_i.R - cd.R)\| + \|(z_i.G - cd.G)\| + \|(z_i.B - cd.B)\| = \min \ \cap$$
$$\|(z_i.R - cd.R)\| + \|(z_i.G - cd.G)\| + \|(z_i.B - cd.B)\| < \delta. \tag{4}$$

Thus it is possible to dynamically apply different δ for different scenarios and lighting conditions. Subsequently, the color for a part is computed by averaging over the pixels that were assigned a specific color. This computation reduces the problem of reflection, since colors that are out of range do not contribute to the resulting part color.

As shown in Figure 18(a) another problem, that needed to be resolved, is based on the fact that there is a partial displacement between the color information and the depth information. Since the color detection algorithms rely on the assumption that for each pixel in the depth information the image data corresponds to that exact same spot in the real world, this displacement leads to miscalculations of a part's color. As the Kinect is attached to the flange of a robot,

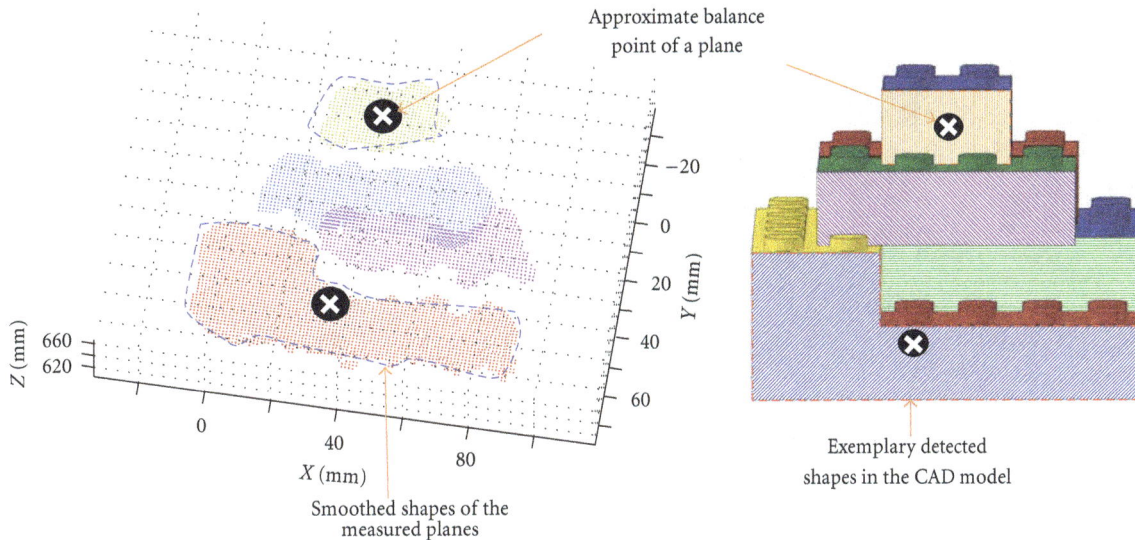

Figure 19: Model-based approach.

however, the information about the current position of the robot's tool center point (TCP) can be utilized for the exact determination of a viewpoint's displacement within the assembly area. The resulting image containing the mapping of both depth and color data is shown in Figure 18(b).

3.3. Generalized, Model-Based Approach. As a modification to the presented approach, the elimination of assumptions for the matching process offers to broaden the range of detected parts without additional engineering efforts. Due to the fact that the model needs to be verified, it is apparent that there is already a CAD model present for the desired assembly group. The superior goal is to use this model in order to match CAD and Kinect data without the need for additional assumptions regarding the presence of an object. This task is often achieved by photogrammetry methods [45, 46], but normally relies on different conditions regarding time and accuracy [47–49].

As a first step to realize this approach, the CAD model has to be analyzed in regard to different layers that can be perceived from different viewpoints. The result of this step is a "virtual model" of the assembly group that should be built. The model that is recorded with the Kinect forms the "perceived model". For this generalized approach it can be stated, that the point of view from which an assembly group is looked at in the virtual and the perceived perspective have to match to achieve the best results. But since it is always possible to correct the perspective for both the virtual and perceived model, there is no need for an exact calibration between these two perspectives. The only requirement is that the basic "sides," from which an assembly group is looked at, match. As previously stated, the four viewpoints correspond to the four sides of the rectangle that defines the maximum assembly space (see Figure 4). Thus for each of those sides the visible layers need to be extracted from the virtual model. The fact that the sides are either parallel to the x-y-layer or the y-z-layer of the modeling environment simplifies

this task, inasmuch as that for the different viewpoints the polygons with the minimum or maximum x or z value, respectively, form the foremost layer for each height, which is given in the y-coordinate. As an example for viewpoint 1 the maximum x values for a given height y define the visible planes. The polygons themselves are extracted from the underlying CAD model. All of the different polygons that are adjacent to each other comprise a region. In order to utilize these regions for further processing, their shape and balance point are determined.

Once the different layers have been identified in the virtual model, the same steps as in the approach mentioned above—from recording the sensor data to the RANSAC algorithm—are performed. Hence the results of these steps are the identified regions for the perceived data. For these regions the different shapes and balance points are extracted from those results. The shape herein is defined by the outmost data points for a specific region. Figure 19 shows this process. The process is illustrated exemplarily for two out of four planes within the assembly. In the left of Figure 19 the planes, found by the RANSAC algorithm within the measured data, are presented. For both planes their smoothed shapes and balance points are shown. The right side illustrates both planes within the underlying virtual model as well as their ideal shapes and ideal balance points.

As a last step, this information about shapes and balance points is used to match the regions of the virtual and the perceived model. This is realized by first comparing the number of balance points for the virtual and perceived model. If there is a difference between those numbers then it becomes apparent that there is an error. After that the Gauss algorithm (cp. [50]) is applied to match the balance points of the different models. The algorithm completes when either the distance between the balance points is small enough or the Gauss algorithm exceeds a certain number of iterations. Next the balance points of the two models can be mapped by taking the smallest difference between two points as the main

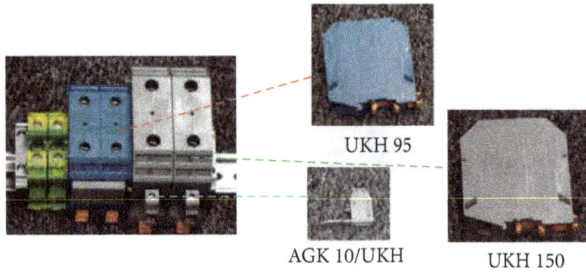

FIGURE 20: Final assembly.

FIGURE 21: Mapped depth and color data for the switch-cabinet scenario (one viewpoint).

criteria. If the numbers of balance points were different, only points that have a minimum distance are considered. In the following the regions that belong to a balance point in the perceived model are projected onto the region of the corresponding balance point in the virtual model. By calculating the overlapping areas it is possible to determine if the two models match or if it is more likely that there is a difference between them.

4. Industrial Application

In order to apply the achieved results in an industrial application, the concepts already tested with the demonstrator cell were simulated, in collaboration with automation technology manufacturer Phoenix Contact, on a scenario for the assembly of switch cabinets. This approach addresses the customer-driven switch cabinet assembly system, which Phoenix Contact operates in addition to the final assembly line in its mass-production-style manufacturing system. The assembly of switch cabinets consists of components (control units, terminals, etc.) in configurations predefined by the customer that are mounted onto top-hat sections and passed on as completed modules to switch-cabinet production (module assembly). Additionally some of these components need to be equipped with specific clip combinations. Figure 20 shows an example of this system.

A model of each customer-specific and mass-production-style switch cabinet is available in a CAD format. Currently this data is used to derive assembly instructions which are printed out and handed to a human worker. Finally the manually assembled switch cabinet is analyzed by an image recognition system and the results are compared to the mounting requirements.

Since the target assembly is available electronically, it is feasible to cognitively automate the process, hence achieving economic advantages. The main challenges involved are as follows:

(i) Construction of a continuous flow of information from the CAD system to the manufacturing system.

(ii) Robust system components and joining processes.

(iii) Logistical concepts and components for efficient stock placement during the assembly process.

(iv) Automated verification of the final assembly and man-machine interaction in case of errors.

The switch-cabinet scenario bears strong similarities to the demonstrator-cell scenario, that is, in regard to the customer-defined CAD models and requirements regarding an automated verification of the final assembly.

Hence, in order to verify the capabilities of the Assembly Group Analysis, the perception and verification process using the Kinect was simulated for a given scenario. Mapped depth and color data for the final assembly indicating the different parts that it consists of is given in Figure 21.

For the verification process, three different cases are taken into consideration. In the first case, the assembly was built correctly. In the second case, one of the larger feed-through modular terminal blocks UKH 95 has not been placed onto the mounting rail. In the third case only one of the smaller pick-off terminal blocks AGK 10/UKH has not been connected to the UKH 150. Figure 22 presents the resulting Kinect data for the different cases. For the first erroneous case the missing UKH 95 results in a smaller region of the matching depth data. In comparison with the Hubelino scenario, this might be the case, if two adjacent bricks that have the same color are present in the desired CAD model, but only one of them has been placed in the real scenario. For the second error case the AGK 10/UKH defines its own region, since it is not directly connected to any other component. A comparable Hubelino scenario would be that a single brick, that is, one that does not comprise a layer with any other brick, has not been placed into the assembly group.

Through transferring the described approaches in cognition as well as in 3D Assembly Group Analysis onto this scenario, an autonomous assembly process can be established. Particularly, within the focus of customer-defined switch-cabinet assemblies, a cognitive production cell is useful as it easily adapts the production process on changing assemblies. Additionally, an automated analysis process relieves the human worker and is less error-prone.

5. Conclusion

In order to resolve the problem that the commissioning and programming of complex robotized assembly takes considerable planning efforts which do not directly contribute to

Final assembly   Error UKH 95   Error AGK 10/UKH

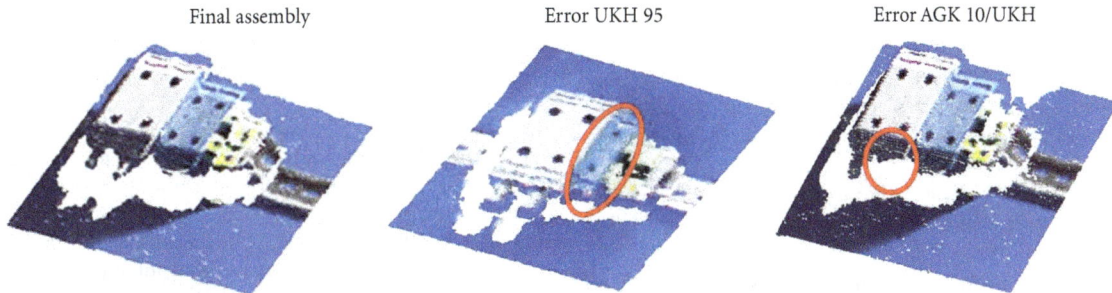

FIGURE 22: Resulting depth data for switch-cabinet scenario.

the added value, the overall approach was to shift planning tasks to the level of execution. Hence, within the Cluster of Excellence "Integrative Production Technology for High-Wage Countries" at RWTH Aachen University, a Cognitive Control Unit had been created, which is able to plan and execute action flows autonomously for a given goal state. The combination with an ergonomic user-centered human machine interface allows for an interaction with a human expert in order to partially transfer planning and implementation tasks to the technical cognitive system as well as to present him/her the opportunity to intervene in case of assembly errors. To test and further enhance the CCU in a near-reality production environment in a variety of different assembly operations, a robotic assembly cell had been set up. The scenario comprises major aspects of an industrial application, while at the same time easily illustrating the potential of a cognitive control system. As a demonstration scenario, the building of an assembly group comprised of Hubelino bricks, which is given to the CCU in the form of a CAD file, was chosen.

In order to verify an assembled group, the CCU requires a recognition process. Additionally, the human operator must be supported in case of assembly errors. This verification process including the feedback to a human expert was realized using the Microsoft Kinect as a perception device. The Kinect was chosen since it merges the capability of a depth sensor and a color camera in a single device, hence providing an innovative approach for research and development for the field of environment recognition.

In the scenario, the Kinect was attached to the flange of an industrial robot, allowing for a positioning at multiple viewpoints in order to create a virtual image of the current assembly group. For each viewpoint the depth data as well as color information is recorded and analyzed to create a discrete model for each side. Afterwards, the discrete models are combined into a complete model of the assembly group. Finally, the superior goal of verifying a real assembly group of Hubelino bricks was achieved by comparing the resulting perceived model with the virtual model from the customer-defined CAD file.

In order to demonstrate the implications in an industrial application, the concepts that were tested with the demonstrator cell were transferred to a scenario for the assembly of switch-cabinets. This scenario bears strong similarities to the demonstrator cell, that is, in regard to the customer-defined CAD models and requirements regarding an automated verification of the final assembly.

Acknowledgment

The authors would like to thank the German Research Foundation (DFG) for its kind support of the research on cognitive automation within the Cluster of Excellence Integrative Production Technology for High-Wage-Countries. No other sponsorship or economic interests were involved which ensured free evaluation of the experiments.

References

[1] E. von Weizsäcker, *Globalisierung der Weltwirtschaft - Herausforderungen und Antworten*, vol. 5 of *Politik und Zeitgeschichte*, 2003, http://www.bpb.de/publikationen/U27MV5,0,Globalisierung_der_Weltwirtschaft_Herausforderungen_und_Antworten.html.

[2] DIHK, "Produktionsverlagerung als Element der Globalisierungsstrategie von Unternehmen: Ergebnisse einer Unternehmensbefragung," 2011, http://www.dihk.de/ressourcen/downloads/produktionsverlagerung.pdf.

[3] DIHK, "Auslandsinvestitionen in der Industrie," 2011, http://www.muenchen.ihk.de/mike/ihk_geschaeftsfelder/starthilfe/Anhaenge/DIHK-Auslandsinvestitionen-2010.pdf.

[4] F. Klocke, "Production technology in high-wage countries: from ideas of today to products of tomorrow," in *Industrial Engineering and Ergonomics: Visions, Concepts, Methods and Tools*, C. Schlick, Ed., Festschrift in Honor of Professor Holger Luczak, Springer, New York, NY, USA, 2009.

[5] S. Kinkel, *Arbeiten in der Zukunft: Strukturen und Trends der Industriearbeit*, Ed. Sigma, Berlin, Germany, 2008.

[6] L. Bainbridge, "Ironies of automation," in *New Technology and Human Error*, J. Rasmussen, K. Duncan, and J. Leplat, Eds., pp. 775–779, John Wiley & Sons, Chichester, UK, 1987.

[7] R. Onken and A. Schulte, *System-Ergonomic Design of Cognitive Automation: Dual-Mode Cognitive Design of Vehicle Guidance and Control Work Systems*, Springer, Berlin, Germany, 2010.

[8] C. Brecher, Ed., *Integrative Produktionstechnik für Hochlohnländer*, Springer, Berlin, Germany, 2011.

[9] M. F. Zäh, M. Beetz, K. Shea et al., "The cognitive factory," in *Changeable and Reconfigurable Manufacturing Systems*, H. A. El-Maraghy, Ed., Springer, Berlin, Germany, 2009.

[10] S. Karim, L. Sonenberg, and A. H. Tan, "A hybrid architecture combining reactive plan execution and reactive learning," in

Proceedings of the 9th Biennial Pacific Rim International Conference on Artificial Intelligence (PRICAI '06), China, 2006.

[11] E. Gat, "On three-layer architectures," in *Artificial Intelligence and Mobile Robots*, D. Kortenkamp, R. Bonnasso, and R. Murphy, Eds., pp. 195–211, AAAI Press, Menlo Park, Calif, USA, 1998.

[12] S. J. Russell and P. Norvig, *Artificial Intelligence: A Modern Approach*, Pearson Education, Upper Saddle River, NJ, USA, 2003.

[13] K. Paetzold, "On the importance of a functional description for the development of cognitive technical systems," in *International Design Conference*, Dubrovnik, Croatia, 2006.

[14] P. Adelt, J. Donath, J. Gausemeier et al., "Selbstoptimierende Systeme des Maschinenbaus," in *HNI-Verlagsschriftenreihe*, J. Gausemeier, F. Rammig, and W. Schäfer, Eds., Westfalia Druck GmbH, Paderborn, Germany, 2009.

[15] M. Zäh and M. Wiesbeck, "A model for adaptively generating assembly instructions using state-based graphs," in *Manufacturing Systems and Technologies for the New Frontier*, The 41st CIRP Conference on Manufacturing Systems, Tokyo, Japan, 2008.

[16] M. Zäh, M. Wiesbeck, F. Engstler et al., "Kognitive Assistenzsysteme in der manuellen Montage," in *wt Werkstatttechnik*, vol. 97, no. 9, pp. 644–650, 2007.

[17] H. Ding, S. Kain, F. Schiller, and O. A. Stursberg, "Control architecture for safe cognitive systems," in *10. Fachtagung Entwurf komplexer Automatisierungssysteme*, Magdeburg, Germany, 2008.

[18] S. Kammel, J. Ziegler, B. Pitzer et al., "Team AnnieWAY's autonomous system for the 2007 DARPA Urban Challenge," *Journal of Field Robotics*, vol. 25, no. 9, pp. 615–639, 2008.

[19] H. J. Putzer, *Ein uniformer Architekturansatz für Kognitive Systeme und seine Umsetzung in ein operatives Framework*, Dr. Köster, Berlin, Germany, 2004.

[20] D. Ewert, E. Hauck, A. Gramatke, and S. Jeschke, "Cognitive assembly planning using state graphs," in *Proceedings of the 3rd International Conference on Applied Human Factors and Ergonomics*, Miami, Fla, USA, 2010.

[21] T. Kempf, W. Herfs, and C. Brecher, "Cognitive Control Technology for a Self-optimizing Robot Based Assembly Cell," in *ASME Design Engineering Technical Conferences and Computers and Information in Engineering Conferences (DETC '08)*, New York, NY, USA, August 2008.

[22] M. Mayer, B. Odenthal, M. Faber et al., "Cognitive engineering for direct human-robot cooperation in self-optimizing assembly cells," in *First International Conference on Human Centered Design (HCD '09)*, M. Kurosu, Ed., San Diego, Calif, USA, July 2009.

[23] C. Schlick, B. Odenthal, M. Mayer et al., "Design and evaluation of an augmented vision system for self-optimizing assembly cells," in *Industrial Engineering and Ergonomics: Visions, Concepts, Methods and Tools*, C. Schlick, Ed., Festschrift in Honor of Professor Holger Luczak, Springer, New York, NY, USA, 2009.

[24] T. B. Sheridan, *Humans and Automation: System Design and Research Issues*, Human Factors and Ergonomics Soc, Santa Monica, Calif, USA, 2001.

[25] M. Mayer, B. Odenthal, and M. Grandt, "Task-oriented process planning for cognitive production systems using MTM," in *Proceedings of the 2nd International Conference on Applied Human Factors and Ergonomic*, Louisville, Ky, USA, 2008.

[26] V. Gazzola, G. Rizzolatti, B. Wicker, and C. Keysers, "The anthropomorphic brain: the mirror neuron system responds to human and robotic actions," *NeuroImage*, vol. 35, no. 4, pp. 1674–1684, 2007.

[27] E. Drumwright, "Toward a vocabulary of primitive task programs for humanoid robots," in *Proceedings of the International Conference on Development and Learning (ICDL '06)*, Bloomington, Ind, USA, 2006.

[28] B. Odenthal, M. Mayer, N. Jochems, and C. Schlick, "Cognitive engineering for human-robot interaction—the effect of subassemblies on assembly strategies," in *Advances in Human Factors, Ergonomics, and Safety in Manufacturing and Service Industries*, pp. 1–10, CRC Press, Boca Raton, Fla, USA, 2011.

[29] M. Mayer, C. Schlick, D. Ewert et al., "Automation of robotic assembly processes on the basis of an architecture of human cognition," *Production Engineering Research and Development*, vol. 5, no. 4, pp. 423–431, 2011.

[30] T. Kempf, *Ein kognitives Steuerungsframework für robotergestützte Handhabungsaufgaben*, Apprimus, Aachen, Germany, 1st edition, 2010.

[31] B. Odenthal, M. Mayer, W. Kabuß, B. Kausch, and C. Schlick, "Investigation of error detection in assembled workpieces using an augmented vision system," in *Proceedings of the 17th World Congress on Ergonomics (IEA '09)*, Beijing, China, August 2009.

[32] B. Odenthal, M. Mayer, W. Kabuß, and C. Schlick, "Einfluss der Bildschirmposition auf die Fehlererkennung in einem Montagebauteil," in *Neue Arbeits- und Lebenswelten gestalten: Vom 24. - 26. März*, M. Schütte, Ed., pp. 203–206, GfA-Press, Dortmund, Germany, 2010.

[33] B. Odenthal, M. Mayer, W. Kabuß, B. Kausch, and C. Schlick, "An empirical study of disassembling using an augmented vision system," in *3rd International Conference on Digital Human Modeling (ICDHM '11)*, Orlando, Fla, USA, 2011.

[34] C. Beder, B. Bartczak, and R. Koch, "A comparison of PMD-cameras and stereo-vision for the task of surface reconstruction using patchlets," in *IEEE Computer Society Conference on Computer Vision and Pattern Recognition (CVPR '07)*, June 2007.

[35] C. Sa, "Time of Flight Camera Technology," *Technology*, 2009.

[36] D. Schauer, *Integration of a 3D time of flight camera system into a robot system: integration, validation and comparison*, VDM Verlag Dr. Müller, Saarbrücken, Germany, 2011.

[37] Microsoft, Ed, "Programming Guide: Getting Started with the Kinect for Windows SDK Beta from Microsoft Research," 2011.

[38] D. Laing, "Kinect - Could this technology have a future in industrial automation?" http://blog.vdcresearch.com/industrial_automation/2010/11/kinect-could-this-technology-have-a-future-in-industrial-automation.html.

[39] D. L. Andrews, *Structured light and its applications: an introduction to phase-structured beams and nanoscale optical forces*, Academic Press, Amsterdam, The Netherlands, 2008.

[40] Z. Song, *Use of structured light for 3D reconstruction*, Hong Kong, 2008.

[41] K. Clague and M. Agullo, *LEGO software power tools*, Syngress, Rockland, Mass, USA, 2002.

[42] Z. Yaniv, "Random Sample Consensus (RANSAC) Algorithm, A Generic Implementation," Information Systems Journal, 2010, http://isiswiki.georgetown.edu/zivy/writtenMaterial/RANSAC.pdf.

[43] R. Szeliski, "Computer vision: Algorithms and applications," Computer Vision, 2011, http://www.worldcat.org/oclc/700473658.

[44] C. Stephens, "Lego Color List - official names, numbers, CYMK, RGB, Pantone values," 2011, http://isodomos.com/Color-Tree/Lego-List.html.

[45] C. Zhang, N. Xi, and Q. Shi, "Object-orientated registration method for surface inspection of automotive windshields," in *IEEE/RSJ International Conference on Intelligent Robots and Systems (IROS '08)*, pp. 3553–3558, September 2008.

[46] X. Liang, J. Liang, J. Liu, and C. Guo, "A rapid inspection method for large water turbine blade," in *IEEE International Conference on Mechatronics and Automation (ICMA '10)*, pp. 712–716, August 2010.

[47] F. Bosché, "Automated recognition of 3D CAD model objects in laser scans and calculation of as-built dimensions for dimensional compliance control in construction," *Advanced Engineering Informatics*, vol. 24, no. 1, pp. 107–118, 2010.

[48] L. Yue and X. Liu, "Application of 3D optical measurement system on quality inspection of turbine blade," in *16th IEEE International Conference on Industrial Engineering and Engineering Management (IE and EM '09)*, pp. 1089–1092, October 2009.

[49] A. Tellaeche, R. Arana, A. Ibarguren, and J. Martínez-Otzeta, "Automatic quality inspection of percussion cap mass production by means of 3D machine vision and machine learning techniques," in *Lecture Notes in Computer Science, Hybrid Artificial Intelligence Systems*, M. Graña Romay, E. Corchado, and M. Garcia Sebastian, Eds., pp. 270–277, Springer, Berlin, Germany, 2010.

[50] B. Muralikrishnan and J. Raja, *Computational Surface and Roundness Metrology*, Springer, 2009.

5

Fuzzy Interpolation and Other Interpolation Methods Used in Robot Calibrations

Ying Bai,[1] Nailong Guo,[2] and Gerald Agbegha[3]

[1] *Johnson C. Smith University, Charlotte, NC 28216, USA*
[2] *Benedict College, Columbia, SC 29204, USA*
[3] *Georgia Gwinnett College, Atlanta, GA 30043, USA*

Correspondence should be addressed to Ying Bai, ybai@jcsu.edu

Academic Editor: G. Muscato

A novel interpolation algorithm, fuzzy interpolation, is presented and compared with other popular interpolation methods widely implemented in industrial robots calibrations and manufacturing applications. Different interpolation algorithms have been developed, reported, and implemented in many industrial robot calibrations and manufacturing processes in recent years. Most of them are based on looking for the optimal interpolation trajectories based on some known values on given points around a workspace. However, it is rare to build an optimal interpolation results based on some random noises, and this is one of the most popular topics in industrial testing and measurement applications. The fuzzy interpolation algorithm (FIA) reported in this paper provides a convenient and simple way to solve this problem and offers more accurate interpolation results based on given position or orientation errors that are randomly distributed in real time. This method can be implemented in many industrial applications, such as manipulators measurements and calibrations, industrial automations, and semiconductor manufacturing processes.

1. Introduction

A suitable interpolation method is important to fit the target pose errors based on the pose errors of the neighboring grid points around the target. In recent years, many advanced interpolation algorithms have been designed and developed by different researchers [1–5]. Jakobsson et al. developed a technique for interpolation with quotients of two radial basis function expansions to approximate functions with poles [6]. Duan et al. constructed a bivariate rational interpolation method using both function values and partial derivatives of the function that was interpolated as the interpolation data. They developed a new rational interpolation with a biquadratic denominator to create a space surface using only values of the interpolated function and designed a bivariate rational Hermite interpolation to create a space surface using both function values and the first-order partial derivatives of the function and presented a weighted rational cubic spline interpolation using two kinds of rational cubic splines with quadratic denominator [7–11].

Luo et al. developed a range-restricted C^1 interpolation local scheme to scattered data. C^μ-rational spline function classes over triangles and quadrilaterals were investigated [12]. Hu and Tan presented an adaptive osculatory rational interpolation for image processing that preserves the contours or edges [13]. Zhao and Tan introduced block-based inverse differences to extend the point-based Thiele-type interpolation to the block-based Thiele-like blending rational interpolation. Also, a bivariate analogy and numerical examples were given to show the effectiveness of their method [14]. Sarfraz and Hussein developed a smooth curve interpolation scheme for positive, monotonic, and convex data by using piecewise rational cubic functions [15].

Goodman and Meek presented a planar interpolation method using a pair of rational spirals to solve planar and two-point G^2 Hermite interpolation problem [16]. Hussain and Sarfraz used a C^1 piecewise rational cubic function to visualize the data arranged over a rectangular grid [17]. Bejancu Built a new treatment of univariate semicardinal interpolation for natural cubic splines, and the solution was obtained as a Lagrange series with suitable localization and polynomial reproduction properties [18]. Zhu and Wang applied the Noether-type theorem of piecewise algebraic curves on cross-cut partitions and used interpolation along a piecewise algebraic curve [19]. Maleknejad and Derili used box spline quasiinterpolants based on local linear functionals of point evaluator and integral type to reproduce the whole spline space [20].

Among those interpolation methods, two of them are very popular and widely implemented in most industrial and manufacturing processes, trilinear and cubic spline interpolation algorithms [21–26].

Both linear and cubic spline interpolation methods can achieve satisfactory interpolation results for a common measurement and calibration process [27]. Generally, the linear interpolation method is based on the assumption that the error distribution is approximately linear, and the interpolated errors are obtained from three plans that are constructed based on 8 neighboring errors on the grid points around the target cubic cell [28]. The cubic spline interpolation technique also assumes that the error of the target pose is located on a cubic curve that is constructed by the pose errors of 8 neighboring grid points around the target [28, 29]. In essence, both methods approximate a spatial error surface based on the errors of known points and assume that the error of the target point is located on that surface. Consequently, the target pose error is estimated by utilizing the equations of the error surface. However, since the actual pose errors are randomly distributed with the time and locations in the measured machine workspace, and therefore it is impossible to pinpoint a pose on the error surface at any given moment accurately, the result is that the traditional interpolation techniques may not provide an accurate estimation of the pose errors.

The fuzzy error interpolation technique utilizes a fuzzy inference system to estimate machine or manipulator pose errors, which is consistent with the random distributed nature of the pose errors. These pose errors can be considered as a fuzzy set at any given moment of time. The fuzzification process takes into account a range of errors rather than only a crisp error value. Therefore, the fuzzy error interpolation technique has the potential to improve the error estimation and compensation results for the target.

Fuzzy interpolation techniques have been rapidly developed and implemented in many academic and industrial fields in recent years [30–42]. Different strategies of fuzzy interpolation have been developed and applied in real applications. Triantafilis et al. and Dragicevic et al. reported approaches of using fuzzy interpolation methods to estimate the soil layer and geographical distributions for GIS database [43, 44]. Song et al. described a fuzzy logic methodology for four-dimensional (4D) systems with optimal global performance using enhanced cell state space [45]. Li et al. reported a multidimensional fuzzy interpolation neural network to perform the fuzzy interpolations for a multidimensional system [46]. Chang et al. reported to use fuzzy interpolation methods to obtain the trajectory data for a multijoint animation robot [47]. Bai et al. developed a robot calibration algorithm to calibrate parallel machine tools using fuzzy interpolation techniques [48]. A control algorithm combined with Lagrange fuzzy interpolation, which was reported by Cheng Wang and Shanzhen Xu can be used to improve control effect and enhance control precision

effectively compared with traditional fuzzy control algorithm [49]. Bai et al. developed a mind assistant system using fuzzy interpolation technique to support the judgment of emotion state for elderly living alone [48]. A supervisory semiactive, nonlinear control system combined with a fuzzy interpolation algorithm is reported by Kim et al. to improve the controllability of a MISO controller [50].

A comparison between trilinear, cubic spline, and fuzzy interpolation methods used in accurate measurements and compensations for machine or manipulator calibration are discussed in this paper. The simulation results show that the fuzzy interpolation outperform other interpolation methods.

The remainder of the paper is organized into the following four sections. The principles of the two popular traditional interpolation techniques, trilinear and cubic spline, are outlined in Section 2. Section 3 discusses the fuzzy error interpolation method. Results from a simulation study are given in Section 4 to illustrate the effectiveness of the fuzzy error interpolation technique. The conclusion is provided in Section 5.

2. Trilinear and Cubic Spline Methods

The trilinear and cubic spline interpolation methods are designed to construct a surface based on the known errors of neighboring points. The target pose error is then derived by using an error surface equation. The operation principles of the trilinear and spline interpolation methods are discussed in this section.

2.1. Trilinear Interpolation. Trilinear interpolation is a computational process of linearly interpolating points within a 3D box given values at the vertices of the box, and it is the most common application in interpolating within cells of a volumetric dataset [29, 30]. The whole process can be simplified to perform three consecutive linear interpolations along three coordinate axes: x, y, and z, respectively.

Refer to Figures 1 and 2, and assume that it is a unit cube with the lower-left-base vertex at the origin. The coordinate values at each vertex will be denoted by C_{000}, C_{100}, $C_{010}, \ldots C_{111}$. Let x_d, y_d, and z_d be the differences between the target of x, y, z and the smaller coordinate related to the cubic lattice $[x]$, $[y]$, and $[z]$, the error values at 8 corners of the cubic lattice are V_{000}, V_{100}, $V_{010}, \ldots$ and so forth $\ldots V_{111}$, which is

$$x_d = x - [x], \qquad y_d = y - [y], \qquad z_d = z - [z]. \quad (1)$$

We can first perform the linear interpolation along the z-axis (pushing the front face of the cube to the back), which is

$$
\begin{aligned}
V_{00} &= V_{000}(1 - z_d) + V_{100} z_d, \\
V_{10} &= V_{010}(1 - z_d) + V_{110} z_d, \\
V_{01} &= V_{001}(1 - z_d) + V_{101} z_d, \\
V_{11} &= V_{011}(1 - z_d) + V_{111} z_d.
\end{aligned}
\quad (2)
$$

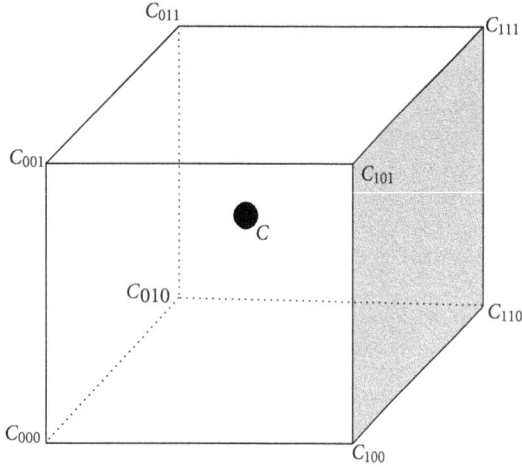

FIGURE 1: Eight corner points.

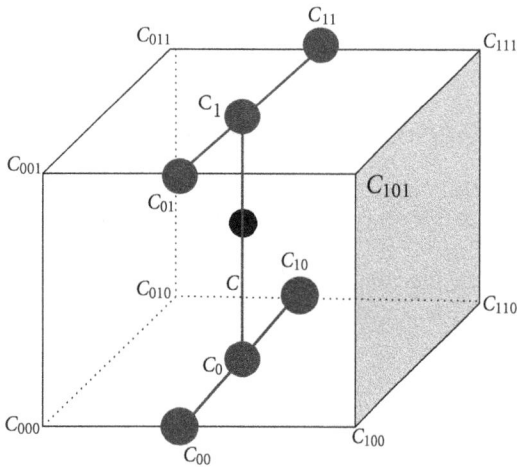

FIGURE 2: 3D trilinear interpolation.

Then we interpolate these values along y-axis, as we were pushing the top edge to the bottom, giving:

$$V_0 = V_{00}(1 - y_d) + V_{10}y_d,$$
$$V_1 = V_{01}(1 - y_d) + V_{11}y_d. \tag{3}$$

Finally, we interpolate these values along x-axis (walking through a line), and this gives us a predicted error value for the target point:

$$t_p = V_0(1 - x_d) + V_1 x_d. \tag{4}$$

The above operations can be illustrated by the following sequence: first we perform linear interpolation between C_{000} and C_{100} to find V_{00}, C_{001} and C_{101} to find V_{01}, C_{011} and C_{111} to find V_{11}, C_{010} and C_{110} to find V_{10}. Then we do interpolation between C_{00} and C_{10} to find V_0, C_{01} and C_{11} to find V_1. Finally, we calculate the error value C via linear interpolation of C_0 and C_1. In practice, a trilinear interpolation is identical to three successive linear interpolations, or two bilinear interpolations combined with a linear interpolation.

Combining (1) through (4), we can obtain the following equation to interpolate the error value V_{xyz} at the target position $[x, y, z]$ assumed that the cubic lattice is a unit one:

$$\begin{aligned} V_{xyz} = {} & V_{000}(1 - x)(1 - y)(1 - z) + V_{100}x(1 - y)(1 - z) \\ & + V_{010}(1 - x)y(1 - z) + V_{001}(1 - x)(1 - y)z \\ & + V_{101}x(1 - y)z + V_{011}(1 - x)yz \\ & + V_{110}xy(1 - z) + V_{111}xyz. \end{aligned} \tag{5}$$

In general, the box will not be of unit size nor will it be aligned at the origin. Simple translation and scaling (possibly of each axis independently) can be used to transform into then out of this simplified situation.

As illustrated in Figures 1 and 2, the trilinear interpolation technique is based on two assumptions. First, the pose error of the target e_p must be located on three error surfaces, which is built based on errors of 8 neighboring grid points around a cubic cell. Secondly, the error surface has to be constructed prior to the application of the trilinear interpolation technique. However, these assumptions have their drawbacks. Pose errors on each cell are randomly distributed and the error curving surfaces, e_x, e_y, and e_z, are also randomly distributed at any given moment. One can consider the $e_x(x, y, z, \alpha, \beta, \gamma)$ as a fourth-dimensional function value based on the pose $[x, y, z, \alpha, \beta, \gamma]$ inside each cell. The same consideration is applied to $e_y(x, y, z, \alpha, \beta, \gamma)$ and $e_z(x, y, z, \alpha, \beta, \gamma)$. Therefore, the compensation accuracy of bilinear interpolation is limited by these assumptions.

2.2. Cubic Spline Interpolation. The cubic spline method is to estimate a cubic surface $S(x, y, z)$ based on the position errors of the neighboring grid points around the target. This method assumes that both the 1st- and the 2nd-order derivatives ($S'(x, y, z)$ and $S''(x, y, z)$) of the interpolated points are existing, and the function $S''(x, y, z)$ is a trilinear surface on each cubic cell [43]. To simplify our discussion, consider the one-dimensional situation. Since the function $S''(x)$ is a linear function at the interval of each cell in the x-direction, the error function $S(x)$ should be a cubic curve. If a and b are two neighboring points in the x direction, we define two values M_a and M_b as

$$M_a = S''(a), \qquad M_b = S''(b). \tag{6}$$

A linear equation can be derived as follows:

$$S''(x) = \frac{(b - x)M_a + (x - a)M_b}{b - a}. \tag{7}$$

After quite a bit of manipulation, this result is in the cubic polynomial [11]:

$$\begin{aligned} S(x) = {} & \frac{(b - x)^3 M_a + (x - a)^3 M_b}{6(b - a)} \\ & + \frac{(b - x)S(a) + (x - a)S(b)}{b - a} \\ & - \frac{(b - a)[(b - x)M_a + (x - a)M_b]}{6}. \end{aligned} \tag{8}$$

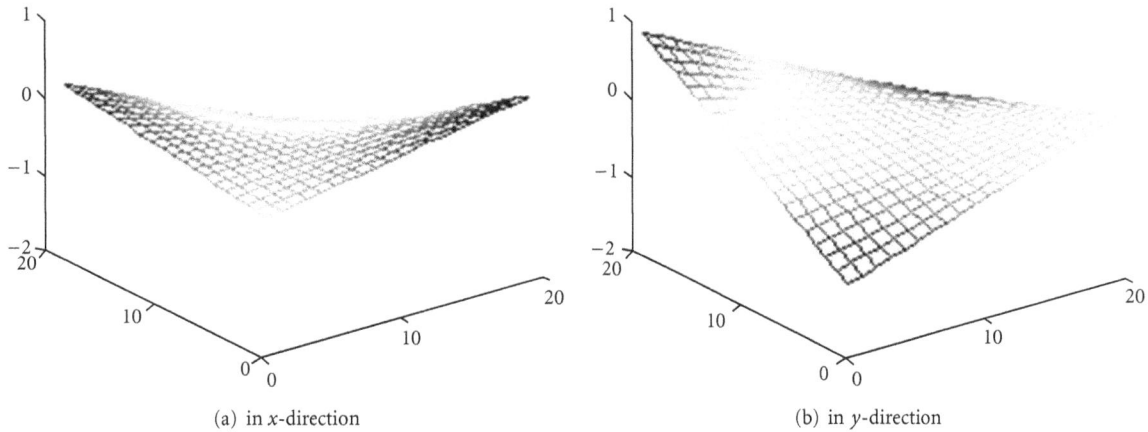

(a) in x-direction   (b) in y-direction

FIGURE 3: Error surfaces in x- and y-directions on a cell.

By using the tridiagonal matrix, M_a and M_b can be derived. By substituting these 2nd-order derivatives back to (8), any point's error in x-direction can be interpolated.

An example of a 2D error surfaces in x direction, $e_x(x, y)$, and in y-direction, $e_y(x, y)$, for a cell is shown in Figure 3. The error surfaces are estimated based on neighboring grid position errors around the target position using the cubic spline technique.

Compared with trilinear interpolation method, the cubic spline method uses a more arbitrary shaped surface to approximate the error. Therefore, it provides more accurate position compensation results for known positions around a target. For unknown interpolated data, such as random noises, the interpolation results may not as good as desired since the interpolated data are randomly distributed noises.

3. Fuzzy Error Interpolation Method

From the structure of the trilinear interpolation technique, it can be observed that the method assumes that the position error on the target point $\mathbf{P}(\mathbf{x}, \mathbf{y}, \mathbf{z})$ must be located on the intersection of the three plans that are built based on errors of the 8 neighbouring grid points $\mathbf{P_1}$–$\mathbf{P_8}$. However in the real world, this assumption may not hold. The compensation accuracy of using this interpolation technique is limited by this assumption. For the cubic spline interpolation method, the assumption is that all interpolated data should be definite or with little degree of uncertainty. However, this assumption cannot be satisfied when the interpolated data are random noises, and therefore the interpolation results may not be as good as desired.

In order to solve this problem and to improve the measurement and compensation accuracy, a dynamic online fuzzy interpolation method is introduced. The traditional fuzzy inference system uses predefined membership functions and control rules to construct lookup tables and then picks up the associated control output from the lookup table as the fuzzy inference system works in an application. This kind of system is often called an offline fuzzy inference system because all inputs and outputs have been defined prior to the application process. This offline fuzzy system may not meet accuracy requirements in certain applications based on the following reasons First, the pose error of the target is estimated based on errors of 8 neighbouring grid points, and these neighbouring errors are randomly distributed. The offline fuzzy output membership functions are defined based on the errors range, say the neighbouring errors' range. However, this range estimation is not as good as the one deduced from the actual errors obtained on 8 grid points. Second, since each cell needs one lookup table for the offline fuzzy system, it needs a large memory space to save a great number of lookup tables, which is both space and time consuming, and therefore not suitable for real-time processing. For example, in our study, the robot workspace is divided into $40 \times 40 \times 40$ small cubic cells, and each cell is $20 \times 20 \times 20$ mm^3. Assume that one lookup table is for one cubic cell, and this needs about 64000 lookup tables! By using an online dynamic fuzzy inference system, one can estimate the target pose error by combining the output membership functions, which are obtained from real errors on the neighbouring grid points, with the control rules in real-time after this online fuzzy system is implemented. Therefore, we do not need any offline lookup tables at all. This means that one cannot determine the output membership functions until the fuzzy inference system is applied to a real process, and this is based on the real errors on the grid points, not a range.

The definition of this dynamic online fuzzy inference algorithm is shown in Figure 4. Each small cube, which is surrounded by 8 neighboring grid points, is defined as a cubic cell. Furthermore this cubic cell is divided into 8 equal smaller cubic cells, which are also shown in Figure 4(a).

The pose error at each grid point is defined as $P_1, P_2, P_3, P_4, P_5, P_6, P_7$, and P_8. For the fuzzy inference system, the interpolation method is divided into three dimensions separately, so the inputs to the fuzzy inference system are e_x, e_y, and e_z. The outputs are ee_x, ee_y, ee_z, ee_α, ee_β, and ee_γ, which are shown in Figure 4(b).

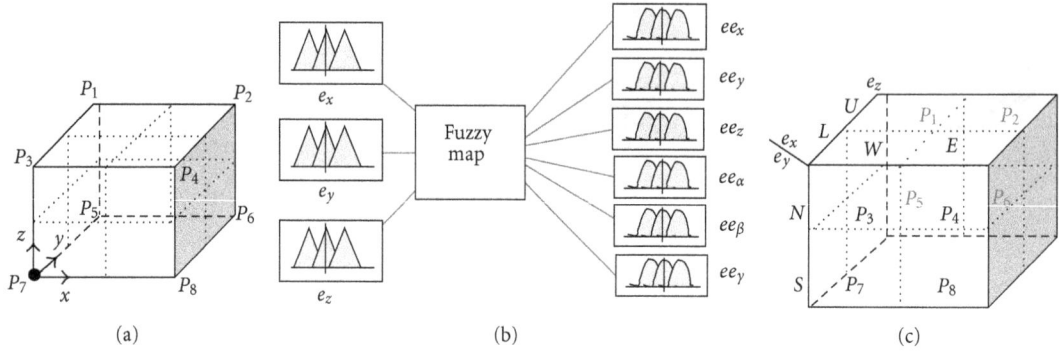

FIGURE 4: Definition of the fuzzy interpolation inference system.

The control rule is shown in Figure 4(c), which is straightforward and based on the human being knowledge. It is worth to note that each P_i should be considered as a combination of three position and three orientation error components on each grid point.

The distance between the neighboring grid points of each cell on the workspace is 20 mm in x-, y- and z-directions for our current study, which is a standard interval for a small-size calibration workspace. Totally, the workspace includes 20 by 20 by 20 cells, which is equivalent to a 400 by 400 by 400 mm^3 space. This is a typical workspace of most popular manipulators implemented in semiconductor manufacturing operations [46, 50, 51]. The input membership functions for x-, y-, and z-directions, and the predefined output membership functions are shown in Figure 5.

The predefined output membership functions are used as default ones, and the actual output membership function will be obtained by shifting the default one based on the actual error values on the grid points. For each cell, 8 output membership functions are implemented, and each one is associated with the error at one grid point. In Figure 5(b), only 4 position output membership functions are shown here because of the space limitation. In a real application, total 8 orientation and 8 position membership functions should be utilized.

The Gaussian-bell waveforms are selected as the shape of the membership functions for three inputs. As shown in Figure 5(a), the ranges of inputs are between -10 and 10 mm (20 mm interval on grid points). The reason for this selection is that the Gaussian-bell waveform has a smooth curve and therefore can make measurements more accurate [43, 44]. W and E represent the inputs located at different areas in the x-direction, N and S represent those in the y-direction, and L and U represent those in the z-direction. Unlike the traditional fuzzy inference system, in which all membership functions should be determined to produce the lookup table prior to the implementation of the fuzzy system, in this study, the output membership functions will not be defined until the implementation of the fuzzy error interpolation to compensate the pose errors. So the output membership functions will be determined during the application of the fuzzy inference system online or dynamically. Figure 5(b) shows an example of the output membership functions,

which are related to the simulated random errors at neighboring grid points. Each P_{xi}, P_{yi}, and P_{zi} corresponds to the pose error at the ith grid point, respectively. During the design stage, all output membership functions should be initialized to a gaussian waveform with a mean of 0 and a range that is close to the actual possible output range which can be estimated based on the different manipulators for the different applications. These output membership functions will be determined online based on the errors of the neighboring grid points around the target point in the workspace during the compensation process.

The control rules shown in Figure 4(c) can be interpreted as follows after the output membership functions are determined:

(i) If e_x is W, e_y is N and e_z is U, then ee_x is P_{x1},

ee$_y$ is P_{y1} and ee_z is P_{z1}, and ee_α is α_1, ee_β is β_1

and ee_γ is γ_1. (P$_1$)

(ii) If e_x is W, e_y is N and e_z is L, then ee_x is P_{x3},

ee$_y$ is P_{y3} and ee_z is P_{z3}, and ee_α is α_3, ee_β is β_3

and ee_γ is γ_3. (P$_3$)

(iii) If e_x is W, e_y is S and e_z is U, then ee_x is P_{x5},

ee$_y$ is P_{y5} and ee_z is P_{z5}, and ee_α is α_5, ee_β is β_5

and ee_γ is γ_5. (P$_5$)

(iv) If e_x is W, e_y is S and e_z is L, then ee_x is P_{x7},

ee$_y$ is P_{y7} and ee_z is P_{z7}, and ee_α is α_7, ee_β is β_7

and ee_γ is γ_7. (P$_7$)

(v) If e_x is E, e_y is N and e_z is U, then ee_x is P_{x2},

ee$_y$ is P_{y2} and ee_z is P_{z2}, and ee_α is α_2, ee_β is β_2

and ee_γ is γ_2. (P$_2$)

(vi) If e_x is E, e_y is N and e_z is L, then ee_x is P_{x4},

ee$_y$ is P_{y4} and ee_z is P_{z4}, and ee_α is α_4, ee_β is β_4

and ee_γ is γ_4. (P$_4$)

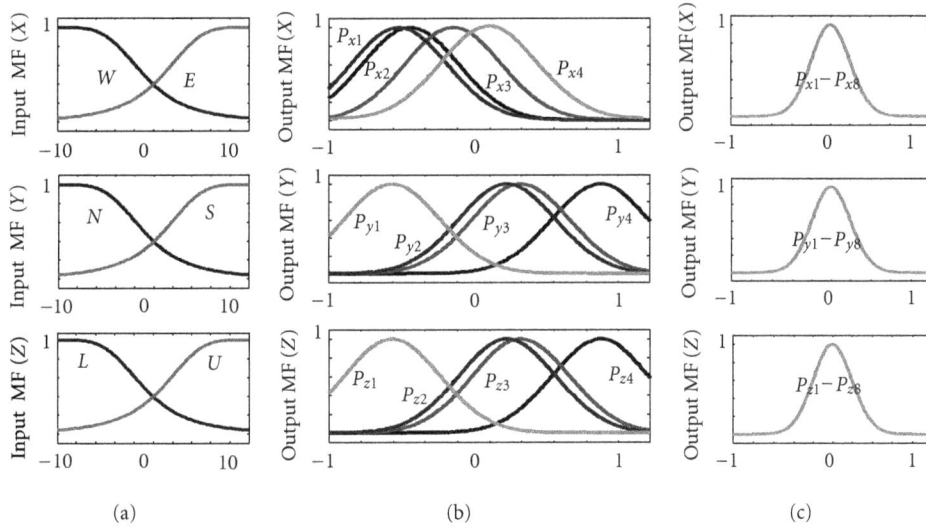

FIGURE 5: Input and output membership functions.

(vii) If e_x is E, e_y is S and e_z is U, then ee_x is P_{x6},

ee_y is P_{y6} and ee_z is P_{z6}, and ee_α is α_6, ee_β is β_6

and ee_γ is γ_6.(P_6)

(viii) If e_x is E, e_y is S and e_z is L, then ee_x is P_{x8},

ee_y is P_{y8} and ee_z is P_{z8}, and ee_α is α_8, ee_β is β_8

and ee_γ is γ_8.(P_8)

$$(9)$$

The control rules are straightforward, and they are based on the human knowledge. The error on P_1 grid point should carry larger weight if the target position (input) is located inside the NWU area on a cell. Similar consideration should be given for errors on all other grid points.

The input error variables can be expressed as a label set L, with E being a linguistic input variable:

$$L(E) = \{NWU, NWL, NEU, NEL, SWU, SWL, SEU, SEL\}. \tag{10}$$

Assume that u_i is the membership function, U_i the universe of discourse, and m the number of contributions, the traditional output of the fuzzy inference system can be represented as

$$u = \frac{\sum_{i=1}^{m}(u_i \times U_i)}{\sum_{i=1}^{m} u_i}, \tag{11}$$

where u is the current crisp output of the fuzzy inference system, and (11) is obtained by using the center-of-gravity method (COG). In this study, both u_i and U_i in the output membership functions are randomly distributed variables, and the actual values of these variables depend upon the

position errors of 8 neighboring grid points around the target position. These relationships can be expressed as

$$u_{xi} = F_{xi}(P_{x1}, P_{x2}, P_{x3}, P_{x4}, P_{x5}, P_{x6}, P_{x7}, P_{x8}),$$
$$U_{xi} = Q_{xi}(P_{x1}, P_{x2}, P_{x3}, P_{x4}, P_{x5}, P_{x6}, P_{x7}, P_{x8}), \tag{12}$$

where F_{xi} is the membership function of the input pose in the x-direction, and it is a predetermined membership function as shown in Figure 5(a). Q_{xi} is the real error output membership function, which is a randomly distributed function, and it gives the error output contributions in the x direction. This membership function is determined by the real pose errors at the 8 neighboring grid points in the x-direction: P_{x1}–P_{x8}. This membership function determines the degree to which the current pose input belongs to each different real error output based on the 8 control rules defined in (9) in the x direction, and it is equivalent to the universe of discourse or a weighing factor. Substituting (12) into (11), one obtains:

$$u_i = \frac{\sum_{i=1}^{m} F_{xi}(P_{x1}, P_{x2}, \ldots P_{x8}) \times Q_{xi}(P_{x1}, P_{x2}, \ldots P_{x8})}{\sum_{i=1}^{m} F_{xi}(P_{x1}, P_{x2}, P_{x3}, P_{x4}, P_{x5}, P_{x6}, P_{x7}, P_{x8})}. \tag{13}$$

Here u_x represents the final error output of the fuzzy interpolation method in the x direction. In (13), Q_{xi} will not be determined until the fuzzy error interpolation technique is applied in an actual compensation process, which means that this fuzzy inference system is an online process. The final crisp output of the fuzzy error interpolation system is determined by the neighboring pose errors of 8 grid points. Similar calculations can be implemented for the error outputs in the y- and z-directions as well as three orientations.

The advantage of using the online fuzzy inference system is that the control output has the real-time control ability, but the drawback is that this type of control has a relative longer response time because of the calculation performed

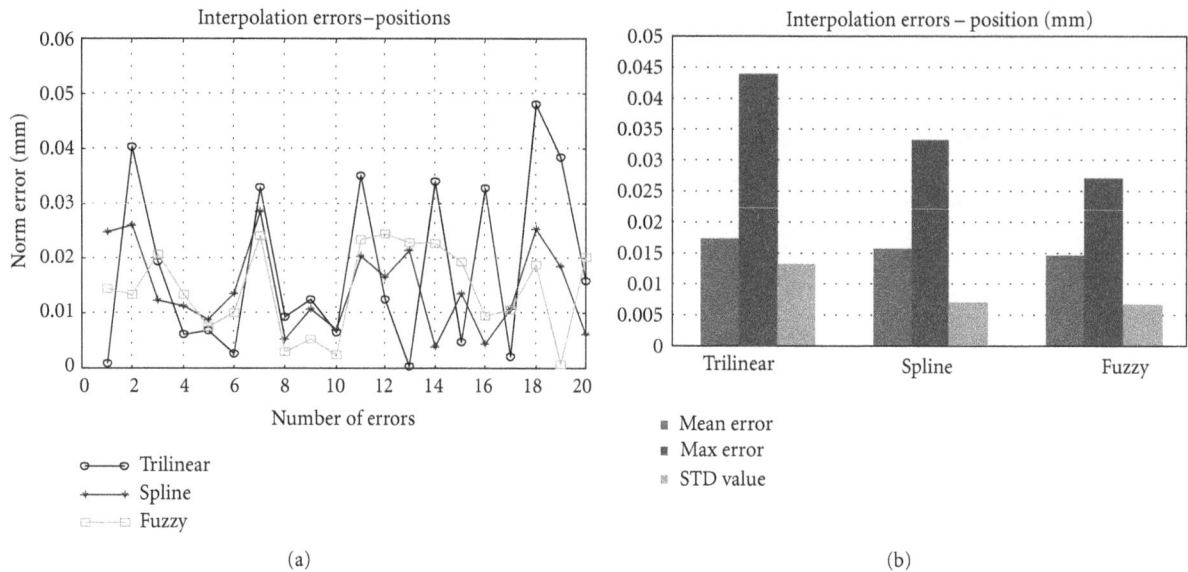

(a)

(b)

FIGURE 6: Simulated interpolation results—position errors.

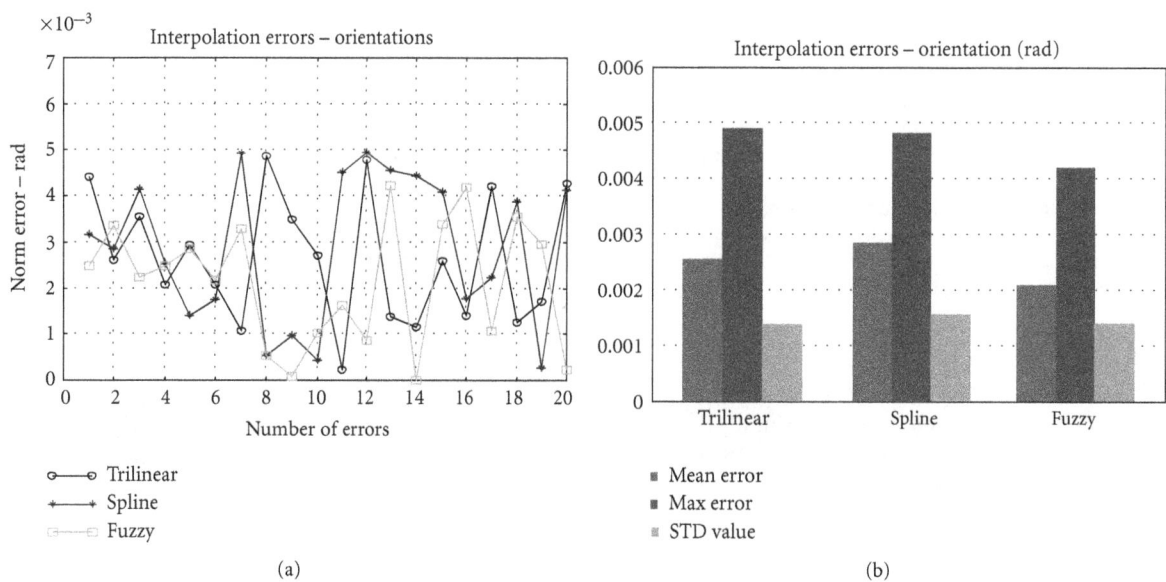

(a)

(b)

FIGURE 7: Simulated interpolation results—orientation errors.

in the fuzzy inference system. This shortcoming becomes of little importance as the availability of high-speed CPUs for the controllers.

4. Simulation Results

Extensive simulation studies have been performed with a PUMA 560 robot in order to illustrate the effectiveness of the proposed fuzzy error interpolation technique in comparison to the trilinear and cubic spline interpolation methods. The simulated position error is a uniformly distributed random noise U [−0.05, 0.05] mm, and the simulated orientation error is also a uniformly distributed random noise U

[−0.001, 0.001] radian degrees. Figure 6(a) shows a comparison of interpolated position errors using three interpolation techniques: trilinear, cubic spline, and fuzzy. Figure 6(b) shows a histogram comparison among three interpolation methods. Figures 7(a) and 7(b) show the comparisons of three interpolation techniques for orientation compensation results.

It can be found that the fuzzy interpolation method has more accurate compensation result for both position and orientation errors compared with both trilinear and cubic spline methods. The max position error of the fuzzy interpolation method is about 0.026 mm, which is about 41% smaller compared with the error obtained from the trilinear method (0.044 mm) and 25% smaller with respect to the error interpolated from the cubic spline method (0.033 mm). For the mean position errors, the fuzzy interpolation method also outperforms the other two methods. Similar comparison results can be obtained from the orientation errors shown in Figure 7.

The numbers in the horizontal axes in Figures 6 and 7 are the number of errors in 20 cubic cells.

The FIA method provided in this paper has better performances compared with other interpolation methods. One possible shortcoming of this method is that it may need high-speed computer and large memory space to process and store predefined data on all grid points. However, this disadvantage can be easily overcome by using high-speed CPUs and huge memory spaces in today's computers.

5. Conclusions

A comparison of fuzzy error interpolation technique with tri-linear and cubic spline interpolation methods used for high accuracy measurement and calibration of robots is discussed and analyzed in this paper. The simulation results show that the measurement and calibration results can be greatly improved when a fuzzy interpolation method is adopted. By using this fuzzy error interpolation algorithm, both position and orientation errors, especially for the random-distributed errors, can be significantly reduced and suppressed, and therefore the measurement and calibration accuracy can be greatly improved. This algorithm can be conveniently implemented in the real manufacturing process to reduce the production cost and operation times. The key technology used in this algorithm is the dynamic and online process in which the output membership functions are determined online based on the real position and orientation errors of the grid points around the target.

References

[1] M. Krajnc, "Interpolation scheme for planar cubic G2 spline curves," *Acta Applicandae Mathematicae*, vol. 113, no. 2, pp. 129–143, 2011.

[2] A. Cuyt and W.-S. Lee, "Sparse interpolation of multivariate rational functions," *Theoretical Computer Science*, vol. 412, no. 16, pp. 1445–1456, 2011.

[3] R. Feng and X. Zhou, "A kind of multiquadric quasi-interpolation operator satisfying any degree polynomial reproduction property to scattered data," *Journal of Computational and Applied Mathematics*, vol. 235, no. 5, pp. 1502–1514, 2011.

[4] G. Allasia and C. Bracco, "Two interpolation operators on irregularly distributed data in inner product spaces," *Journal of Computational and Applied Mathematics*, vol. 235, no. 7, pp. 1763–1774, 2011.

[5] N. Giménez, J. Heintz, G. Matera, and P. Solernó, "Lower complexity bounds for interpolation algorithms," *Journal of Complexity*, vol. 27, no. 2, pp. 151–187, 2011.

[6] S. Jakobsson, B. Andersson, and F. Edelvik, "Rational radial basis function interpolation with applications to antenna design," *Journal of Computational and Applied Mathematics*, vol. 233, no. 4, pp. 889–904, 2009.

[7] Q. Duan, Y. Zhang, and E. H. Twizell, "A bivariate rational interpolation and the properties," *Applied Mathematics and Computation*, vol. 179, no. 1, pp. 190–199, 2006.

[8] Q. Duan, H. Zhang, A. Liu, and H. Li, "A bivariate rational interpolation with a bi-quadratic denominator," *Journal of Computational and Applied Mathematics*, vol. 195, no. 1-2, pp. 24–33, 2006.

[9] Q. Duan, L. Wang, and E. H. Twizell, "A new C2 rational interpolation based on function values and constrained control of the interpolant curves," *Applied Mathematics and Computation*, vol. 161, no. 1, pp. 311–322, 2005.

[10] Q. Duan, S. Li, F. Bao, and E. H. Twizell, "Hermite interpolation by piecewise rational surface," *Applied Mathematics and Computation*, vol. 198, no. 1, pp. 59–72, 2008.

[11] Q. Duan, L. Wang, and E. H. Twizell, "A new weighted rational cubic interpolation and its approximation," *Applied Mathematics and Computation*, vol. 168, no. 2, pp. 990–1003, 2005.

[12] Z. Luo and X. Peng, "A C1-rational spline in range restricted interpolation of scattered data," *Journal of Computational and Applied Mathematics*, vol. 194, no. 2, pp. 255–266, 2006.

[13] M. Hu and J. Tan, "Adaptive osculatory rational interpolation for image processing," *Journal of Computational and Applied Mathematics*, vol. 195, no. 1-2, pp. 46–53, 2006.

[14] Q. J. Zhao and J. Tan, "Block-based Thiele-like blending rational interpolation," *Journal of Computational and Applied Mathematics*, vol. 195, no. 1-2, pp. 312–325, 2006.

[15] M. Sarfraz and M. Z. Hussain, "Data visualization using rational spline interpolation," *Journal of Computational and Applied Mathematics*, vol. 189, no. 1-2, pp. 513–525, 2006.

[16] T. N. T. Goodman and D. S. Meek, "Planar interpolation with a pair of rational spirals," *Journal of Computational and Applied Mathematics*, vol. 201, no. 1, pp. 112–127, 2007.

[17] M. Z. Hussain and M. Sarfraz, "Positivity-preserving interpolation of positive data by rational cubics," *Journal of Computational and Applied Mathematics*, vol. 218, no. 2, pp. 446–458, 2008.

[18] A. Bejancu, "Semi-cardinal interpolation and difference equations: from cubic B-splines to a three-direction box-spline construction," *Journal of Computational and Applied Mathematics*, vol. 197, no. 1, pp. 62–77, 2006.

[19] C. G. Zhu and R. H. Wang, "Lagrange interpolation by bivariate splines on cross-cut partitions," *Journal of Computational and Applied Mathematics*, vol. 195, no. 1-2, pp. 326–340, 2006.

[20] K. Maleknejad and H. Derili, "Numerical solution of Hammerstein integral equations by using combination of spline-collocation method and Lagrange interpolation," *Applied Mathematics and Computation*, vol. 190, no. 2, pp. 1557–1562, 2007.

[21] T. M. Lehmann, C. Gönner, and K. Spitzer, "Survey: interpolation methods in medical image processing," *IEEE Transactions on Medical Imaging*, vol. 18, no. 11, pp. 1049–1075, 1999.

[22] R. P. Woods, *Handbook of Medical Image Processing and Analysis*, Academic Press, New York, NY, USA, 2000, Edited by I. N. Bankman.

[23] http://www.cgg-journal.com/1999-3/02/Cg_97h.htm.

[24] R. Huang, K. L. Ma, P. McCormick, and W. Ward, "Visualizing industrial CT volume data for nondestructive testing applications," in *Proceedings of the 14th IEEE Visualization*, pp. 547–554, Seattle, Wash, USA, October 2003.

[25] http://www.freepatentsonline.com/20030033050.pdf.

[26] http://www.freepatentsonline.com/5726896.pdf.

[27] M. A. Penna, "Camera calibration: a quick and easy way to determine the scale factor," *IEEE Transactions on Pattern Analysis and Machine Intelligence*, vol. 13, no. 12, pp. 1240–1245, 1991.

[28] G. Farin, *Curves and Surfaces for Computer Aided Geometric Design—A Practical Guide*, Academic Press, 3rd edition, 1993.

[29] I. J. Schoenberg, "Cardinal spline interpolation," in *Proceedings of the Regional Conference Series in Applied Mathematics, Society for Industrial and Applied Mathematics*, Philadelphia, Pa, USA, 1973.

[30] Z. Huang and Q. Shen, "Fuzzy interpolation and extrapolation: a practical approach," *IEEE Transactions on Fuzzy Systems*, vol. 16, no. 1, pp. 13–28, 2008.

[31] Y. Bai and H. Zhuang, "On the comparison of model-based and modeless robotic calibration based on the fuzzy interpolation technique," in *Proceedings of the IEEE Conference on Robotics, Automation and Mechatronics*, pp. 892–897, December 2004.

[32] Y. Bai and D. Wang, "Improve the position measurement accuracy using a dynamic on-line fuzzy interpolation technique," in *Proceedings of the IEEE International Symposium on Computational Intelligence for Measurement Systems and Applications*, pp. 227–232, 2003.

[33] N. Bizon, I. Gabriel, and M. Oproescu, "Fuzzy interpolation of the average signal steps," in *Proceedings of the International Symposium on Signals, Circuits and Systems (ISSCS '09)*, pp. 1–4, July 2009.

[34] T. K. Yin, "A characteristic-point-based fuzzy inference classifier by a closeness matrix," *IEEE Transactions on Fuzzy Systems*, vol. 13, no. 5, pp. 673–687, 2005.

[35] N. Bizon, I. Gabriel, and M. Oproescu, "Fuzzy interpolation of the average signal steps," in *Proceedings of the International Symposium on Signals, Circuits and Systems (ISSCS '09)*, pp. 1–4, July 2009.

[36] D. Li, Y. Yue, C. Maple, V. Schetinin, and H. Qiu, "Multidimensional fuzzy interpolation neural network," in *Proceedings of the IEEE International Conference on Automation and Logistics (ICAL '09)*, pp. 186–190, Shenyang, China, August 2009.

[37] Z. Huang and Q. Shen, "Preserving piece-wise linearity in fuzzy interpolation," in *Proceedings of the IEEE International Conference on Fuzzy Systems (FUZZ-IEEE '09)*, pp. 575–580, August 2009.

[38] C. Wang and S. Xu, "Lagrange fuzzy interpolating controlling and simulation on semi-active suspension of vehicle," in *Proceedings of the 2nd International Conference on Intelligent Computing Technology and Automation (ICICTA '09)*, vol. 2, pp. 78–81, Changsha, Hunan, China, 2009.

[39] H. C. Chen and W. J. Wang, "Two-stage interpolation algorithm based on fuzzy logics and edges features for image zooming," *Eurasip Journal on Advances in Signal Processing*, vol. 2009, Article ID 372180, 2009.

[40] http://en.wikipedia.org/wiki/Trilinear_interpolation..

[41] http://local.wasp.uwa.edu.au/~pbourke/miscellaneous/ Interpolation.

[42] G. Agbegha, N. Guo, and Y. Bai, "Compare fuzzy interpolation algorithm with other interpolation methods used in industrial applications," in *Proceedings of the Hawaii University Engineering and Mathematics International Conference*, Honolulu, Hawaii, USA, 2011.

[43] J. Triantafilis, W. T. Ward, I. O. A. Odeh, and A. B. McBratney, "Creation and interpolation of continuous soil layer classes in the lower Namoi Valley," *Soil Science Society of America Journal*, vol. 65, no. 2, pp. 403–413, 2001.

[44] S. Dragicevic, D. J. Marceau, and C. Marois, "Space, time, and dynamics modeling in historical GIS databases: a fuzzy logic approach," *Environment and Planning B*, vol. 28, no. 4, pp. 545–562, 2001.

[45] F. Song, S. M. Smith, and C. G. Rizk, "Fuzzy logic controller design methodology for 4D systems with optimal global performance using enhanced cell state space based best estimate directed search method," in *Proceedings of theIEEE International Conference on Systems, Man, and Cybernetics 'Human Communication and Cybernetics'*, October 1999.

[46] D. Li, Y. Yue, C. Maple, V. Schetinin, and H. Qiu, "Multidimensional fuzzy interpolation neural network," in *Proceedings of the IEEE International Conference on Automation and Logistics (ICAL '09)*, pp. 186–190, Shenyang, China, August 2009.

[47] S. R. Chang, C. H. Lee, and U. Y. Huh, "The natural motion using fuzzy interpolation," in *Proceedings of the International Conference on Control, Automation and Systems (ICCAS '07)*, pp. 425–428, Seoul, Republic of Korea, October 2007.

[48] Y. Bai, H. Zhuang, and D. Wang, "Calibration of parallel machine tools using fuzzy interpolation method," in *Proceedings of the IEEE International Conference on Technologies for Practical Robot Applications (TePRA '08)*, pp. 56–61, Woburn, Mass, USA, November 2008.

[49] C. Wang and S. Xu, "Lagrange fuzzy interpolating controlling and simulation on semi-active suspension of vehicle," in *Proceedings of the 2nd International Conference on Intelligent Computing Technology and Automation (ICICTA '09)*, vol. 2, pp. 78–81, Changsha, Hunan, China, 2009.

[50] Y. Kim, R. Langari, and S. Hurlebaus, "Supervisory semiactive nonlinear control of a building-magnetorheological damper system," in *Proceedings of the American Control Conference (ACC '08)*, pp. 2540–2545, Seattle, Wash, USA, June 2008.

[51] C.-C. Hsu and Y. Y. Chien, "An intelligent fuzzy affective computing system for elderly living alone," in *Proceedings of the 5th International Conference on Hybrid Intelligent Systems (HIS '09)*, vol. 1, pp. 293–297, Shenyang, China, 2009.

6

Medical Robots: Current Systems and Research Directions

Ryan A. Beasley

Department of Engineering Technology and Industrial Distribution, Texas A&M University,
3367 TAMU, College Station, TX 77843, USA

Correspondence should be addressed to Ryan A. Beasley, beasley@entc.tamu.edu

Academic Editor: Farrokh Janabi-Sharifi

First used medically in 1985, robots now make an impact in laparoscopy, neurosurgery, orthopedic surgery, emergency response, and various other medical disciplines. This paper provides a review of medical robot history and surveys the capabilities of current medical robot systems, primarily focusing on commercially available systems while covering a few prominent research projects. By examining robotic systems across time and disciplines, trends are discernible that imply future capabilities of medical robots, for example, increased usage of intraoperative images, improved robot arm design, and haptic feedback to guide the surgeon.

1. Introduction

Medical robotics is causing a paradigm shift in therapy. The most widespread surgical robot, Intuitive Surgical's da Vinci system, has been discussed in over 4,000 peer-reviewed publications, was cleared by the United States' Food and Drug Administration (FDA) for multiple categories of operations, and was used in 80% of radical prostatectomies performed in the U.S. for 2008, just nine years after the system went on the market [1–3]. The rapid growth in medical robotics is driven by a combination of technological improvements (motors, materials, and control theory), advances in medical imaging (higher resolutions, magnetic resonance imaging, and 3D ultrasound), and an increase in surgeon/patient acceptance of both laparoscopic procedures and robotic assistance. New uses for medical robots are created regularly, as in the initial stages of any technology-driven revolution.

In 1979, the Robot Institute of America, an industrial trade group, defined a robot as "a reprogrammable, multifunctional manipulator designed to move materials, parts, tools, or other specialized devices through various programmed motions for the performance of a variety of tasks." Such a definition leaves out tools with a single task (e.g., stapler), anything that cannot move (e.g., image analysis algorithms), and nonprogrammable mechanisms (e.g., purely manual laparoscopic tools). As a result, robots are generally indicated for tasks requiring programmable motions, particularly where those motions should be quick, strong, precise, accurate, untiring, and/or via complex articulations. The downsides generally include high expense, space needs, and extensive user training requirements. The greatest impact of medical robots has been in surgeries, both radiosurgery and tissue manipulation in the operating room, which are improved by precise and accurate motions of the necessary tools. Through robot assistance, surgical outcomes can be improved, patient trauma can be reduced, and hospital stays can be shortened, though the effects of robot assistance on long-term results are still under investigation.

Medical robots have been reviewed in various papers since the 1990s [4–7]. Many such reviews are domain-specific, for example, focusing on surgical robots, urological robots, spine robots, and so forth [8–13]. For an overview of the basic science behind medical robots (e.g., kinematics, degrees of freedom, ergonomics, and telesurgery) along with a discussion of urologic robotic systems, see Challacombe and Stoianovici [14]. Similarly focused on surgery, Kenngott et al. provide a recent Medline metareview on the outcomes of laparoscopic robot-assisted surgeries (urologic, gynecologic, and abdominal) [15], while Gomes covers market drivers and roadblocks [16], and Okamura et al. explore big picture issues like societal drivers, quantitative diagnosis, and system adaptation/learning [17]. The most recent coverage of medical robots across various domains was by Najarian et al. and the articles collected by Rosen et al. [18, 19].

This paper provides an overview of the impact of robots in multiple medical domains. This work builds on top of the aforementioned papers by providing an updated review of various robotic systems, covering system improvements (technical and regulatory) and changes in manufacturers due to corporate buyouts. Furthermore, to the author's knowledge this work covers more breadth in the medical domains benefiting from robot assistance than any other single paper, and thus provides a big picture view of how robots are improving the medical field. Though primarily focused on commercially available medical robotic systems and the history that describes their evolution, this paper also covers multiple next-generation systems and discusses their potential impacts on the future of the medical field.

2. Neurological

Brain surgery involves accessing a buried target surrounded by delicate tissue, a task that benefits from the ability for robots to make precise and accurate motions based on medical images [18, 20, 21]. Thus, the first published account investigating the use of a robot in human surgery was in 1985 for brain biopsy using a computed tomography (CT) image and a stereotactic frame [22]. In that work, an industrial robot defined the trajectory for a biopsy by keeping the probe oriented toward the biopsy target even as the surgeon manipulated the approach. This orientation was determined by registering a preoperative CT with the robot via fiducials on a stereotactic frame attached to the patient's skull. That project was discontinued after the robot company was bought out, due to safety concerns of the new owning company, which specified that the robot arm (54 kg and capable of making 0.5 m/s movements) was only designed to operate when separated by a barrier from people. Then in 1991, the Minerva robot (University of Lausanne, Switzerland) was designed to direct tools into the brain under real-time CT guidance. Real-time image guidance allows tracking of targets even as the brain tissue swells, sags, or shifts due to the operation. Minerva was discontinued in 1993 due to the limitation of single-dimensional incursions and its need for real-time CT [23].

The currently available neurosurgery robots exhibit a purpose similar to historical systems, namely, image-guided positioning/orientation of cannulae or other tools (Figure 1). The NeuroMate (by Renishaw, previously by Integrated Surgical Systems, previously by Innovative Medical Machines International) has a Conformité Européenne (CE) mark and is currently used in the process for FDA clearance (the previous generation was granted FDA clearance in 1997) [24]. In addition to biopsy, the system is marketed for deep brain stimulation, stereotactic electroencephalography, transcranial magnetic stimulation, radiosurgery, and neuroendoscopy. Li et al. report in-use accuracy as submillimeter for a frame-based configuration, the same level of application accuracy as bone-screw markers with infrared tracking, and an accuracy of 1.95 mm for the frameless configuration [25].

Another robotic system, Pathfinder (Prosurgics, formerly Armstrong Healthcare Ltd.), has been cleared by the FDA for neurosurgery (2004) [26]. Using the system, the surgeon specifies a target and trajectory on a pre-operative medical image, and the robot guides the instrument into position with submillimeter accuracy [27]. Reported uses of the system include guiding needles for biopsy and guiding drills to make burr holes [28].

Renaissance (Mazor Robotics, the first generation system was named SpineAssist) has FDA clearance (2011) and CE mark for spinal surgery, and a CE mark for brain operations (2011) [29]. The device consists of a robot the size of a soda can that mounts directly onto the spine and provides tool guidance based on planning software for various procedures including deformity corrections, biopsies, minimally invasive surgeries, and electrode placement procedures. Renaissance includes an add-on for existing fluoroscopy C-arms that provides 3D images for intraoperative verification of implant placement. Studies show increased implant accuracy and provide evidence that the Renaissance/SpineAssist may allow significantly more implants to be placed percutaneously [30].

3. Orthopedics

The expected benefit of robot assistance in orthopedics is accurate and precise bone resection [31, 32]. Through good bone resection, robotic systems (Figure 2) can improve alignment of implant with bone and increase the contact area between implant and bone, both of which may improve functional outcomes and implant longevity [5]. Orthopedic robots have so far targeted the hip and knee for replacements or resurfacing (the exception being the Renaissance system in Section 2 and its use on the spine). Initial systems required the bones to be fixed in place, and all systems use bone screws or pins to localize the surgical site.

The initial robot assistance for orthopedics came via Robodoc (Curexo Technology Corp, originally by Integrated Surgical Systems), first used in 1992 for total hip replacement [5, 33]. Robodoc has received a CE mark (1996), and FDA clearance for total hip replacement (1998) and total knee replacement (2009) [34]. The robot is used in conjunction with OrthoDoc, a surgical planner, with which the surgeon plans bone milling is based on preoperative CT. During the procedure, the patient's leg is clamped to the robot's pedestal, and a second clamp locates the femoral head to automatically halt the robot if the leg moves. The Robodoc then performs the milling automatically based on the surgical plan. Many initial attempts in surgical robotics involved such autonomous motions, which generated concerns about patient and doctor safety. To address those concerns, Robodoc has force sensing on all axes, as well as a six-axis force sensor at the wrist [35]. The force sensing is used for safety monitoring, to allow the surgeon to manually direct the robot arm and to vary the velocity of tool motion as a function of the forces experienced during the milling operation.

Though no longer for sale, CASPAR (Computer Assisted Surgical Planning and Robotics) was another robotic system for knee and hip surgery, introduced in 1997 by OrtoMaquet,

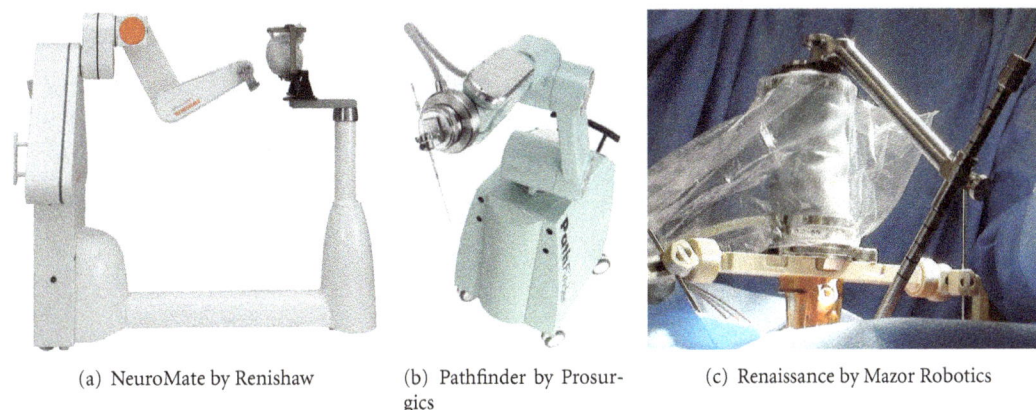

(a) NeuroMate by Renishaw   (b) Pathfinder by Prosurgics   (c) Renaissance by Mazor Robotics

FIGURE 1: Neurosurgery robots for image-guided tool positioning/orientation. The NeuroMate image is ©2012 Renishaw. The Pathfinder image is ©2012 Prosurgics. The Renaissance image is ©2011 Mazor Robotics Ltd. All rights reserved with nonexclusive permission.

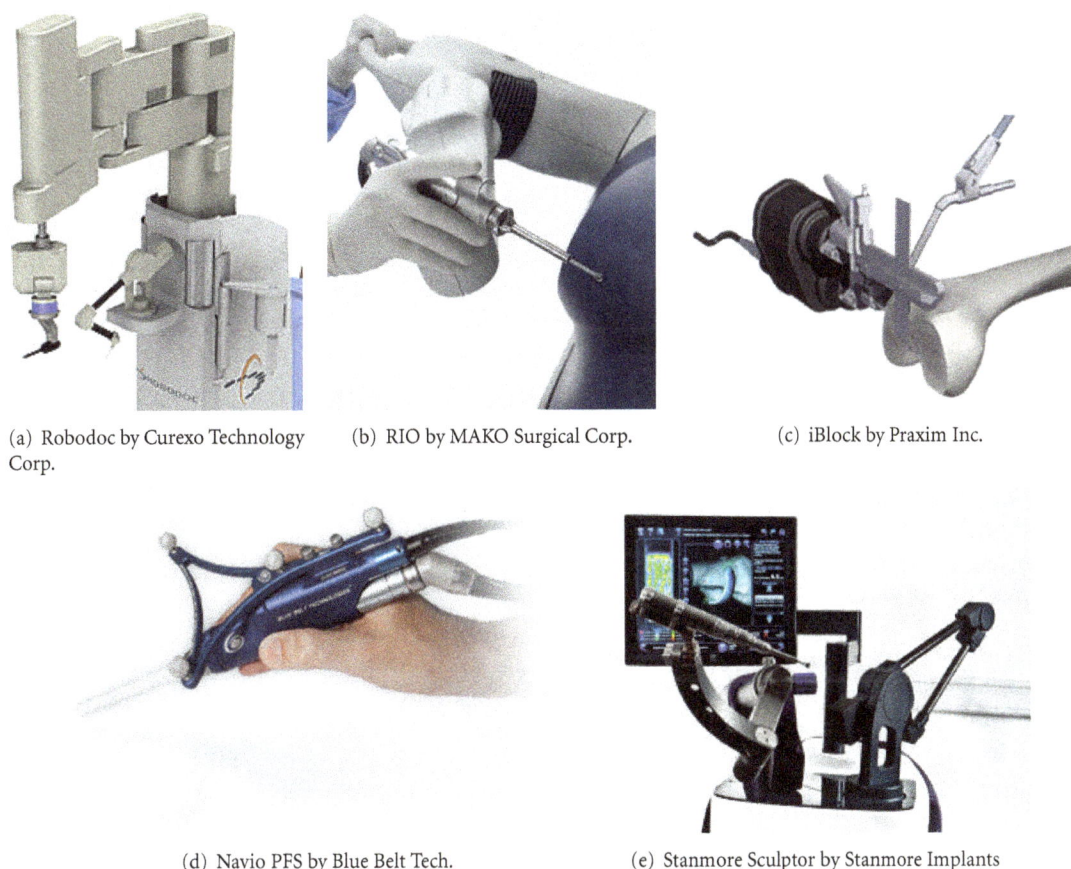

(a) Robodoc by Curexo Technology Corp.   (b) RIO by MAKO Surgical Corp.   (c) iBlock by Praxim Inc.

(d) Navio PFS by Blue Belt Tech.   (e) Stanmore Sculptor by Stanmore Implants

FIGURE 2: Orthopedic robots for accurate bone resection. The Robodoc image is ©2012 Curexo Technology Corp. The RIO image is ©2012 MAKO Surgical Corp. The iBlock image is ©2012 Praxim Inc. The Navio PFS image is ©2012 Blue Belt Tech. The Stanmore Sculptor image is ©2012 Stanmore Implants.

acquired by Getinge in 2000, acquired and discontinued by Universal Robot Systems (URS) in 2001. The robot was a direct competitor to Robodoc. It automatically performed bone drilling from a preoperative plan based on CT data.

In 2008, the RIO robotic arm (MAKO Surgical Corp, previous generation called the Tactile Guidance System)

was released and received FDA clearance. The RIO is used for implantation of medial and lateral unicondylar knee components, as well as for patellofemoral arthroplasty [36, 37]. As part of the trend away from autonomous robot motions, both the RIO and the surgeon simultaneously hold the surgical tool, with which the surgeon moves about the surgical site. The arm is designed to be low friction and low inertia, so that the surgeon can easily move the tool, backdriving the arm's joint motors in the process [19]. The arm's purpose is to act as a haptic device during the milling procedure, resisting motions outside of the planned cutting envelope by pushing back on the surgeon's hand. Unlike other orthopedic systems, the RIO does not require the bone to be fixed in place, instead relying on a camera system to track bone pins and tools intraoperatively and instantaneously registering the planned cutting envelope to the patient in the operating room. With this configuration, the system has promise for use as a surgical training tool.

Further reducing robotic influence on the cutting tool, the iBlock (Praxim Inc., an Orthopaedic Synergy Inc. company, previous generation the Praxiteles, FDA clearance 2010) is an automated cutting guide for total knee replacement [38]. The iBlock is mounted directly to the bone, preventing any relative motion between the robot and the bone and aligns a cutting guide that the surgeon uses to manually perform planar cuts based on a preoperative plan. Koulalis et al. report reduced surgical time and increased cut accuracy compared with freehand navigation of cutting blocks [39].

The Navio PFS (Blue Belt Technologies, CE mark 2012) does not require a CT scan for unicondylar knee replacement, instead it uses intraoperative planning [40, 41]. The drill tool is tracked during the procedure, and the drill bit is retracted when it would leave the planned cutting volume. Limited information is available on the system due to its recent development.

The Stanmore Sculptor (Stanmore Implants, previous generation the Acrobot Sculptor by Acrobot Company Ltd.) is a synergistic system similar to the RIO, with active constraints to keep the surgeon in the planned workspace [42]. The company's "Savile Row" system tailors a personalized unicondylar knee implant to the patient, incorporates the 3D model of that implant into the surgical planning interface, and uses active constraints with the Stanmore Sculptor to ensure proper preparation of the bone surface. The system does not currently have FDA clearance, but has been in use in Europe since 2004.

4. General Laparoscopy

Prior to the 1980s, surgical procedures were performed through sizable incisions through which the surgeon could directly access the surgical site. In the late 1980s, camera technology had improved sufficiently for laparoscopy (a.k.a. minimally invasive surgery), in which one or more small incisions are used to access the surgical site with tools and camera [43]. Laparoscopy significantly reduces patient trauma in comparison with traditional "open" procedures, thereby reducing morbidity and length of hospital stay, but at the cost of increased complexity of the surgical task. Compared with open surgery, in laparoscopy the surgeon's feedback from the surgical site is impaired (reduced visibility and cannot manually palpate the tissue) and tool control is

reduced ("mirror-image" motions due to fulcrum effect and loss of degrees of freedom in tool orientation) [16, 44, 45].

Robot assistance for soft-tissue surgery was first done in 1988 using an industrial robot to actively remove soft tissue during transurethral resection of the prostate [5]. As with neurosurgery, the researchers deemed use of an industrial robot in the operating room to be unsafe. The experience provided the impetus for a research system, Probot, with the same purpose [46].

4.1. Zeus. Commercial robotic systems for laparoscopy started with Computer Motion's Aesop (discontinued, FDA clearance 1993) for holding endoscopes [47]. Aesop was clamped to the surgical table or to a cart, and either moved the endoscope under voice control or allowed the endoscope to be manually positioned. In 1995, Computer Motion combined two tool-holding robot arms with Aesop to create the Zeus system (discontinued, FDA clearance 2001) [48]. The Zeus's tool arms were teleoperated, following motions the surgeon made with instrument controls (a.k.a. "master" arms or joysticks) at the surgeon console. Technically, the Zeus is not a robot because it does not follow programmable motions, but rather is a remote computer-assisted telemanipulator with interactive robotic arms. To improve precision in tool motion, the Zeus filters out hand tremor, and can scale large hand motions by the surgeon down to short and precise motions by the tool. As described by Marescaux et al., the Zeus was used in the Lindbergh Operation, the first surgery was (cholecystectomy) performed with the surgeon and patient being separated by a distance of several thousand kilometers [49].

4.2. da Vinci. Meanwhile, Intuitive Surgical Inc. was developing the da Vinci (initial FDA clearance 1995, Figure 3(a)). Like the Zeus, the da Vinci is a teleoperated system, wherein the surgeon manipulates instrument controls at a console and the robot arms follow those motions with motion scaling and tremor reduction. Also like the Zeus, the da Vinci was initially offered with three arms to hold two tools and an endoscope, which are mounted to a single bedside cart.

The da Vinci system provides several technical enhancements over the Zeus. The grasper tools have two degrees of freedom inside the patient, the EndoWrist (Figure 3(b)), an enhanced articulation that increases the ease of suturing and other complex manipulations. The console puts increased emphasis on surgeon ergonomics and incorporates a separate video screen for each eye to display 3D video from the 3D endoscope. The motions of the surgeon's hands are mapped to motions of the operational ends of the tools, providing a more intuitive control than the "mirror-image" laparoscopic mapping. In 2003, Intuitive Surgical began selling a fourth arm for the da Vinci, and Intuitive Surgical and Computer Motion were merged (discontinuing the Zeus).

The da Vinci system is the only surgical robot with over a thousand systems installed worldwide and has been sold in four models so far: Standard (1999), S (2006), Si (2009), and Si-e (2010) [50, 51]. The S model increased the image resolution, redesigned the patient-side manipulators

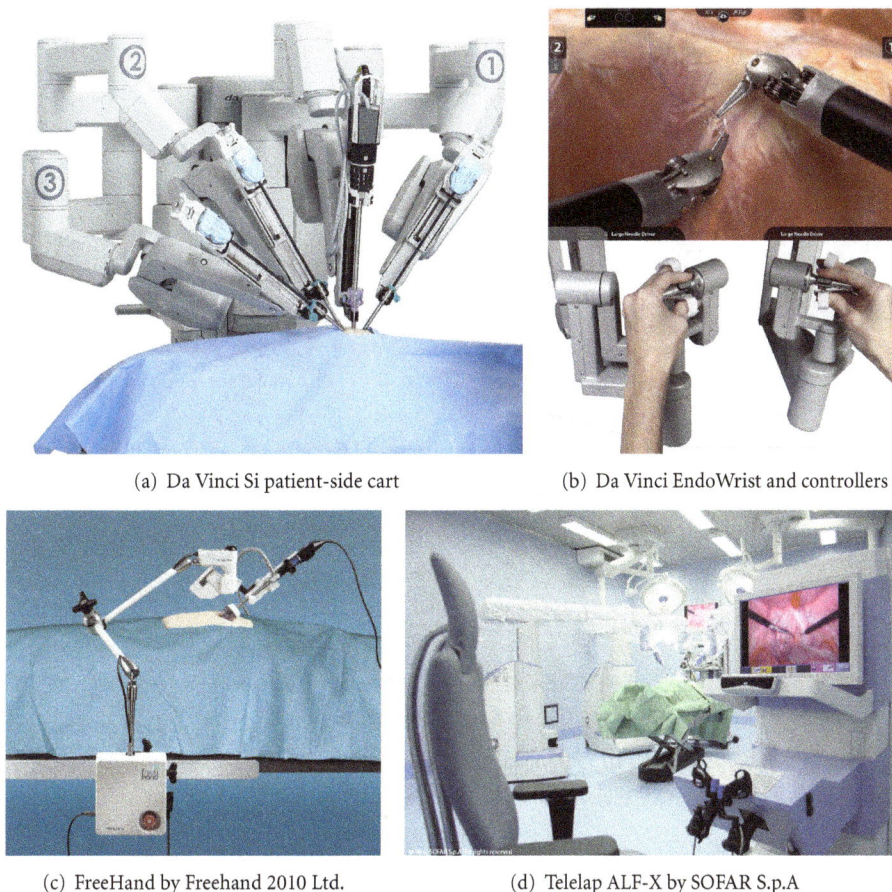

(a) Da Vinci Si patient-side cart

(b) Da Vinci EndoWrist and controllers

(c) FreeHand by Freehand 2010 Ltd.

(d) Telelap ALF-X by SOFAR S.p.A

FIGURE 3: Laparoscopy robots. The da Vinci Si, by Intuitive Surgical Inc. (a) Cart and (b) image mosaic showing the tool tips with EndoWrist articulation, and the instrument controls. (c) FreeHand, a next-generation endoscope holder. (d) A computer model of the Telelap ALF-X, by SOFAR S.p.A. The da Vinci images are ©2012 Intuitive Surgical, Inc. The FreeHand image is ©2012 Freehand 2010 Ltd. The Telelap ALF-X image is ©2012 SOFAR S.p.A. All rights reserved.

to enable multiquadrant access, and shortened setup time. The Si model further improved the visual resolution, refined the instrument controllers, and increased the ergonomics and ease for the surgeon to provide input to the system. The Si-e model is a 3-arm system that is fully upgradeable to the Si model. Continuing the da Vinci focus on improved visualization, the Firefly Fluorescence Imaging add-on product combines fluorescent dye and a special endoscope to identify vasculature beneath the tissue surface.

The da Vinci was initially cleared for general laparoscopy, became commonly used for radical prostatectomy, and is now cleared by the FDA for various procedures [52, 53]. Even so, as with most or even all robotic systems, long-term benefits continue to be uncertain [15, 54]. The enhanced endoscopic visualization and increased tool articulation are commonly considered improvements, but detractors point out the system's expense (between $1 M and $2.3 M), the reduced patient access due to the amount of space the arms take over/around the patient, and the significant amount of training necessary for the best outcomes [55, 56]. To address this last point, the Si model also allows dual console use for training and collaboration, in which both consoles get the

same images and can cooperatively control the instruments [57]. Additionally, the da Vinci Skills Simulator is an add-on case that can be used with an Si or Si-e console to practice operations in a virtual environment [58].

In an attempt to further reduce patient trauma, surgeons are exploring Single-Port Access (SPA), LaparoEndoscopic Single-Site surgery (LESS), and Natural Orifice Transluminal Endoscopic Surgery (NOTES) [59, 60]. To meet this need, Intuitive Surgical has recently developed the Single-Site platform for the da Vinci Si model. The Single-Site platform passes two semirigid tools and the endoscope through a single multichannel port, reducing the number of incisions but preventing EndoWrist articulation [61].

4.3. FreeHand. The FreeHand robot (Freehand 2010 Ltd., previously Freehand Surgical, previously Prosurgics, the previous generation was called EndoAssist, FDA clearance and CE mark 2009) is a next-generation endoscope holder. The arm (Figure 3(c)) is more compact, easier to setup, and cheaper than its predecessor. Furthermore, endoscope motion is controlled by gentle head motions by the surgeon, which are tracked with an optical system.

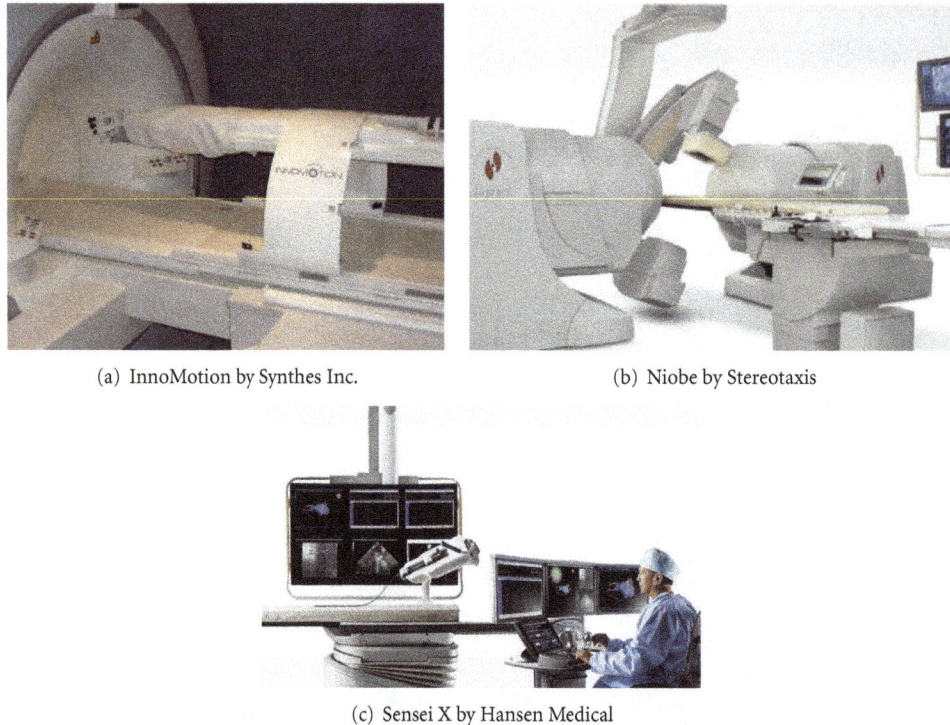

(a) InnoMotion by Synthes Inc.

(b) Niobe by Stereotaxis

(c) Sensei X by Hansen Medical

FIGURE 4: Real-time image guided percutaneous (a) and catheter robots ((b) and (c)). The InnoMotion image is ©2012 Synthes Inc. The Niobe image is ©2012 Stereotaxis. The Sensei X image is ©2012 Hansen Medical.

4.4. Telelap ALF-X. SOFAR S.p.A has developed Telelap ALF-X (CE mark 2011, Figure 3(d)), a four-armed surgical robotic system, to compete with the da Vinci [62]. The system uses eyetracking to control the endoscopic view and to enable activation of the various instruments. Compared to the da Vinci, the system moves the base of the manipulators away from the bed (about 80 cm) and has a realistic tactile-sensing capability due to a patented approach to measure tip/tissue forces from outside the patient, with a sensitivity of 35 grams. The system has been used in animal trials demonstrating a significant reduction in the time for cholecystectomy compared with a "conventional telesurgical system" [62].

5. Percutaneous

Noncatheter percutaneous procedures employ needles, cannulae, and probes for biopsy, drainage, drug delivery, and tumor destruction. During the procedure, accurate targeting can be reduced by soft tissue displacements that occur due to patient breathing, changes in posture, or tissue forces exerted during the insertion. Two options to guide a needle to its target are tissue modeling for needle steering and three-dimensional intraoperative imaging [63]. Unfortunately, tissue modeling is excessively complex [64]. So following the latter approach, InnoMotion (Synthes Inc., previously by Innomedic GmbH, CE mark 2005) is a robot arm designed to operate within a CT or magnetic resonance imaging (MRI) machine [65–67]. For MRI-compatibility, the arm (Figure 4(a)) is pneumatically actuated and joint sensing is via MRI-compatible encoders.

6. Steerable Catheters

Vascular catheterization is used to diagnose and treat various cardiac and vasculature diseases, including direct pressure measurements, biopsy, ablation for atrial fibrillation, and angioplasty for obstructed blood vessels [68–70]. The catheter is inserted into a blood vessel and the portion external to the patient is manipulated to move the catheter tip to the surgical site, while fluoroscopy provides image guidance. Due to the supporting tissue, catheters only require three degrees of freedom, typically: tip flexion, tip rotation, and insertion depth. Possible benefits of robot-steered catheters are shorter procedures, reduced forces exerted on the vasculature by the catheter tip, increased accuracy in catheter positioning, and teleoperation (reducing exposure of the physician to radiation) [71].

The Sensei X (Hansen Medical, FDA clearance and CE mark 2007, previous generation the Sensei, Figure 4(c)) uses two steerable sheaths, one inside the other, to create a tight bend radius [72–74]. The sheaths are steered via a remotely operated system of pulleys. IntelliSense force sensing allows constant estimation of the contact forces by gently pulsing the catheter a short distance in and out of the steerable inner sheath and measuring forces at the proximal end of the catheter. These forces are communicated visually as well as through a vibratory feedback to the surgeon's hand on the "3D joystick". Corindus's CorPath 200 is a direct competitor with the Sensei X, but is not yet commercially available.

The Niobe (Stereotaxis, CE mark 2008, FDA clearance 2009) is a remote magnetic navigation system, in which

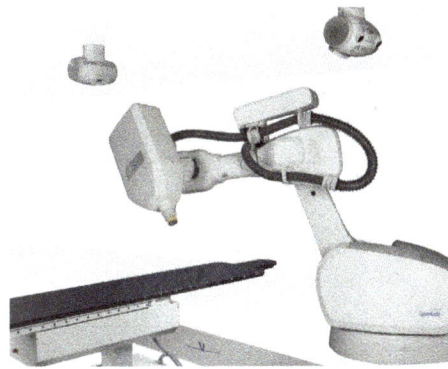

(a) CyberKnife by Accuray Inc.   (b) Novalis with TrueBeam STx by BrainLab Inc. and Varian Medical Systems

FIGURE 5: Radiosurgery robots use X-ray images taken during the treatment to control robotic patient tables, ensuring accurate targeting of the radiosurgery beams. The Cyberknife image and any Accuray trademarks or logos are used with permission from Accuray Incorporated. The image of the Novalis with TrueBeam STx is ©2012 BrianLab Inc.

a magnetic field is used to guide the catheter tip [75]. The magnetic field is generated by two permanent magnets contained in housings on either side of a fluoroscopy table (Figure 4(b)). The surgeon manipulates a joystick to specify the desired orientation of the catheter tip, causing the orientations of the magnets to vary under computer-control, and thereby controlling the magnetic field. A second joystick controls advancement/retraction of the catheter. Chun et al. report significant improvements in surgical outcomes due to advances in the design of magnetically guided catheters [76].

7. Radiosurgery

Radiosurgery is a treatment (not a surgery), in which focused beams of ionizing radiation are directed at the patient, primarily to treat tumors [77, 78]. By directing the beam through the tumor at various orientations, high-dose radiation is delivered to the tumor while the surrounding tissue receives significantly less radiation. Prior to real-time tissue tracking, radiosurgery was practically limited to treating the brain using stereotactic frames mounted to the skull with bone screws. Now that real-time tissue tracking is feasible, systems are commercially available.

The CyberKnife (Accuray Inc., FDA cleared 1999, Figure 5(a)) is a frameless radiosurgery system consisting of a robotic arm holding a linear accelerator, a six degree of freedom robotic patient table called the RoboCouch, and an X-ray imaging system that can take real-time images in two orthogonal orientations simultaneously [79, 80]. The two simultaneous, intraoperative X-ray images are not sufficient to provide good definition of the tumor, but are used to register a high-definition preoperative CT image. The robotic arm can then provide the preplanned radiation dosage with a wide range of orientations. For targets that move during treatment (e.g., due to breathing), the optional synchrony system can optically track the tissue surface, correlate the motion of the tissue surface to the motion of radio-opaque fiducials inserted near the target, and thus continuously predict target motion [81]. The intraoperative

tracking obviates the need for a stereotactic frame, reducing patient trauma and making it practical to fractionate the dosage over longer time periods.

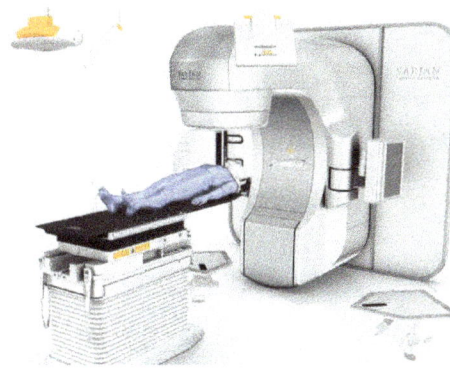

The Novalis with TrueBeam STx (BrainLab Inc. and Varian Medical Systems, previously Novalis and Trilogy, initial FDA clearance 2000, Figure 5(b)) is also a frameless system with a linear accelerator, but with micro-multileaf collimators for beam shaping [82–84]. Similar to CyberKnife, intraoperative X-rays are compared with a CT, and skin-mounted fiducials are optically tracked in real-time. The delivery system also includes cone beam CT. The patient is moved into position on top of a six degree of freedom robotic couch. The major differences between Cyberknife and Novalis are that the Cyberknife radiation source has more degrees of freedom to be oriented around the patient while the Novalis can shape the radiation beam and claim reduced out-of-field dosage [85, 86].

8. Emergency Response

Few medical robot systems are suitable for use outside of the operating room, despite significant research funding on medical devices for disaster response and battlefield medicine. Typical goals for such research include improved extraction of patients from dangerous environments, rapid diagnosis of injuries, and semiautonomous delivery of life-saving interventions. Current Emergency Response robots are little more than single-motor systems, but those systems can be controlled by health monitors to minimize the necessary attention by Emergency Responders. Such a feedback control makes it more likely that such systems will be autonomous, for example, automated external defibrillators.

The AutoPulse Plus (ZOLL Medical Corp., previously by Revivant) is an automated, portable device that combines the functions of the AutoPulse (FDA clearance 2008, Figure 6(a)) cardiopulmonary resuscitation device and the E Series monitor/defibrillator (FDA clearance 2010) [87, 88]. Consisting of a half-backboard containing a battery-powered motor that actuates a chest band, the AutoPulse rhythmically tightens

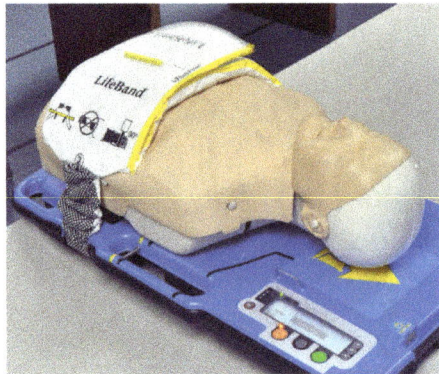

(a) AutoPulse by ZOLL Medical Corp.

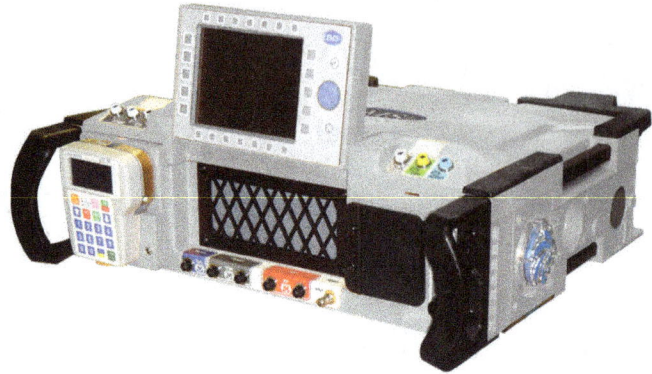

(b) LS-1 by Integrated Medical Systems Inc.

FIGURE 6: Commercially available Emergency Response robots perform simple actuations compared to robots in other medical disciplines, but their actions are tightly coupled with patient measurements. The AutoPulse image is ©2012 ZOLL Medical Corp. The LS-1 image is ©2012 Integrated Medical Systems Inc.

the band to perform chest compressions. The tightness of the band during compressions is a function of the patient's resting chest size, to adjust for interpatient variability. Meanwhile, the E Series monitor/defibrillator measures the rate and depth of chest compressions in real time and filters cardiopulmonary resuscitation artifacts from the electrocardiogram signal. If combined with an automatic battery-powered ventilator, for example, the SAVe (AutoMedx Inc., FDA clearance 2007), basic cardiopulmonary emergency response treatments could be automated while on battery power.

The LS-1 "suitcase intensive care unit" (Integrated Medical Systems Inc., previous generation called MedEx 1000, previous generation called LSTAT, FDA clearance 2008, Figure 6(b)) takes an inclusive approach to portable life support [89]. The system contains a ventilator with oxygen and carbon dioxide monitoring, electrocardiogram, invasive and noninvasive blood pressure monitoring, fluid/drug infusion pumps, temperature sensing, and blood oxygen level measurement. The LS-1 is battery powered and can be powered by facility or vehicular electrical sources. The system is FDA-cleared for remote control of its diagnostic and therapeutic capabilities.

9. Prosthetics and Exoskeletons

Microprocessor-controlled prosthetics have been available since 1993, specifically the Intelligent Prosthesis knee (Chas. A. Blatchford & Sons, Ltd.). Several microprocessor-controlled prosthetics exist today, predominantly for knee prosthetics, hand prosthetics, and exoskeletons. For example, one current generation knee prosthetic is the C-leg (Otto Bock, FDA clearance 1999, CE mark) which is designed to automatically adjust the swing phase dynamics and improve stability during the stance phase by controlling knee flexion [90]. An example of hand prosthetic is the i-limb ultrahand (Touch Bionics, previous version i-limb hand, FDA clearance and CE mark), the first commercially available hand prosthesis with five individually powered digits, controlled via

myoelectric signals generated by muscles in the remaining portion of the patient's limb [91]. For wheelchair users, the ReWalk (Argo Medical Technologies, FDA clearance 2011, CE mark 2010) is one walking assistance exoskeleton that allows users to stand, walk, and climb stairs and is controlled with a wrist-mounted remote and a posture detection sensor [92]. Significant research on exoskeletons is ongoing, such as the research on upper-limb exoskeletons by Rosen and Perry [93]. For further information in the area of prosthetics and exoskeletons see the works by Kazerooni [94] and Bogue [95].

10. Assistive and Rehabilitation Systems

Assistive robotic systems are designed to allow people with disabilities more autonomy, and they cover a wide range of everyday tasks. In 1992, Handy 1 (Rehab Robotics, Ltd.) became the first commercial assistive robot [96]; it interacts with different trays for tasks such as eating, shaving, and painting, and it is controlled by a single switch input to select the desired action. One task-specific system is the Neater Eater (Neater Solutions Ltd.), a modular device that scoops food from a plate to a person's mouth, and can be controlled manually or via head or foot switches. More general systems rely on arms with many degrees of freedom, such as Exact Dynamics' iARM, a robotic arm with a two-fingered grasper, that attaches to electric wheelchairs and can be controlled via keypad, joystick, or single button.

Rehabilitation systems can be similar to assistive systems, but are designed to facilitate recovery by delivering therapy and measuring the patient's progress, often following a stroke [97]. The Mobility System (Myomo, Inc.) is a wearable robotic device that moves the patient's arm in response to his/her muscle signals, thus creating feedback to facilitate muscle reeducation. The InMotion (Interactive Motion Technologies, based on the MIT-MANUS research platform) is a robotic arm that moves, guides, or perturbs the patient's arm within a planar workspace, while recording motions, velocities, and forces to evaluate progress [98]. For information on research efforts, see Dallaway et al. [99] for an overview of thirty assistive and rehabilitation systems, such as the MASTER II system that uses a rail-mounted robotic arm to make manually controlled, remote-controlled, or preprogrammed motions for various domestic and office tasks [100]. Difficulties in developing rehabilitation robots and potential future uses are investigated by Ceccarelli [101].

11. Current Research and Development in Medical Robotics

Many more medical robots are currently being researched [16, 19, 21]. Such research will lead to the new capabilities of future commercial systems. This section discusses just a few systems of note.

11.1. RAVEN and MiroSurge. Two prominent academic robot-assisted surgical systems are currently used for research into endoscopic telesurgery: RAVEN II and Miro-Surge. The RAVEN II (University of Washington and UC Santa Cruz) is a teleoperated laparoscopic system that was designed to maximize surgical performance based on objective clinical measurements [102–104]. The system has two patient-side arms that are cable-driven with 7 degrees of freedom each. The arm kinematics are based on a spherical mechanism such that the tool always passes through a remote center (e.g., the insertion point for minimally invasive surgery). The length and angles of the links were optimized to maximize performance throughout the workspace. The arms are lighter, smaller, and less expensive than current robotic systems for laparoscopy. The instrument controllers are haptic devices, allowing force feedback on the operator's hands based on tool forces or virtual fixtures (e.g., forbidden regions) defined with respect to patient anatomy (see [105–107] for the impact of haptics on surgery). Teleoperation experiments have been conducted with the RAVEN, including routing the data transmission through an unmanned aircraft. In February 2012, five systems were provided to various other surgical robotics research labs to spur collaboration and further development efforts.

In another endoscopic research effort, the German Aerospace Center (DLR) is developing MiroSurge to be highly versatile with respect to the number of surgical domains, arm-mounting locations, number of robots, different control modes (e.g., control of position versus control of force), and the ability to integrate with other technologies [108]. The expectation is for a base robot system to hold specialized instruments, such as DLR's MICA instrument (which is itself a robotic tool with 3 degrees of freedom and force sensing) [109]. By using a general robotic base to hold a specialized robotic instrument that has its own motors, sensors, and control electronics, the same base system can be specialized for various procedures just by switching the instrument. The base robot, the DLR Miro, masses 10 kg with a 3 kg payload and has serial kinematics that resemble the kinematics of the human arm, with joint ranges and link lengths optimized based on certain medical procedures

[110]. Unlike the RAVEN II, the MIRO arm does not have a remote center of motion, and thus must be controlled to direct the instrument through any insertion point, but is more easily able to handle moving insertion points (e.g., through the chest wall during respiration).

11.2. Amadeus. Titan Medical Inc. is currently developing Amadeus, a four-armed laparoscopic surgical robot system, to compete with Intuitive Surgical's da Vinci system. The Amadeus uses snakelike multiarticulating arms for improved maneuverability, and the system is being designed to facilitate teleoperation for long-distance surgery. Human trials are planned for late 2013.

11.3. NeuroArm and MrBot. At least two renowned research systems are investigating improved MR-compatible robots. The neuroArm (University of Calgary, MacDonald Dettwiler and Associates, IMRIS) is a two-armed, MRI-guided neurosurgical robot actuated via piezoelectric motors [111, 112]. The neuroArm end effectors are equipped with 3 degrees of freedom optical force sensors and are accurate to tens of micrometers. The MrBot (Johns Hopkins University) is a parallel linkage arm designed for MRI-guided access of the prostate gland, actuated by novel pneumatic stepper motors for reduced MR interference [113].

11.4. TraumaPod. TraumaPod (highly collaborative, led by SRI International) is a semi-autonomous telerobotic surgical system designed to be rapidly deployable [114]. The surgical cell consists of a surgical robot (da Vinci for Phase I testing), Scrub Nurse Subsystem, Tool Rack System, Supply Dispensing System, Patient Imaging System (a movable X-ray tube), predecessor of the aforementioned LS-1 ("suitcase intensive care unit"), and Supervisory Controller System. The TraumaPod has demonstrated successful teleoperation of a bowel closure and shunt placement on a phantom without a human in the surgical cell. That success implies the potential for increased automation in the operating room, though challenges were reported in sterilization, anesthesia, and robustness.

11.5. HeartLander. The heart has long been a target for surgical robots and various systems continue to investigate how best to treat cardiac diseases, particularly while the heart is beating (e.g., see Section 6) [115]. The HeartLander (HeartLander Surgical) is a minimally invasive robot that uses suction to crawl around the surface of the heart [116, 117]. The system is designed for intrapericardial drug delivery, cell transplantation, epicardial atrial ablation, and other such procedures.

11.6. Robots In Vivo. Various groups are expanding and exploring the da Vinci system's approach to enhance surgery by increasing the dexterity of the tool inside the patient. One such example is the University of Nebraska's laparoscopic robotic system for research into single-site surgeries [118]. The system has two arms with six degrees of freedom each, and those arms are fully inserted into the abdomen.

The expectation is that, by increasing the tool's dexterity inside the patient, fewer incisions will be needed to insert instruments because multiple tools/arms can pass through a single incision and then spread out inside the patient. Further, miniaturizing the robotics reduces the difficulty in working with (and around) the system in the operating room. A limited number of animal trials (colon resections) have been performed to demonstrate feasibility.

Swallowable capsules take patient trauma reduction to an extreme, but current systems are limited to diagnostic uses. Core temperature measurement has been FDA cleared since 1990, by CorTemp (HQ Inc., formerly HTI Technologies). More recently, capsule endoscopy systems (PillCam by Given Imaging with FDA clearance 2001, and EndoCapsule by Olympus with FDA clearance 2007) consist of a forward-looking wide-angle camera taking regularly timed pictures, a battery, and lights, all contained in a capsule [119, 120]. SmartPill (SmartPill Corp., FDA clearance 2006) utilizes multiple sensors to measure pressure, pH level, gastric emptying time, and bowel emptying time [121]. Sayaka (RF Co Ltd.) is a novel design, not FDA cleared, using a lateral-facing camera that rotates inside the capsule to image the entire tract and is designed without a battery, instead relying on an externally applied magnetic field for inductive power supply [122]. For the future, many enhancements have been proposed, including biopsy, real-time localization of the capsule, drug delivery, ultrasonic imaging, increasing motility by electrically inducing peristalsis, and utilizing an active locomotion system involving treads or legs [123].

In a more dramatic approach to *in vivo* robotics, micro/ nanotechnology is a multibillion dollar area of research [124, 125], including investigation for various medical robotic uses such as inexpensive directable drug delivery vehicles, radio-controlled biomolecules, tissue micromanipulation platforms, artificial mechanical white blood cells, and many other therapeutic approaches that may benefit from robots working at the cellular level [126–129]. Construction of functional systems is an ongoing area of research, particularly with respect to generating and powering motion. Many current prototypes are propelled and guided via magnetic fields, though some utilize external electrical energy sources [130, 131]. To the author's knowledge, clinical trials have not begun for any medical micro/nanorobot.

12. Discussion

Medical robotics is a young and relatively unexplored field made possible by technical improvements over the past couple of decades. Currently available systems have been available for too short time to allow long-term studies. Nor are the benefits potentially provided by medical robots fully understood. Medical robots have only passed through a few technological generations and the technology continues to change and leap into new areas. Yet by looking at the current market and representative research systems, educated guesses can be made about the impacts of robots on near-future medicine.

In surgical robotics, there has been a trend away from autonomous or even semiautonomous motions, and toward synergistic manipulation and virtual fixtures. Thus, the robot acts as a guidance tool, providing information (and possibly a physical nudge) to keep the surgeon on target. Such use requires accurate localization of the tissues in the surgical site, even as the tissues are manipulated during surgery. Improved imaging systems (e.g., Explorer, an intraoperative soft tissue tracker by Pathfinder Therapeutics [132]) or robot compatibility with MRI or CT will provide that localization. In particular, MRI-guided robots will benefit from intraoperative 3D images with excellent soft tissue contrast and accurate registration between the tool and the tissue, thus allowing precise virtual fixtures, "snap-to" and "stand-off" behaviors. Further, such imaging will allow modeling and rapid prototyping of patient-specific templates/jigs/implants.

The physical designs for medical robots will continue to improve, reducing expense and size, while minimizing or compensating for nonidealities such as flexion, for example, the CRIGOS robot [133]. With better physical designs, semiautonomous behavior will likely become more useful. "Macros" may become commonplace, wherein the surgeon presses a button and the robot performs a preprogrammed motion, such as passing a suture needle between graspers, or the Sensei's autoretract feature [13].

Robots will see more use for medical training purposes, bolstered by improved tissue-modeling capabilities, by the increasing objectivity in healthcare assessment, by advances in computer simulations, and as a result of increased data mining arising naturally from improved data connectivity between devices and between institutions. Some such systems are already available, such as the aforementioned da Vinci Skills Simulator, the Virtual I.V. Simulator by Laerdal, and the EndoscopyVR Surgical Simulator by CAE. For the same reasons, robotics will continue to make possible new medical procedures and treatments, such as new Single-Port Access procedures.

Even as robots are developed for new medical areas, other tools may encroach on medical needs currently filled by robots. Medical robots must develop a firm basis in improved medical outcomes, or risk being displaced by pharmaceuticals, tissue engineering, gene therapy, and rapid innovation in manual tools (e.g., the SPIDER Surgical System by TransEnterix, and the EndoStitch by Covidien). To that end, improvements in medical robotics must address and solve real problems in healthcare, ultimately providing a clear improvement in quality of life when compared with the alternatives.

References

[1] Intuitive Surgical Incorporation, "Webpage on da Vinci clinical evidence," March 2012, http://www.intuitivesurgical .com/company/clinical-evidence.

[2] Intuitive Surgical Incorporation, "Webpage on da Vinci regulatory approval," March 2012, http://www.intuitivesurgical .com/company/regulatory-clearance.html.

[3] H. Lavery, D. Samadi, and A. Levinson, "Not a zero-sum game: the adoption of robotics has increased overall prostatectomy utilization in the united states," in *Proceedings of the American Urological Association Annual Meeting, Poster Session*, Washington, DC, USA, 2011.

[4] R. D. Howe and Y. Matsuoka, "Robotics for surgery," *Annual Review of Biomedical Engineering*, vol. 1, pp. 211–240, 1999.

[5] B. Davies, "A review of robotics in surgery," *Proceedings of the Institution of Mechanical Engineers H*, vol. 214, no. 1, pp. 129–140, 2000.

[6] R. H. Taylor and D. Stoianovici, "Medical robotics in computer-integrated surgery," *IEEE Transactions on Robotics and Automation*, vol. 19, no. 5, pp. 765–781, 2003.

[7] A. R. Lanfranco, A. E. Castellanos, J. P. Desai, and W. C. Meyers, "Robotic surgery: a current perspective," *Annals of Surgery*, vol. 239, no. 1, pp. 14–21, 2004.

[8] P. Berkelman, J. Troccaz, and P. Cinquin, "Body-supported medical robots: a survey," *Journal of Robotics and Mechatronics*, vol. 16, pp. 513–519, 2004.

[9] L. Guo, X. Pan, Q. Li, F. Zheng, and Z. Bao, "A survey on the gastrointestinal capsule micro-robot based on wireless and optoelectronic technology," *Journal of Nanoelectronics and Optoelectronics*, vol. 7, no. 2, pp. 123–127, 2012.

[10] C. Stüer, F. Ringel, M. Stoffel, A. Reinke, M. Behr, and B. Meyer, "Robotic technology in spine surgery: current applications and future developments," *Intraoperative Imaging*, vol. 109, pp. 241–245, 2011.

[11] S. Badaan and D. Stoianovici, "Robotic systems: past, present, and future," in *Robotics in Genitourinary Surgery*, pp. 655–665, Springer, New York, NY, USA, 2011.

[12] I. Singh, "Robotics in urological surgery: review of current status and maneuverability, and comparison of robot-assisted and traditional laparoscopy," *Computer Aided Surgery*, vol. 16, no. 1, pp. 38–45, 2011.

[13] G. P. Moustris, S. C. Hiridis, K. M. Deliparaschos, and K. M. Konstantinidis, "Evolution of autonomous and semi-autonomous robotic surgical systems: a review of the literature," *International Journal of Medical Robotics and Computer Assisted Surgery*, vol. 7, no. 4, pp. 375–392, 2011.

[14] B. Challacombe and D. Stoianovici, "The basic science of robotic surgery," in *Urologic Robotic Surgery in Clinical Practice*, pp. 1–23, 2009.

[15] H. Kenngott, L. Fischer, F. Nickel, J. Rom, J. Rassweiler, and B. Muller-Stich, "Status of robotic assistance: a less traumatic and more accurate minimally invasive surgery?" *Langenbeck's Archives of Surgery*, vol. 397, no. 3, pp. 1–9, 2012.

[16] P. Gomes, "Surgical robotics: reviewing the past, analysing the present, imagining the future," *Robotics and Computer-Integrated Manufacturing*, vol. 27, no. 2, pp. 261–266, 2011.

[17] A. M. Okamura, M. J. Matarić, and H. I. Christensen, "Medical and health-care robotics," *IEEE Robotics and Automation Magazine*, vol. 17, no. 3, pp. 26–37, 2010.

[18] S. Najarian, M. Fallahnezhad, and E. Afshari, "Advances in medical robotic systems with specific applications in surgery—a review," *Journal of Medical Engineering and Technology*, vol. 35, no. 1, pp. 19–33, 2011.

[19] J. Rosen, B. Hannaford, and R. Satava, Eds., *Surgical Robotics: Systems Applications and Visions*, Springer, New York, NY, USA, 2011.

[20] N. Nathoo, M. C. Çavuşoğlu, M. A. Vogelbaum, and G. H. Barnett, "In touch with robotics: neurosurgery for the future," *Neurosurgery*, vol. 56, no. 3, pp. 421–431, 2005.

[21] T. Haidegger, L. Kovacs, G. Fordos, Z. Benyo, and P. Kazanzides, "Future trends in robotic neurosurgery," in *Proceedings of the 14th Nordic-Baltic Conference on Biomedical Engineering and Medical Physics (NBC '08)*, pp. 229–233, Springer, June 2008.

[22] Y. S. Kwoh, J. Hou, E. A. Jonckheere, and S. Hayati, "A robot with improved absolute positioning accuracy for CT guided stereotactic brain surgery," *IEEE Transactions on Biomedical Engineering*, vol. 35, no. 2, pp. 153–160, 1988.

[23] D. Glauser, H. Fankhauser, M. Epitaux, J. L. Hefti, and A. Jaccottet, "Neurosurgical robot Minerva: first results and current developments," *Journal of Image Guided Surgery*, vol. 1, no. 5, pp. 266–272, 1995.

[24] T. R. K. Varma and P. Eldridge, "Use of the NeuroMate stereotactic robot in a frameless mode for functional neurosurgery," *International Journal of Medical Robotics and Computer Assisted Surgery*, vol. 2, no. 2, pp. 107–113, 2006.

[25] Q. H. Li, L. Zamorano, A. Pandya, R. Perez, J. Gong, and F. Diaz, "The application accuracy of the NeuroMate robot—a quantitative comparison with frameless and frame-based surgical localization systems," *Computer Aided Surgery*, vol. 7, no. 2, pp. 90–98, 2002.

[26] P. Morgan, T. Carter, S. Davis et al., "The application accuracy of the pathfinder neurosurgical robot," in *International Congress Series*, vol. 1256, pp. 561–567, Elsevier, Amsterdam, The Netherlands, 2003.

[27] G. Deacon, A. Harwood, J. Holdback et al., "The pathfinder image-guided surgical robot," *Proceedings of the Institution of Mechanical Engineers H*, vol. 224, no. 5, pp. 691–713, 2010.

[28] J. Brodie and S. Eljamel, "Evaluation of a neurosurgical robotic system to make accurate burr holes," *International Journal of Medical Robotics and Computer Assisted Surgery*, vol. 7, no. 1, pp. 101–106, 2011.

[29] L. Joskowicz, R. Shamir, Z. Israel, Y. Shoshan, and M. Shoham, "Renaissance robotic system for keyhole cranial neurosurgery: in-vitro accuracy study," in *Proceedings of the Simposio Mexicano en Ciruga Asistida por Computadora y Procesamiento de Imgenes Mdicas (MexCAS '11)*, 2011.

[30] D. P. Devito, L. Kaplan, R. Dietl et al., "Clinical acceptance and accuracy assessment of spinal implants guided with spineassist surgical robot: retrospective study," *Spine*, vol. 35, no. 24, pp. 2109–2115, 2010.

[31] M. Yang, J. Jung, J. Kim et al., "Current and future of spinal robot surgery," *Korean Journal of Spine*, vol. 7, no. 2, pp. 61–65, 2010.

[32] J. E. Lang, S. Mannava, A. J. Floyd et al., "Robotic systems in orthopaedic surgery," *Journal of Bone and Joint Surgery B*, vol. 93, no. 10, pp. 1296–1299, 2011.

[33] W. L. Bargar, A. Bauer, and M. Börner, "Primary and revision total hip replacement using the robodoc system," *Clinical Orthopaedics and Related Research*, vol. 354, pp. 82–91, 1998.

[34] A. P. Schulz, K. Seide, C. Queitsch et al., "Results of total hip replacement using the Robodoc surgical assistant system: clinical outcome and evaluation of complications for 97 procedures," *International Journal of Medical Robotics and Computer Assisted Surgery*, vol. 3, no. 4, pp. 301–306, 2007.

[35] P. Kazanzides, J. Zuhars, B. Mittelstadt, and R. H. Taylor, "Force sensing and control for a surgical robot," in *Proceedings of the IEEE International Conference on Robotics and Automation*, pp. 612–617, May 1992.

[36] A. D. Pearle, P. F. O'Loughlin, and D. O. Kendoff, "Robot-assisted unicompartmental knee arthroplasty," *Journal of Arthroplasty*, vol. 25, no. 2, pp. 230–237, 2010.

[37] A. D. Pearle, D. Kendoff, V. Stueber, V. Musahl, and J. A. Repicci, "Perioperative management of unicompartmental knee arthroplasty using the MAKO robotic arm system

(MAKOplasty)," *American Journal of Orthopedics*, vol. 38, no. 2, pp. 16–19, 2009.

[38] C. Plaskos, P. Cinquin, S. Lavallée, and A. J. Hodgson, "Praxiteles: a miniature bone-mounted robot for minimal access total knee arthroplasty," *The International Journal of Medical Robotics and Computer Assisted Surgery*, vol. 1, no. 4, pp. 67–79, 2005.

[39] D. Koulalis, P. F. O'Loughlin, C. Plaskos, D. Kendoff, M. B. Cross, and A. D. Pearle, "Sequential versus automated cutting guides in computer-assisted total knee arthroplasty," *Knee*, vol. 18, no. 6, pp. 436–442, 2010.

[40] G. Brisson, T. Kanade, A. DiGioia, and B. Jaramaz, "Precision freehand sculpting of bone," in *Proceedings of the 7th International Conference on Medical Image Computing and Computer-Assisted Intervention (MICCAI '04)*, pp. 105–112, September 2004.

[41] G. Brisson, *The Precision Freehand Sculptor: a Robotic Tool for Less Invasive Joint Replacement Surgery*, ProQuest, 2008.

[42] P. L. Yen and B. L. Davies, "Active constraint control for image-guided robotic surgery," *Proceedings of the Institution of Mechanical Engineers H*, vol. 224, no. 5, pp. 623–631, 2010.

[43] A. G. Harrell and B. T. Heniford, "Minimally invasive abdominal surgery: lux et veritas past, present, and future," *American Journal of Surgery*, vol. 190, no. 2, pp. 239–243, 2005.

[44] G. Dogangil, B. L. Davies, and F. Rodriguez Y Baena, "A review of medical robotics for minimally invasive soft tissue surgery," *Proceedings of the Institution of Mechanical Engineers H*, vol. 224, no. 5, pp. 653–679, 2010.

[45] C. Kuo and J. Dai, "Robotics for minimally invasive surgery: a historical review from the perspective of kinematics," in *Proceedings of the International Symposium on History of Machines and Mechanisms*, pp. 337–354, Springer, 2009.

[46] S. J. Harris, F. Arambula-Cosio, and Q. Mei, "The probot—an active robot for prostate resection," *Proceedings of the Institution of Mechanical Engineers H*, vol. 211, no. 4, pp. 317–325, 1997.

[47] G. H. Ballantyne, "Robotic surgery, telerobotic surgery, telepresence, and telementoring: review of early clinical results," *Surgical Endoscopy and Other Interventional Techniques*, vol. 16, no. 10, pp. 1389–1402, 2002.

[48] G. T. Sung and I. S. Gill, "Robotic laparoscopic surgery: a comparison of the da Vinci and Zeus systems," *Urology*, vol. 58, no. 6, pp. 893–898, 2001.

[49] J. Marescaux, J. Leroy, M. Gagner et al., "Transatlantic robot-assisted telesurgery," *Nature*, vol. 413, no. 6854, pp. 379–380, 2001.

[50] P. Mozer, J. Troccaz, and D. Stoinaovici, "Robotics in urology: past, present, and future," in *Atlas of Robotic Urologic Surgery*, L. Su, Ed., Current Clinical Urology, ch. 1, pp. 3–13, Springer, New York, NY, USA, 2011.

[51] K. Shah and R. Abaza, "Comparison of intraoperative outcomes using the new and old generation da Vinci robot for robot-assisted laparoscopic prostatectomy," *British Journal of Urology International*, vol. 108, no. 10, pp. 1642–1645, 2011.

[52] J. Bodner, H. Wykypiel, G. Wetscher, and T. Schmid, "First experiences with the da Vinci operating robot in thoracic surgery," *European Journal of Cardio-Thoracic Surgery*, vol. 25, no. 5, pp. 844–851, 2004.

[53] A. Tewari, A. Srivasatava, and M. Menon, "A prospective comparison of radical retropubic and robot-assisted prostatectomy: experience in one institution," *British Journal of Urology International*, vol. 92, no. 3, pp. 205–210, 2003.

[54] S. Maeso, M. Reza, J. A. Mayol et al., "Efficacy of the da Vinci surgical system in abdominal surgery compared with that of laparoscopy: a systematic review and meta-analysis," *Annals of Surgery*, vol. 252, no. 2, pp. 254–262, 2010.

[55] R. E. Link, S. B. Bhayani, and L. R. Kavoussi, "A prospective comparison of robotic and laparoscopic pyeloplasty," *Annals of Surgery*, vol. 243, no. 4, pp. 486–491, 2006.

[56] A. Amodeo, A. Linares Quevedo, J. V. Joseph, E. Belgrano, and H. R. H. Patel, "Robotic laparoscopic surgery: cost and training," *Minerva Urologica e Nefrologica*, vol. 61, no. 2, pp. 121–128, 2009.

[57] W. Jeong, F. Petros, and C. Rogers, *Robotic Surgery: Basic Instrumentation and Troubleshooting*, ch. 72, Wiley-Blackwell, Hoboken, NJ, USA, 2012.

[58] M. A. Lerner, M. Ayalew, W. J. Peine, and C. P. Sundaram, "Does training on a virtual reality robotic simulator improve performance on the da Vinci surgical system?" *Journal of Endourology*, vol. 24, no. 3, pp. 467–472, 2010.

[59] K. Cleary and T. M. Peters, "Image-guided interventions: technology review and clinical applications," *Annual Review of Biomedical Engineering*, vol. 12, pp. 119–142, 2010.

[60] M. E. Hagen, O. J. Wagner, I. Inan et al., "Robotic single-incision transabdominal and transvaginal surgery: initial experience with intersecting robotic arms," *International Journal of Medical Robotics and Computer Assisted Surgery*, vol. 6, no. 3, pp. 251–255, 2010.

[61] M. Kroh, K. El-Hayek, S. Rosenblatt et al., "First human surgery with a novel single-port robotic system: cholecystectomy using the da Vinci Single-Site platform," *Surgical Endoscopy*, vol. 25, no. 11, pp. 3566–3573, 2011.

[62] M. Stark, T. Benhidjeb, S. Gidaro, and E. Morales, "The future of telesurgery: a universal system with haptic sensation," *Journal of the Turkish-German Gynecological Association*, vol. 13, no. 1, pp. 74–76, 2012.

[63] S. DiMaio and S. Salcudean, "Needle steering and model-based trajectory planning," in *Proceedings of the 6th International Conference on Medical Image Computing and Computer-Assisted Intervention (MICCAI '03)*, pp. 33–40, 2003.

[64] H. Delingette, "Toward realistic soft-tissue modeling in medical simulation," *Proceedings of the IEEE*, vol. 86, no. 3, pp. 512–523, 1998.

[65] A. Melzer, B. Gutmann, T. Remmele et al., "Innomotion for percutaneous image-guided interventions," *IEEE Engineering in Medicine and Biology Magazine*, vol. 27, no. 3, pp. 66–73, 2008.

[66] M. Li, A. Kapoor, D. Mazilu, and K. A. Horvath, "Pneumatic actuated robotic assistant system for aortic valve replacement under MRI guidance," *IEEE Transactions on Biomedical Engineering*, vol. 58, no. 2, pp. 443–451, 2011.

[67] S. Zangos, A. Melzer, K. Eichler et al., "MR-compatible assistance system for biopsy in a high-field-strength system: initial results in patients with suspicious prostate lesions," *Radiology*, vol. 259, no. 3, pp. 903–910, 2011.

[68] H. J. Swan, W. Ganz, J. Forrester, H. Marcus, G. Diamond, and D. Chonette, "Catheterization of the heart in man with use of a flow-directed balloon-tipped catheter," *The New England Journal of Medicine*, vol. 283, no. 9, pp. 447–451, 1970.

[69] M. R. Franz, D. Burkhoff, and H. Spurgeon, "In vitro validation of a new cardiac catheter technique for recording monophasic action potentials," *European Heart Journal*, vol. 7, no. 1, pp. 34–41, 1986.

[70] J. M. Gore, R. J. Goldberg, D. H. Spodick, J. S. Alpert, and J. E. Dalen, "A community-wide assessment of the use of pulmonary artery catheters in patients with acute myocardial infarction," *Chest*, vol. 92, no. 4, pp. 721–727, 1987.

[71] D. Steven, H. Servatius, T. Rostock et al., "Reduced fluoroscopy during atrial fibrillation ablation: benefits of robotic guided navigation," *Journal of Cardiovascular Electrophysiology*, vol. 21, no. 1, pp. 6–12, 2010.

[72] V. Y. Reddy, P. Neuzil, Z. J. Malchano et al., "View-synchronized robotic image-guided therapy for atrial fibrillation ablation: experimental validation and clinical feasibility," *Circulation*, vol. 115, no. 21, pp. 2705–2714, 2007.

[73] K. R. J. Chun, B. Schmidt, B. Köktürk et al., "Catheter ablation—new developments in robotics," *Herz*, vol. 33, no. 8, pp. 586–589, 2008.

[74] C. V. Riga, C. D. Bicknell, D. Wallace, M. Hamady, and N. Cheshire, "Robot-assisted antegrade in-situ fenestrated stent grafting," *CardioVascular and Interventional Radiology*, vol. 32, no. 3, pp. 522–524, 2009.

[75] S. Ernst, F. Ouyang, C. Linder et al., "Initial experience with remote catheter ablation using a novel magnetic navigation system," *Circulation*, vol. 109, no. 12, pp. 1472–1475, 2004.

[76] J. K. R. Chun, S. Ernst, S. Matthews et al., "Remote-controlled catheter ablation of accessory pathways: results from the magnetic laboratory," *European Heart Journal*, vol. 28, no. 2, pp. 190–195, 2007.

[77] L. Leksell, "Stereotactic radiosurgery," *Journal of Neurology Neurosurgery and Psychiatry*, vol. 46, no. 9, pp. 797–803, 1983.

[78] R. Schulz and N. Agazaryan, *Shaped-Beam Radiosurgery: State of the Art*, Springer, New York, NY, USA, 2011.

[79] J. R. Adler Jr., S. D. Chang, M. J. Murphy, J. Doty, P. Geis, and S. L. Hancock, "The cyberknife: a frameless robotic system for radiosurgery," *Stereotactic and Functional Neurosurgery*, vol. 69, no. 1–4, pp. 124–128, 1997.

[80] G. J. Gagnon, N. M. Nasr, J. J. Liao et al., "Treatment of spinal tumors using cyberKnife fractionated stereotactic radiosurgery: pain and quality-of-life assessment after treatment in 200 patients," *Neurosurgery*, vol. 64, no. 2, pp. 297–306, 2009.

[81] M. Hoogeman, J. B. Prevost, J. Nuyttens, J. Poll, P. Levendag, and B. Heijmen, "Clinical accuracy of the respiratory tumor tracking system of the cyberknife: assessment by analysis of log files," *International Journal of Radiation Oncology *Biology* Physics*, vol. 74, no. 1, pp. 297–303, 2009.

[82] J. P. Rock, S. Ryu, F. F. Yin, F. Schreiber, and M. Abdulhak, "The evolving role of stereotactic radiosurgery and stereotactic radiation therapy for patients with spine tumors," *Journal of Neuro-Oncology*, vol. 69, no. 1–3, pp. 319–334, 2004.

[83] R. E. Wurm, S. Erbel, I. Schwenkert et al., "Novalis frameless image-guided noninvasive radiosurgery: initial experience," *Neurosurgery*, vol. 62, no. 5, pp. A11–A17, 2008.

[84] Z. Chang, T. Liu, J. Cai, Q. Chen, Z. Wang, and F. Yin, "Evaluation of integrated respiratory gating systems on a novalis tx system," *Journal of Applied Clinical Medical Physics*, vol. 12, no. 3, article 3495, 2011.

[85] A. Liu, N. Agazaryan, C. Yu, H. Han, T. Schultheiss, and J. Wong, "A multi-center consortium study of competing platforms for intracranial stereotactic irradiation," *International Journal of Radiation Oncology *Biology* Physics*, vol. 72, supplement 1, pp. S213–S213, 2008.

[86] M. Abacioglu, "Advances in technology in radiation oncology," *Oncology*, vol. 2, no. 1, pp. 11–14, 2012.

[87] H. R. Halperin, N. Paradis, J. P. Ornato et al., "Cardiopulmonary resuscitation with a novel chest compression device in a porcine model of cardiac arrest: improved hemodynamics and mechanisms," *Journal of the American College of Cardiology*, vol. 44, no. 11, pp. 2214–2220, 2004.

[88] A. Hallstrom, T. D. Rea, M. R. Sayre et al., "Manual chest compression vs use of an automated chest compression device during resuscitation following out-of-hospital cardiac arrest: a randomized trial," *Journal of the American Medical Association*, vol. 295, no. 22, pp. 2620–2628, 2006.

[89] R. Palmer, "Integrated diagnostic and treatment devices for enroute critical care of patients within theater," in *Proceedings of the RTO Human Factors and Medicine Panel Symposium*, Amsterdam, The Netherlands, October 2010.

[90] R. Seymour, B. Engbretson, K. Kott et al., "Comparison between the C-leg microprocessor-controlled prosthetic knee and non-microprocessor control prosthetic knees: a preliminary study of energy expenditure, obstacle course performance, and quality of life survey," *Prosthetics and Orthotics International*, vol. 31, no. 1, pp. 51–61, 2007.

[91] O. Otr, H. A. Reinders-Messelink, R. M. Bongers, H. Bouwsema, and C. K. Van Der Sluis, "The i-LIMB hand and the DMC plus hand compared: a case report," *Prosthetics and Orthotics International*, vol. 34, no. 2, pp. 216–220, 2010.

[92] K. Low, "Robot-assisted gait rehabilitation: from exoskeletons to gait systems," in *Proceedings of the Defense Science Research Conference and Expo (DSR '11)*, pp. 1–10, August 2011.

[93] J. Rosen and J. C. Perry, "Upper limb powered exoskeleton," *International Journal of Humanoid Robotics*, vol. 4, no. 3, pp. 529–548, 2007.

[94] H. Kazerooni, "Exoskeletons for human performance augmentation," in *Springer Handbook of Robotics*, B. Siciliano and O. Khatib, Eds., Springer, New York, NY, USA, 2008.

[95] R. Bogue, "Exoskeletons and robotic prosthetics: a review of recent developments," *Industrial Robot*, vol. 36, no. 5, pp. 421–427, 2009.

[96] M. J. Topping and J. K. Smith, "The development of Handy 1. A robotic system to assist the severely disabled," *Technology and Disability*, vol. 10, no. 2, pp. 95–105, 1999.

[97] M. Hillman, "Rehabilitation robotics from past to present—a historical perspective," in *Advances in Rehabilitation Robotics*, pp. 25–44, Springer, New York, NY, USA, 2004.

[98] A. Waldner, C. Werner, and S. Hesse, "Robot assisted therapy in neurorehabilitation," *Europa Medicophysica*, vol. 44, supplement 1, pp. 1–3, 2008.

[99] J. L. Dallaway, R. D. Jackson, and P. H. A. Timmers, "Rehabilitation robotics in Europe," *IEEE Transactions on Rehabilitation Engineering*, vol. 3, no. 1, pp. 35–45, 1995.

[100] M. Busnel, R. Cammoun, F. Coulon-Lauture, J. M. Détriché, G. Le Claire, and B. Lesigne, "The robotized workstation "MASTER" for users with tetraplegia: description and evaluation," *Journal of Rehabilitation Research and Development*, vol. 36, no. 3, pp. 217–229, 1999.

[101] M. Ceccarelli, "Problems and issues for service robots in new applications," *International Journal of Social Robotics*, vol. 3, no. 3, pp. 299–312, 2011.

[102] M. J. H. Lum, D. C. W. Friedman, G. Sankaranarayanan et al., "The RAVEN: design and validation of a telesurgery system," *International Journal of Robotics Research*, vol. 28, no. 9, pp. 1183–1197, 2009.

[103] A. Simorov, R. Otte, C. Kopietz, and D. Oleynikov, "Review of surgical robotics user interface: what is the best way to

control robotic surgery?" *Surgical Endoscopy*, vol. 26, no. 8, pp. 2117–2125, 2012.

[104] C. Kuo, J. Dai, and P. Dasgupta, "Kinematic design considerations for minimally invasive surgical robots: an overview," *The International Journal of Medical Robotics and Computer Assisted Surgery*, vol. 8, no. 2, pp. 127–145, 2012.

[105] C. Wagner, N. Stylopoulos, and R. Howe, "The role of force feedback in surgery: analysis of blunt dissection," in *Proceedings of the 10th Symposium on Haptic Interfaces for Virtual Environment and Teleoperator Systems*, vol. 2002, Citeseer, 2002.

[106] M. Tavakoli, R. V. Patel, and M. Moallem, "Haptic interaction in robot-assisted endoscopic surgery: a sensorized end-effector," *The International Journal of Medical Robotics and Computer Assisted Surgery*, vol. 1, no. 2, pp. 53–63, 2005.

[107] A. M. Okamura, "Haptic feedback in robot-assisted minimally invasive surgery," *Current Opinion in Urology*, vol. 19, no. 1, pp. 102–107, 2009.

[108] U. Hagn, R. Konietschke, A. Tobergte et al., "DLR MiroSurge: a versatile system for research in endoscopic telesurgery," *International Journal of Computer Assisted Radiology and Surgery*, vol. 5, no. 2, pp. 183–193, 2010.

[109] S. Thielmann, U. Seibold, R. Haslinger et al., "MICA—a new generation of versatile instruments in robotic surgery," in *Proceedings of the 23rd IEEE/RSJ International Conference on Intelligent Robots and Systems (IROS '10)*, pp. 871–878, October 2010.

[110] R. Konietschke, T. Ortmaier, H. Weiss, G. Hirzinger, and R. Engelke, "Manipulability and accuracy measures for a medical robot in minimally invasive surgery," in *Advances in Robot Kinematics*, 2004.

[111] G. Sutherland, P. McBeth, and D. Louw, "Neuroarm: an mr compatible robot for microsurgery," in *International Congress Series*, vol. 1256, pp. 504–508, Elsevier, Amsterdam, The Netherland, 2003.

[112] M. J. Lang, A. D. Greer, and G. R. Sutherland, "Intraoperative robotics: NeuroArm," *Intraoperative Imaging*, vol. 109, pp. 231–236, 2011.

[113] D. Stoianovici, D. Song, D. Petrisor et al., "'MRI Stealth' robot for prostate interventions," *Minimally Invasive Therapy and Allied Technologies*, vol. 16, no. 4, pp. 241–248, 2007.

[114] P. Garcia, J. Rosen, C. Kapoor et al., "Trauma pod: a semi-automated telerobotic surgical system," *International Journal of Medical Robotics and Computer Assisted Surgery*, vol. 5, no. 2, pp. 136–146, 2009.

[115] S. G. Yuen, P. M. Novotny, and R. D. Howe, "Quasiperiodic predictive filtering for robot-assisted beating heart surgery," in *Proceedings of the IEEE International Conference on Robotics and Automation (ICRA '08)*, pp. 3875–3880, May 2008.

[116] N. Patronik, C. Riviere, S. El Qarra, and M. Zenati, "The heartlander: a novel epicardial crawling robot for myocardial injections," in *International Congress Series*, vol. 1281, pp. 735–739, Elsevier, Amsterdam, The Netherland, 2005.

[117] D. Moral Del Agua, N. A. Wood, and C. N. Riviere, "Improved synchronization of heartlander locomotion with physiological cycles," in *Proceedings of the 37th Annual Northeast Bioengineering Conference (NEBEC '11)*, April 2011.

[118] T. Wortman, A. Meyer, O. Dolghi et al., "Miniature surgical robot for laparoendoscopic single-incision colectomy," *Surgical Endoscopy*, vol. 26, pp. 727–731, 2012.

[119] Y. Hayashi, H. Yamamoto, T. Yano, and K. Sugano, "Review: diagnosis and management of mid-gastrointestinal bleeding by double-balloon endoscopy," *Therapeutic Advances in Gastroenterology*, vol. 2, no. 2, pp. 109–117, 2009.

[120] A. Van Gossum, M. M. Navas, I. Fernandez-Urien et al., "Capsule endoscopy versus colonoscopy for the detection of polyps and cancer," *The New England Journal of Medicine*, vol. 361, no. 3, pp. 264–270, 2009.

[121] D. Cassilly, S. Kantor, L. C. Knight et al., "Gastric emptying of a non-digestible solid: assessment with simultaneous Smart-Pill pH and pressure capsule, antroduodenal manometry, gastric emptying scintigraphy," *Neurogastroenterology and Motility*, vol. 20, no. 4, pp. 311–319, 2008.

[122] C. Mc Caffrey, O. Chevalerias, C. O'Mathuna, and K. Twomey, "Swallowable-capsule technology," *IEEE Pervasive Computing*, vol. 7, no. 1, pp. 23–29, 2008.

[123] A. Moglia, A. Menciassi, and P. Dario, "Recent patents on wireless capsule endoscopy," *Recent Patents on Biomedical Engineering*, vol. 1, no. 1, pp. 24–33, 2008.

[124] M. C. Roco, "Nanotechnology: convergence with modern biology and medicine," *Current Opinion in Biotechnology*, vol. 14, no. 3, pp. 337–346, 2003.

[125] M. Copot, A. Popescu, I. Lung, and A. Moldovanu, "Achievements and perspectives in the field of nanorobotics," *The Romanian Review Precision Mechanics, Optics and Mechatronics*, vol. 19, no. 36, pp. 61–66, 2009.

[126] R. A. Freitas, "What is nanomedicine?" *Nanomedicine: Nanotechnology, Biology, and Medicine*, vol. 1, no. 1, pp. 2–9, 2005.

[127] L. Zhang, J. J. Abbott, L. Dong, B. E. Kratochvil, D. Bell, and B. J. Nelson, "Artificial bacterial flagella: fabrication and magnetic control," *Applied Physics Letters*, vol. 94, no. 6, Article ID 064107, 2009.

[128] G. Kósa, M. Shoham, and M. Zaaroor, "Propulsion of a swimming micro medical robot," in *Proceedings of the IEEE International Conference on Robotics and Automation (ICRA '05)*, pp. 1327–1331, April 2005.

[129] G. Dogangil, O. Ergeneman, J. J. Abbott et al., "Toward targeted retinal drug delivery with wireless magnetic microrobots," in *Proceedings of the IEEE/RSJ International Conference on Intelligent Robots and Systems (IROS '08)*, pp. 1921–1926, September 2008.

[130] H. Li, J. Tan, and M. Zhang, "Dynamics modeling and analysis of a swimming microrobot for controlled drug delivery," *IEEE Transactions on Automation Science and Engineering*, vol. 6, no. 2, pp. 220–227, 2009.

[131] T. Ebefors, J. Mattsson, E. Kalvesten, and G. Stemme, "A walking silicon micro-robot," in *Proceedings of the 10th International Conference on Solid-State Sensors and Actuators (Transducers '99)*, pp. 1202–1205, 1999.

[132] P. Dumpuri, L. W. Clements, B. M. Dawant, and M. I. Miga, "Model-updated image-guided liver surgery: preliminary results using surface characterization," *Progress in Biophysics and Molecular Biology*, vol. 103, no. 2-3, pp. 197–207, 2010.

[133] G. Brandt, A. Zimolong, L. Carrât et al., "CRIGOS: a compact robot for image-guided orthopedic surgery," *IEEE Transactions on Information Technology in Biomedicine*, vol. 3, no. 4, pp. 252–260, 1999.

7

Advances in Haptics, Tactile Sensing, and Manipulation for Robot-Assisted Minimally Invasive Surgery, Noninvasive Surgery, and Diagnosis

Abbi Hamed,[1] **Sai Chun Tang,**[2] **Hongliang Ren,**[3] **Alex Squires,**[4] **Chris Payne,**[5]
Ken Masamune,[6] **Guoyi Tang,**[7] **Javad Mohammadpour,**[4] **and Zion Tsz Ho Tse**[4]

[1] *Department of Advanced Robotics, Chiba Institute of Technology, 2-17-1 Tsudanuma, Narashino, Chiba 285-0016, Japan*
[2] *Department of Radiology, Brigham and Women's Hospital, Harvard Medical School, 221 Longwood Avenue, Boston,*
MA 02115, USA
[3] *Department of Bioengineering, National University of Singapore, Singapore 117575*
[4] *Driftmier Engineering Center, College of Engineering, The University of Georgia, Athens, GA 30602, USA*
[5] *Department of Mechanical Engineering, Imperial College London, London SW7 2AZ, UK*
[6] *Advanced Therapeutic and Rehabilitation Engineering Laboratory, Faculty of Engineering, The University of Tokyo, Suite No. 83A3,*
Building No. 2, Hongo, Bunkyo-ku, Tokyo 113-8656, Japan
[7] *Advanced Material Institute, Graduate School at Shenzhen, Tsinghua University, Shenzhen, Guangdong 518055, China*

Correspondence should be addressed to Zion Tsz Ho Tse, ziontse@uga.edu

Academic Editor: Yangmin Li

The developments of medical practices and medical technologies have always progressed concurrently. The relatively recent developments in endoscopic technologies have allowed the realization of the "minimally invasive" form of surgeries. The advancements in robotics facilitate precise surgeries that are often integrated with medical image guidance capability. This in turn has driven the further development of technology to compensate for the unique complexities engendered by this new format and to improve the performance and broaden the scope of the procedures that can be performed. Medical robotics has been a central component of this development due to the highly suitable characteristics that a robotic system can purport, including highly optimizable mechanical conformation and the ability to program assistive functions in medical robots for surgeons to perform safe and accurate minimally invasive surgeries. In addition, combining the robot-assisted interventions with touch-sensing and medical imaging technologies can greatly improve the available information and thus help to ensure that minimally invasive surgeries continue to gain popularity and stay at the focus of modern medical technology development. This paper presents a state-of-the-art review of robotic systems for minimally invasive and noninvasive surgeries, precise surgeries, diagnoses, and their corresponding technologies.

1. Introduction

Based on the degree of invasiveness of surgical procedures, there are roughly three main categories: invasive procedures also known as open surgery, minimally invasive procedures, and noninvasive procedures.

Minimally invasive surgery (MIS) is a form of surgery intended to provide great benefits to the patient over conventional open surgery by minimizing unnecessary trauma caused in the process of performing a medical procedure.

Besides less trauma, pain, blood loss, scarring, and better cosmesis, these differences lead to a shorter recovery time for the patient and reduced risk of complications [1]. However, it is also well documented that this approach brings a number of corresponding difficulties to the clinical staff performing it. These difficulties are due to the highly limited workspace, specialized tools requiring further staff training and adaption for use, and greatly reduced visual and touch information.

Despite the aforementioned drawbacks, MIS has continued to gain popularity and to be widely used [2]. A prime factor in this continued adoption is the corresponding development of medical tools and devices intended for use in MIS. Medical robotics has been applied in MIS; robotic platforms are particularly suitable due to their favorable characteristics. Such characteristics include high accuracy, repeatability, and the possibility of designing specialized mechanisms that can be applied to specific procedures and organs. Furthermore, medical robots can incorporate sensors to return touch and force information and can also be combined with medical imaging technology to allow autonomous, semiautonomous, or teleoperated control, which can improve surgical performance and the scope of MIS. In addition, by incorporating emerging imaging techniques and non-invasive modalities, more precise, cost-effective, and portable treatment tools can be made possible. Medical robotics has been in use for approximately 30 years; the first generation of medical robots were used as tool holders and positioners, before the development of active medical robots in the early 1990s [3]. Despite the fact that this technology has been around for three decades, medical robots have not been widely adopted due to limitations of control and the high risk of their applications. The associated regulations and standards, and in particular for active systems, which are intended to be powered during a procedure, are necessarily demanding and as such, the standard development time is very long in this field. There has also been the issue of acceptance from the medical community and surgeons, who still view with distrust any technology that would create a separation of the surgeon from the surgery. A few systems have seen commercial success; these are operated remotely or semiautonomously, where control of the robot remains strictly with the clinician at all times except, in some cases, during a highly restricted function, which can be monitored.

A current trend for medical robotics and devices being developed for use with MIS and non-invasive surgery (NIS) is to further increase the scope of applications with the design of highly function-specific devices that can be combined with medical imaging technology. Another trend for MIS is the development of systems to return touch and force information to aid in surgery or to provide more information for diagnosis purposes. This paper describes the development of robotic systems for MIS, the move towards less autonomous but more form-specialized systems, the rise in research and development of haptics technology in medical robotics, and the trend from MIS towards imaged-guided NIS.

2. MIS Robots and Their Technologies

2.1. From Open Surgery to MIS Robotic Surgery. Most of the first generation of medical robots were designed for MIS tool positioning, in which high accuracy and repeatable motion gave them a significant advantage [19]. Some open surgeries, such as total hip replacement [20], have also benefited from these precise systems, where the main purpose is to augment the performance of human surgeons in precise bone machining procedures. Tool positioners are significant in MIS as these procedures are inherently more difficult to perform; these types of robots essentially reduce the burden

on a surgeon [21]. Orthopedic surgeries were the very first type of medical procedures in which medical robots were developed with the specific functionalities required to play an active role in an operation. The intention for many of the robots developed in the 80s and 90s for orthopedic surgeries was to automate part of the procedure during the surgery. These "active" or automated robots would implement a preoperative plan based on preoperative imaging techniques, such as magnetic resonance imaging (MRI) or computed tomography (CT), and then perform the operations without input from the surgeon. It was perceived that the use of robots in this fashion would improve the overall outcome of a surgery through increased accuracy of implant placement. Furthermore, robots were particularly suitable for this application due to the nondeformability of bones, relative simplicity of the task, and the fact that the minimally invasive approach made knee surgery particularly challenging for surgeons. Commercial robots developed for this task include the ROBODOC (Integrated Systems) and Caspar (Orto Maquet). Development of ROBODOC started in the 1980s and the first human trial in 1992 showed significant improvement of the surgical results gained by using this system over human conducted surgery [22].

2.2. Human-Controlled Robot-Assisted Surgery. Application of automated robotics to soft tissues was found to yield a particular challenge due to the tissue deformability [23]. This means that the imaging registration process to match the robot to the patient from the preoperative images would be insufficient to guarantee positional accuracy within safe limits throughout the procedure. As a solution, either intraoperative imaging or deformable tissue models can be used in conjunction with the preoperative planning to implement real-time controls; however, this is an extremely challenging technical issue [24, 25]. Another issue with automated robotics is that no robotic system can be guaranteed to become 100 percent safe, and robotic artificial intelligence is not sufficiently advanced to the point where the robot can be held responsible if a mistake is made. This is an issue of culpability that concerns manufacturers and has led to the shifting of the MIS technology towards nonautomated telerobotic systems. Telerobotic systems are in fact master-slave platforms, in which the surgeon has direct control of the action of the robotic manipulator. Visual feedback is provided by way of endoscopic sight and results in advantages over minimally invasive procedures through the use of binocular vision, tremor filtering, and motion scaling.

The concept of telesurgery has in fact developed separately from that of the automated surgical robot and was initially driven by the desire to treat critically wounded soldiers near the front lines. The first surgical telerobotic system was developed under the DARPA advanced combat casualty care program in the early 1990s [26]. More advanced and sophisticated examples of these types of systems are the da Vinci robot of Intuitive Surgical Inc. (http://www.intuitivesurgical.com/), Titan Medical Inc's Amadeus System (http://www.titanmedicalinc.com/), and the ZEUS of Computer Motion Inc. (acquired by Intuitive

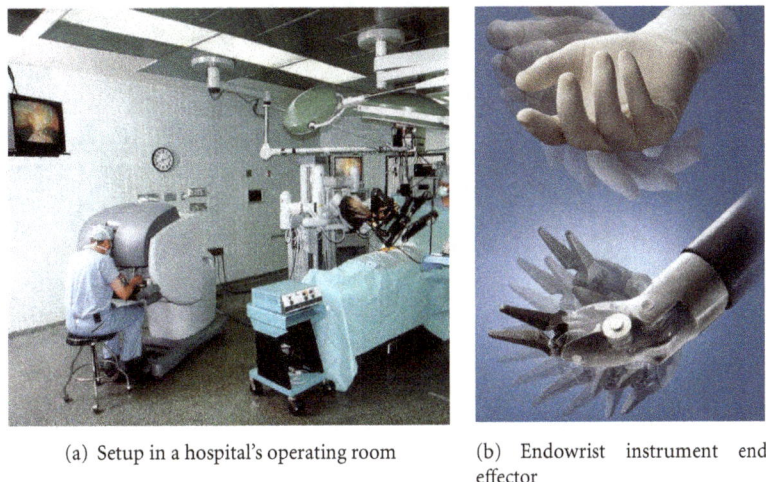

(a) Setup in a hospital's operating room

(b) Endowrist instrument end effector

FIGURE 1: (a) Surgeon operating da Vinci system (by Intuitive Surgical Inc.) for telesurgery. (b) Endowrist is a feature of the robotic system, with multi-DOF at its end effectors. The robotic system provides high dexterity, hand-tremor cancellation, and three-dimensional views for the endoscopic surgery [4].

Surgical in 2003). In the use of conventional MIS techniques in these systems, several robotic arms control endoscopic instruments, and an additional arm guides a laparoscopic camera; the arms are rigid and cable driven. The surgeon operates the robot from a console via two hand masters for the tools and pedal controls or voice control (ZEUS) for the laparoscope arm. With the da Vinci system, the surgeon can view the surgical site through a viewfinder, which generates pseudo-3D images. The da Vinci system shown in Figure 1 also incorporates "endo-wrists," a special feature that provides two extra degrees of freedom (DOF) at the point of intervention and significantly increases the ease of use, especially for cutting and suturing operations [4]. Other examples of minimally invasive procedures performed by the da Vinci system include cardiac surgery, urological surgery, and prostatectomy [27].

Laparoscopic and endoscopic techniques are presently dominated by multiport access methods, which mimic the conventional hand-eye coordination with an instrument-stereovision system. Those techniques require multiple ports in the patient's body for the MIS instrument insertions. Towards even less invasive techniques, surgeons are progressively moving to single-port access (SPA) MIS, with the assistance of robotic arms and computer assistive devices in the operating room. SPA can potentially minimize complications associated with multiport incisions [28–30] although SPA poses significant challenges for the design of endoscopic instruments due to the maneuverability constraints through a single port.

2.3. Surgical Robots with Parallel Mechanisms. Most of the robotic manipulators used for minimally invasive or non-invasive medical interventions are serial structures. The serial structure robots have the advantage of providing a large workspace, high dexterity, and high maneuverability; however, they suffer from low stiffness and poor positioning accuracy. To address the drawbacks associated with serial structures, more attention has recently been paid to parallel structure robots, due to their simplicity, large payload capacity, positional accuracy, and high stiffness. The very first parallel platform was developed by Stewart in 1965 [31], whose platform was composed of a fixed base, a movable platform, and six variable-length actuators connecting with base and platform.

In recent years, several designs of parallel robot structures have been developed for a variety of medical procedures. Brandt et al. developed a compact robotic system for image-guided orthopedic surgery (CRIGOS) that comprised a parallel robot and a computational core for planning of the surgical interventions and control of the parallel platform [32]. Tanikawa and Arai developed a dexterous micromanipulation system based on a parallel mechanism and utilized it for performing microsurgery among a few other tasks [33]. Merlet developed a micro robot (named MIPS) with a parallel manipulator. The 3-DOF system allowed fine positioning of a surgical tool and was used as an active wrist at the tip of an endoscope [34]. In 2003, Shoham et al. developed a miniature robot for surgical procedures (MARS), which was a cylindrical 6-DOF parallel mechanism. MARS was shown to have the capability of being used in a variety of surgical procedures including spine and trauma surgery, where accurate positioning and orientation of a handheld surgical robot is of interest [35]. Maurin et al. developed a 5-DOF parallel robotic platform for CT-guided percutaneous procedures with force feedback and automatic patient-to-image registration of needle [36]. Tsai and Hsu developed a parallel surgical robot for precise skull drilling in stereotactic neurosurgical operations [37]. As mentioned earlier, a major drawback of parallel mechanisms is their restricted workspace compared to serial link robots. In neurosurgical operations, the workspace lies on the surface of the skull located at one side of the robot. Tsai and Hsu analyzed this asymmetric workspace and found the optimal relative positions of the skull and the parallel mechanism to determine the maximum workspace on the skull [37].

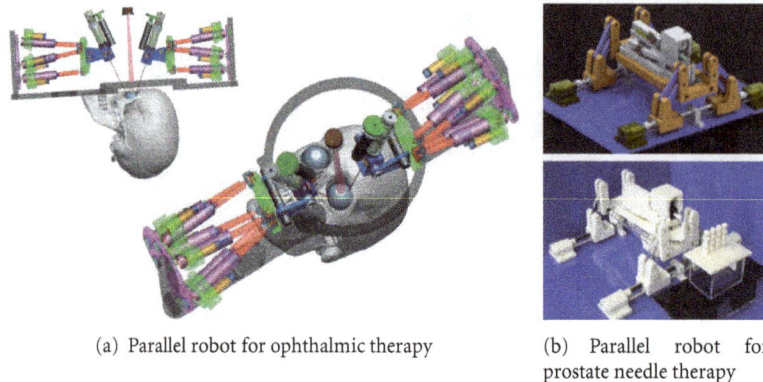

(a) Parallel robot for ophthalmic therapy

(b) Parallel robot for prostate needle therapy

FIGURE 2: (a) The design for a dual-arm robot for use in ophthalmic surgery. Two parallel robots are perched upon a fixation halo, and each can hold modular instruments [5]. (b) The 4-DOF parallel platform manipulator is integrated with a 3-DOF needle driver module to provide 7-DOF needle motion for prostate interventions [6].

More recent developments in the use of parallel structures for medical interventions include the work by Fine et al. [5], where the authors designed a dual-arm ophthalmic surgical robot using a parallel structure mechanism (Figure 2(a)) with high precision ($<5\,\mu$m). Their platform was used in both vascular cannulation and stent deployment in animal models. Fischer and his colleagues at WPI have been developing MRI-compatible parallel mechanism robots for prostate brachytherapy and neurosurgery applications [6, 38]. Figure 2(b) shows a prototype of their surgical manipulator, where the bottom figure illustrates a base platform for prostate interventions. This platform includes the 4-DOF parallel manipulator integrated with a 3-DOF needle driver to pro-vide needle translation, rotation, and stylet retraction, thus in total providing 7-DOF needle motion [6].

Parallel robots are currently being designed and utilized for needle surgery by many other groups, including the dynamic model and control of a needle surgery parallel robot "PmarNeedle" reported by D'Angella et al. in 2011 [39]. A very recent work by Salimi et al. has reported the development and testing of a 4-DOF parallel structure platform that is used to assist in MRI-guided intracardiac interventions, in particular aortic valve implantation [40]. The work of Salimi et al. is the first effort that has been reported in the literature that uses parallel platforms for minimally invasive intracardiac interventions [40].

The compact and lightweight designs of medical robots with parallel mechanisms save necessary space for operation and storage, simplifying the relocation of the robot in the operating room. The compact structure allows easy sterilization by way of covering the robot with a closed drape, and its relatively small workspace can also be an important safety feature. Parallel robots, if designed correctly, can provide higher precision than similar serial robots, and as such they are well suited to applications in ophthalmic [41] and orthopedic surgeries [42], cell manipulations [43], and micropositioning [44, 45].

2.4. Human-Controlled Robots for Microsurgery. One of the first surgical robots developed for eye surgery was the microsurgical manipulator developed at Northwestern University [46]. This telerobotic system is designed to drive a micro pipette through the lumen of a hypodermic needle. The system is controlled via a Trackball master and can target retinal blood vessels to an internal diameter of 20 microns. Since the turn of the century, many more examples of telerobotic systems have appeared in the literature. The Johns Hopkins University steady-hand robot for eye surgery emphasizes on motion scaling over tremor filtering and has a positioning resolution of 5 to 10 microns (see Figure 3(a)) [7]. The RAVEN developed at the University of Washington is a 7-DOF cable actuated manipulator controlled by the PHANToM haptic control interface (Sensable Technologies Inc.), and includes wrist joints like the da Vinci System, and has the potential for haptic feedback [47]. At ETH Zurich, a wireless system has been developed to deliver drugs into the retinal blood vessels [48]; this is a "microbot" and requires only one incision to be made in the sclera wall. A telerobotic system known as Heartlander has been developed at Carnegie Mellon University, which uses suction pads to attach itself to the epicardium and propels itself using small onboard piezoelectric motors [49].

2.5. Wireless MIS Robots. Wireless robotic systems that can reach far inside the human body have become a new branch in minimally invasive robotic systems. Many examples of these systems can be found in applications to gastrointestinal (GI) procedures. The intestine presents a long and convoluted environment in GI endoscopic procedures. By performing the conventional GI procedures, patients complain of pain and discomfort, and clinicians face technical difficulties involved in navigating the instruments. As such, robotic devices such as the PillCam developed by Given Imaging Inc. are short and thin, can be inserted into the GI tract, and maneuvered under their own volition. The PillCam also includes a minicamera, LEDs, and RF transmitter but is transported using the body's own digestive process. The "inchworm" design, developed by Quirini et al. [50] and shown in Figure 3(b), is a quite popular solution although legged microbots have also been employed. Other popular

Advances in Haptics, Tactile Sensing, and Manipulation for Robot-Assisted Minimally Invasive Surgery, Noninvasive Surgery, and Diagnosis

79

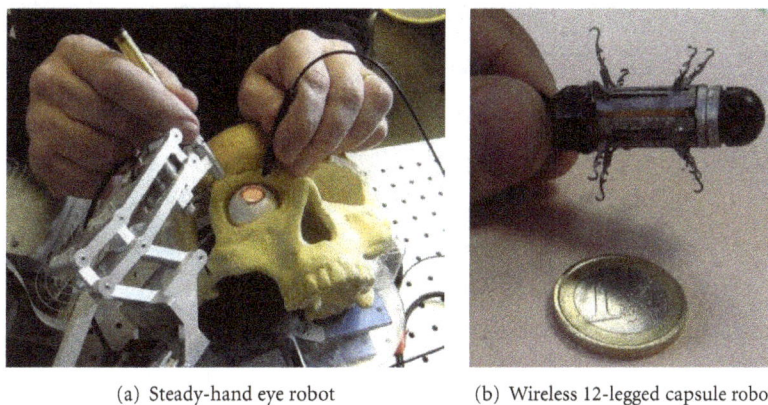

(a) Steady-hand eye robot   (b) Wireless 12-legged capsule robot

FIGURE 3: (a) Steady-hand eye robot developed in Johns Hopkins University for retinal microsurgery. This device minimizes hand tremors by sharing the instrument control between the surgeon and the robotic manipulator, allowing precise and steady motion [7]. (b) Pill-size endoscopic capsule robot developed by Sant'Anna School of Advanced Studies and Vanderbilt University with active control for placing its camera in a desired position and direction in the intestine [8].

actuation mechanisms have also been explored such as the ones driven by external magnetic fields, proposed by Wang and Meng [51], Ciuti et al. [52] and Lien et al. [53], or hybrid driven by internal/external actuators in [54].

3. Use of Haptic and Tactile Sensing in Robotic MIS

Medical haptics is an underdeveloped field of research that is slowly gaining attention [55]. The motivation stems mainly from the rise of teleoperation and from its potential as a research tool [56]. Most research efforts currently focus on bringing haptic feedback to medical tools to aid in operations, especially those in MIS. Haptic systems have become more widely used in surgical training programs to simulate tool behavior for medical trainees [57]. Over the past 15 years or so, medical haptic systems have been used for either medical training via haptic simulation or improving the function of medical tools in minimally invasive surgery through force feedback [58].

3.1. Surgical Simulation and Training. Development of haptic surgical simulators is driven by the limitations of traditional methods of surgical training. Generally, the method of teaching is in two stages: first, the surgeon studies the anatomy from textbooks and other visual aids; this is followed by "hands on" training in the operating room or by cadaver dissection or simulation mannequins [59]. A problem with the traditional form of training is that availability of cadavers or patients to work with is limited and unreliable [57]. Simulators have the ability to generate realistic human anatomical properties and varied morphologies, and since the early 90s virtual reality (VR) simulations have been available for this purpose [60]. It has also been proposed that surgical simulators would aid in diagnosis and treatment planning.

3.2. Haptic Feedback in Telerobotic Surgery. Haptic feedback has become a vital area of research with the rise of teleoperated minimally invasive surgery. Clinically, such feedback can improve a surgeon's sense of telepresence, hopefully leading to an improved performance. Evidence strongly suggests that the ability to confer haptic feedback to present surgical robotic systems such as da Vinci would contribute significantly to safe cardiac surgical procedures using these complex systems [61]. The deficiency of haptic feedback in current robotic systems is a significant handicap in performing the technically more intricate and delicate surgical tasks. Such tasks are inherent in specialties like cardiac surgery [62] and lack of haptic feedback and can lead to unsafe levels of force by the clinician [63]. An area of surgical haptics that provides an interim between autonomous medical robotics and master-slave telesurgery is the use of "virtual fixtures" [64] or "active constraints" [65]. Such robots do not actively drive the tools being used, rather the surgeon's own motive power is used, while the robot can provide controlling forces when the boundary of a predetermined workspace is reached. This synergistic approach adds safety to the robot-assisted surgery and MIS techniques, whilst allowing overall control and judgment to remain with the surgeon [66]. The "Acrobot" developed at Imperial College is an example of this. Another research area that has seen a significant rise in popularity is the development of haptic capabilities for telerobotic systems, such as the da Vinci.

The Black Falcon was created at MIT in the late 90s [67]. Figure 4(a) shows an 8-DOF teleoperator slave for MIS, which includes some novel features to improve surgeon facility during an interventional procedure. The Black Falcon attempts to address some of the main limitations intrinsic to teleoperated MIS. Those restrictions include the discrepancy between tool motions observed via endoscope and the motions of the surgeon's hand, the poor dexterity due to lack of DOF and the lack of force/tactile feedback. The slave system is a 4-DOF wrist with a 2-DOF gripper and

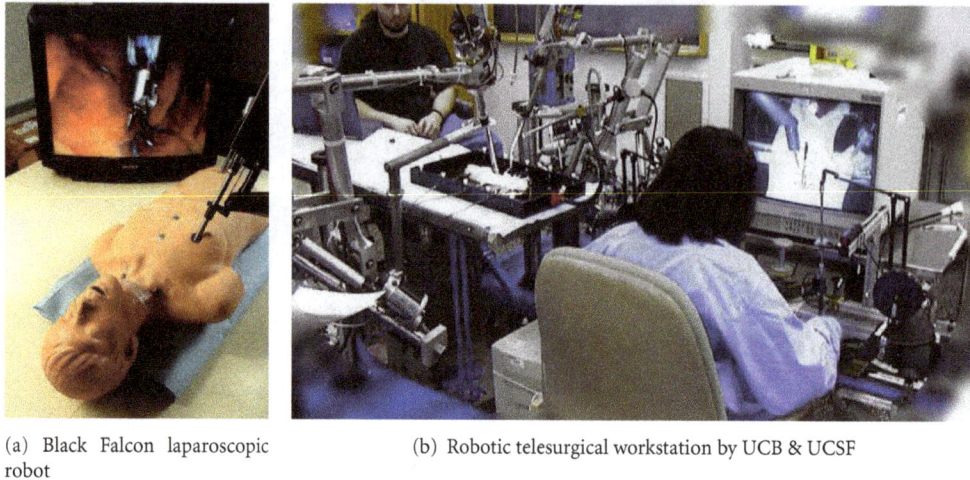

(a) Black Falcon laparoscopic robot

(b) Robotic telesurgical workstation by UCB & UCSF

FIGURE 4: (a) Black Falcon robotic system developed at MIT for laparoscopic surgery [9]. (b) Laparoscopic telesurgical workstation jointly developed by the UC Berkeley and UC San Francisco [10]. Both systems aim to enhance the dexterity and sensation of minimally invasive surgery through using scaled robotic manipulators under the control of the surgeon.

a 3-DOF base positioner. In addition, a modified PHANToM is used as the master system. A dedicated control system called macro-micro is used to implement force reflection. The entire mechanism has careful weight distribution so that the counterbalance is easy to achieve with motor torque. The motors are brushed DC servo motors with planetary gearheads. Transmission is by cable, but the kinematics of the system is decoupled between links due to the cabling scheme.

A robotic telesurgical workstation (RTW) was developed by Çavuşoğlu et al. at the University of California, Berkeley (see Figure 4(b)) [10]. The surgical tasks accomplished with the Berkeley/UCSF RTW are suturing and knot tying. The improved design includes high system bandwidth and good haptic feedback with sufficient fidelity. The slave system comprises two main sections: the gross positioner has 4 DOF, which is conventional for laparoscopic instruments, with an additional 2 DOF forming an "endo-wrist." The whole mechanism is actuated by DC servo motors with cable transmission provided for the wrist. The master workstation is composed of a pair of 6-DOF PHANToMs, one for each slave manipulator. Preliminary testing showed that the addition of force feedback made suturing more successful than without it. It also highlighted some weaknesses in the current mechanical design. Since then, other works have also been carried out at other research groups. Okamura have shown that the absence of haptics in telesurgical systems can lead to an increased duration of operations with increased propensity for error [62]. By fitting strain gauges on the lower shaft of da Vinci needle tools and representing data using visual feedback on the display, they showed significantly reduced tool forces.

At Iwate University, Shimachi et al. report on the addition of a force sensor function to an instrument of the da Vinci using an "adapter frame," through which the instrument is supported by the force sensors [68]. Results show maximum error of 0.5 N due to the frame deformation. Tavakoli and Patel describe how current telerobotic systems still lack the haptic feedback due to the difficulty of creating suitable force sensors and adapting to the complex end effectors [69]. Interpretation of force using visual feedback remains the only solution for the meantime.

3.3. Tactile Devices for Palpation and Characterizing Tissue Stiffness. Enhancing the tactile sensing capability of MIS instruments has become a prime research area [70]. It has been shown that force feedback can significantly improve the performance of robotic surgical systems [1]. Okamura [62] argues that true telemanipulation will require tactile, as well as force feedback. By the end of the 90s, tactile sensing technology was still considered new particularly in the field of medicine [71]. Some of the main challenges with the development of tactile sensing instruments are the limited size and weight permissible, sterility, safety [70], and placement of sensors based on the tool format [72]. Other applications for tactile sensing in medical devices have been developed. The development of tools specifically for quantifying tissue stiffness for use in MIS has been reported. Developing tools to measure tissue stiffness is motivated by the clinicians' inability to directly touch or "palpate" a surface in MIS [71, 73]. This important quantification, although subjective, allows a quick assessment of the health of a tissue.

Design of tactile sensing has received lots of attention with several arrays now available. Tactile data processing and displays are much less well developed. Human tactile sensing is a tremendously complex system and still not fully understood; as such, the challenge of producing comprehensive tactile displays is still insurmountable. A review of tactile displays by Benali-Khoudja concludes that there are no tactile displays that fully incorporate all the physical parameters [74]. Tactile displays are still too large, imprecise, and expensive to be used in MIS [70] and for similar reasons are very rare in training devices [57]. Force feedback is much more established than tactile feedback, and several

Advances in Haptics, Tactile Sensing, and Manipulation for Robot-Assisted Minimally Invasive Surgery, Noninvasive Surgery, and Diagnosis

81

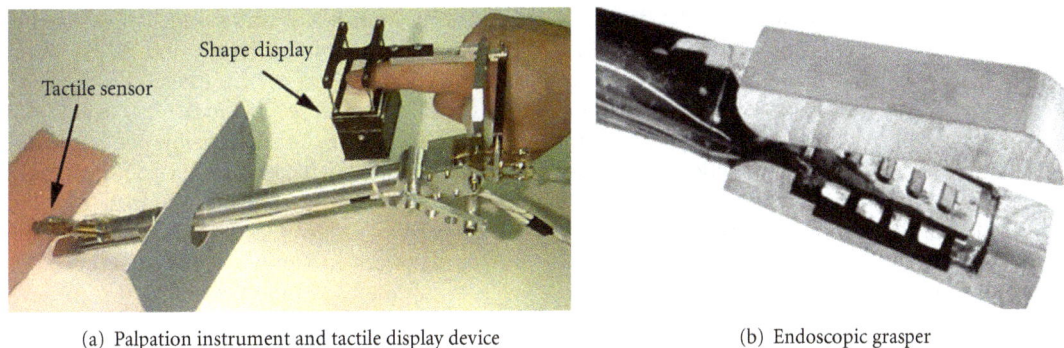

(a) Palpation instrument and tactile display device   (b) Endoscopic grasper

FIGURE 5: (a) A remote palpation and display device developed by Harvard for MIS. Tactile information of the palpated tissue is transmitted from the organ tissue via the endoscopic device to the surgeon's fingertips during MIS [11]. (b) Tactile sensor array made of PVDF piezoelectric films at Concordia University is integrated in the jaws of an endoscopic grasper to measure tissue stiffness characteristics [12].

commercial devices have been developed. One of the most commonly used in both haptic medical systems and training systems is the PHANToM from Sensable Technology Inc., which evolved from haptics research at MIT [75].

One of the first palpation simulations was developed by Langrana et al. to feedback force information using a virtual knee model and a Rutgers Master [76]. The knee model was comprised of nearly 13,500 polygons and included information for the bones, cartilages, and muscles of the locality. Contact forces fed back to the master were calculated in real-time based on Hooke's law. The Rutgers master was a glove incorporating actuators placed proximal to the palm. The actuators were air pistons, with a maximum force of 4 N, which could provide force on the fingers in flexion and extension. Having equipped with spherical joints allowed the adduction and abduction normally. The main drawbacks were that the actuators, although small, still limited the mobility of the hand and due to the absence of wrist feedback, it was impossible to simulate the object weight.

One of the first institutes that dedicated some efforts to the development of haptics for medical devices was the BioRobotics lab at Harvard. In fact, the first haptic medical system reported was a minimally invasive tool devised for laparoscopy by Peine et al. at Harvard [77]. The motivation for the system was to use it in MIS to allow the surgeon to palpate a region to locate arteries and detect blood flow in the same way as would be done in open surgery. The device shown in Figure 5(a) consists of a long endoscope-like probe with a tactile sensor array located at the end. The last portion of the probe is flexible, and a trigger mechanism allows the surgeon to orient the tip to a region of interest. The sensing mechanism is capacitive using an array of 64 force-sensitive elements, which are constructed from copper strips and rubber spacers; forces are measured by determining the change of capacitance between top and bottom layers. Initially, there was no force feedback to the user; tactile information was presented on a visual display. This was later combined with a master device (finger) developed to study human factors in earlier research [60]. The haptic display was tested on a phantom formed of rubber buried 5 mm deep inside a softer foam rubber block. During the test, subjects

were asked to locate the tumor that was randomly moved within the range of ±2 cm. The tests revealed that with force feedback alone, errors were in the range of 13 mm, while with tactile feedback included this error dropped to average of 3 mm. Ottermo et al. developed another sensorized laparoscopic grasper using a commercially available tactile sensing array known as "TactArray" by Pressure Profile Systems [78]. This device had a size of 3.5 cm^2 and comprised a 15×4 array.

A device for "tactile imaging" was developed at Harvard by Wellman et al. [79]. This device was designed to generate stiffness contour maps of the surface area of an anatomical region, in this case specifically the breast, to detect the presence of (tumorous) lumps. Although manual palpation is possible and common for breast examinations, and additionally ultrasound and MRI elastography can generate stiffness quantification, the former is entirely nonquantitative whilst the latter is highly time and resource expensive. The device uses a 16×26 piezoresistive sensor array with a resolution of 1.5 mm. In terms of the target size, the device was found to be twice as accurate as either manual or ultrasound breast exams.

Dargahi et al. [12] have developed a sensorized endoscopic grasper with visual force feedback representation with the aim of allowing stiffness measurements of tissues in MIS. Figure 5(b) shows this device with manually actuated grasper jaws, which can be closed around a tissue so that the jaw surfaces which are equipped with sensors are in contact with the tissue. The sensing material used is polyvinylidene fluoride (PVDF) piezoelectric polymer film. The unique features in construction of the device allow the sensor to measure nonlinear properties such as "softness"; however, it cannot measure static forces.

A haptic flexible endoscope was reported by Petra et al. at the University of Birmingham [80]. The design has a flexible digit to resemble the end of an endoscope created using flexible PVC tubes with varying stiffness through the longitudinal crosssection. The digit is part of a master-slave system and actuated accordingly; the user wears an instrumented glove as the master device. The actuation scheme is quite unique; the flexible tube is divided longitudinally into two chambers.

These are pressurized separately with fluid and the pressure difference causing the tube to bend. In addition, sensing is achieved using a cantilever structure attached along the length of the tubing as this is bent from its natural position strain gauges, which can be implemented to output a signal.

Tavakoli et al. [81] describe a sensorized endoscopic grasper. Strain gauges are mounted on the end effector, and a linear motor and load cell combination is used to actuate the tip and measure forces imparted on it. These are mounted on the exterior (proximal) end of the endoscopic device. The grasper has an optional "wrist" at the distal end to allow angulations inside the work area. The system was evaluated using two PHANToM master devices.

A sensorized laparoscopic grasper has been developed at the Institute of Healthcare Industries in Germany [82]. This device is based on a standard manually actuated 10 mm laparoscopic grasper, in which both jaws are equipped with custom-made hexagonal array of 32 conductive polymer sensors with a spatial resolution of 1.4 mm. The output is displayed graphically as a 2-dimensional color map. A limitation with the system's mechanical design is that not all the tissues are graspable.

At the Canadian Surgical Technologies and Advanced Robotics Lab, a tactile sensing instrument for MIS has been developed, which mounts a flat sensor array on the end of a probe [83]. The sensor is a commercially available capacitive array known as "TactArray" by Pressure Profile Systems comprised of 60 elements. The device was used to assess the difference between robot-conducted and manual-conducted palpation with robotic palpation leading to a 55% decrease in maximum forces applied, a 50% decrease in task completion time, and a 40% increase in detection accuracy.

Yao et al. [84] have developed a tactile enhanced probe called "MicroTactus" for MIS at McGill University. The premise is an arthroscopic "hook" type probe with a combined accelerometer and actuator incorporated in the handle to amplify vibrational forces picked up at the tip. The instrument improved the performance of tear detection in a phantom, especially when it is combined with auditory feedback.

3.4. Instruments for Tactile Display. There has been some fairly recent work conducted on the development of tactile displays for the representation of shape and pressure distribution derived from the same sensing system. Some of the first tactile displays were adapted from Braille machines [85]. Most Braille machines are driven by piezoelectric actuators and have high bandwidth but lack the range to render curved surfaces for a display. At Harvard, the first display developed [86] used a frame that held small pins of around 2 mm diameter, arranged in a 6 × 4 array, which was actuated up to 3 mm into the fingertip using shape memory alloy (SMA) wires. Problems associated with SMA include hysteresis, and directional asymmetry, and delays caused by slow thermal response. The display was mounted on a force-reflecting master.

The very early work on haptic medical devices was completed at the Kernforschungszentrum Karlsruhe GmbH in Germany. A tactile display (see Figure 6(a)) was developed

consisting of an 8 × 8 array, which was based on an SMA spring actuation of the haptic tool [13]. The device was fan cooled and capable of a maximum force of 2.5 N and positional accuracy of 0.1 mm but would provide only 0.1 Hz bandwidth. Other developments around the same time period included a master device for general endoscopic surgery simulators developed at Stanford [87]. This consisted of a mechanical device using a novel system of linear and rotary actuators and transmission. The result was a design with low friction and inertia and high stiffness. The design had 4 DOF, a bandwidth of 100 Hz, and could deliver a force to the load with a mass of up to 1.8 kg at the master end.

Bicchi et al. [88] reported the development of a haptic device at the ARTS lab in the University of Pisa, where they modified an ordinary stiff laparoscope by adding a sensing unit located near the handle. The unit included a force sensor made of an aluminum ring attached to two strain gauges and a position sensor formed by the placement of an LED above a semiconductor. Information was returned to the surgeon via a monitor display. The advantage of this system was that it could provide a small and cheap method of haptic (force) feedback that could be incorporated into any commercial MIS tool. The disadvantage laid in the fact that the sensor placement implies that the tool properties, for example, backlash and friction, affect the feedback. Also, this system would be of very limited or no use with nonrigid devices.

The researchers at Salford University developed a tactile shape display (see [14]) using a 4 × 4 pin array (Figure 6(b)) by choosing pneumatic cylinders for actuation. The advantage of the latter design is that it is generally smaller and more lightweight. The display is small enough to be mounted onto a fingertip and is incorporated as part of a larger hand master that provides vibrotactile, as well as thermal or shear feedback called "tactile glove." A drawback is that the secondary components are bulky and heavy.

4. Noninvasive Robots in Imaged-Guided Therapy

4.1. Imaged-Guided Radiosurgery. A large number of modern non-invasive medical procedures were difficult to perform before the advent of positron emission tomography (PET), CT, MRI, and ultrasound; delicate or critical areas such as brain and heart were simply impossible to image and therefore operate on. Non-invasive principles also exacerbated the loss of information. Surgeons are now able to view the operating site in real time and track the progress of their instruments. Specially adopted robotic systems could be designed to perform intricate and accurate functions using the data returned from the imaging system for control. The CyberKnife manufactured by Accuray Inc. was first developed at Stanford University for neurosurgery with CT/X-ray guidance in the early 1990s (see Figure 7) [89]. It consists of a linear accelerator on a 6-DOF robotic arm and can fire precise radiosurgical beams with the accuracy of better than 2 mm without the use of a stereotactic frame. As no rigidly fixed frame of reference is used, it was possible to extend the

Advances in Haptics, Tactile Sensing, and Manipulation for Robot-Assisted Minimally Invasive Surgery, Noninvasive Surgery, and Diagnosis

83

(A) Actuator pin   (B) Actuator array

(a) SMA-actuated tactile display

(b) Combined tactile & shear feedback array

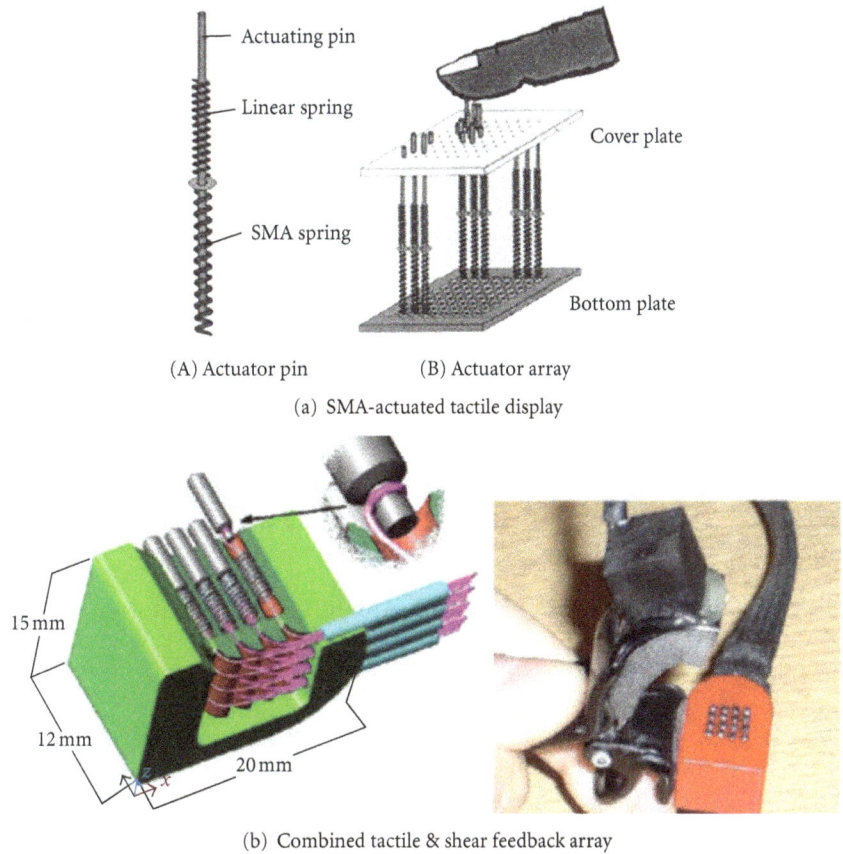

FIGURE 6: (a) An 8×8 actuator array with an SMA-spring actuated principle designed for MIS in the University of Karlsruhe [13]. (b) An integrated tactile and shear feedback 4×4 sensor array developed by the University of Salford for stimulation of finger mechanoreceptor. Actuation principle (left) and the device prototype (right) are shown [14].

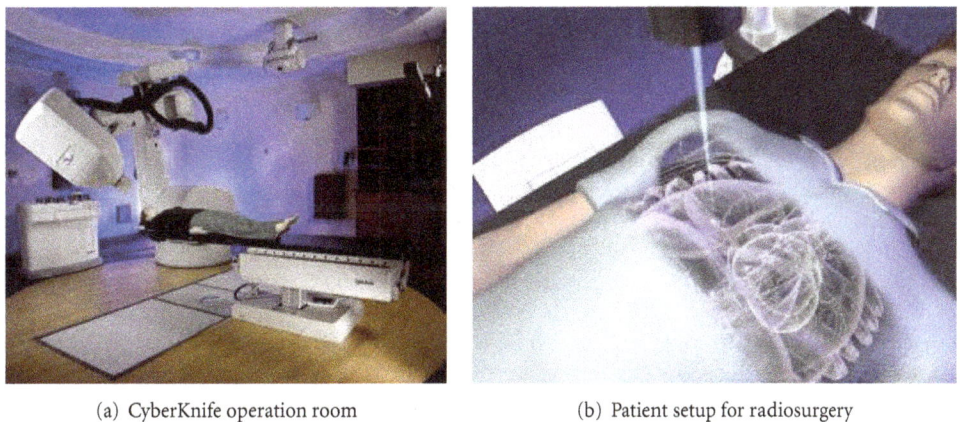

(a) CyberKnife operation room   (b) Patient setup for radiosurgery

FIGURE 7: (a) Patient setup for radiosurgery with the use of fine and high-power radiation beams delivered by a miniature linear accelerator and positioned by a robotically controlled arm in CyberKnife by Accuray Inc. [15, 16]. (b) Cyberknife schematic for non-invasive image-guided therapy.

range of application of the system to other areas such as chest, abdomen, and pelvis. In 1999, the Cyberknife became the first FDA-approved autonomous robotic system for radio-surgery [15]. Concurrent with a rise in sophistication of imaging and medical robotics, a much greater incidence of

non-invasive image-guided robotics applications has become apparent in the literature since the turn of the century.

4.2. Imaged-Guided High-Intensity Focused Ultrasound. Non-invasive thermotherapy using high-intensity focused

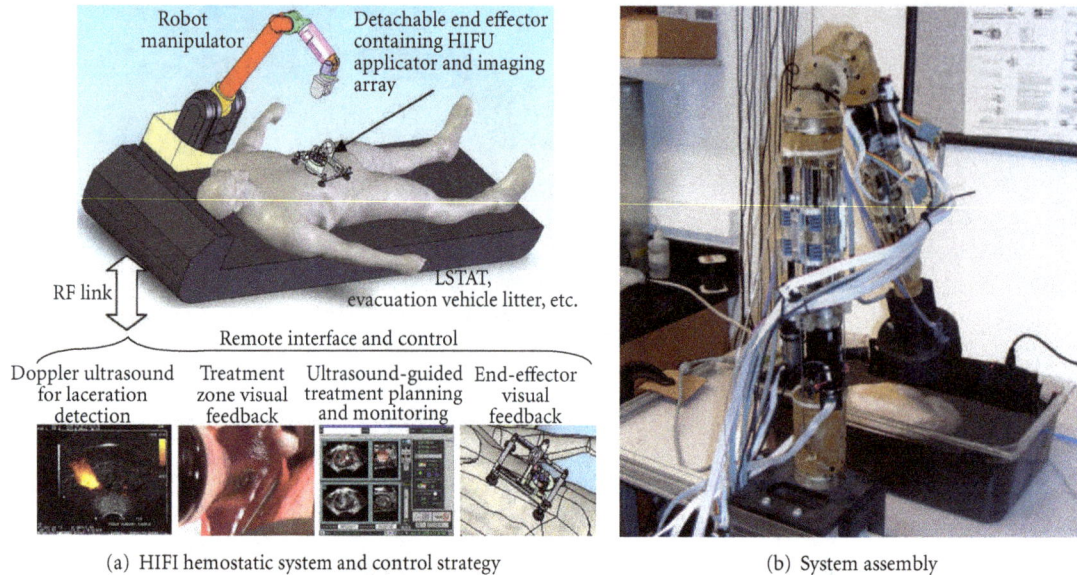

(a) HIFI hemostatic system and control strategy

(b) System assembly

FIGURE 8: (a) Schematics of the LSTAT/CSTAT compatible, remotely operated HIFU hemostatic system (top) and the overview of its control strategy (bottom) [17]. (b) The same system's assembly and experimental setup at the Brigham and Women's Hospital [18].

ultrasound (HIFU) has received increasing interest in the past couple of decades [90–93]. Treatments using HIFU include liver tumors ablations [90], arterial occlusion [91], coagulating benign breast fibroadenomas [92], and brain surgery [93]. HIFU surgery is considered as an alternative to open surgery for ablating some kinds of cancerous tissues, for example, hepatocellular carcinoma (HCC) [90], that are not responsive to other therapeutic strategies, such as radiotherapy and chemotherapy. Comparing to open surgeries that require breaking the overlying tissues, HIFU surgeries do not need general anesthesia, and they have advantages of shorter recovery time and hospital stay and reduced risk of infection and therefore significantly reduce the overall cost.

In HIFU surgeries, ultrasound beams penetrate through soft tissues and are focused at the target to destroy deep-seated tumor tissue by regional heating without overheating or damaging the overlaying or surrounding tissues. The ultrasound beams can be focused by using self-focusing spherical transducers, lenses, or reflectors [94]. The focused ultrasound can also be achieved by a phased array transducer, in which the elements are individually controlled by excitation signals with proper phase differences. The focus can be very precise and small (in the order of 1 mm). Multiple sonications can be used by steering the entire desired target volume through controlling the electrical signal phases applied to each element of the transducer array or properly positioning the transducer by a robotic manipulator, or both.

The use of HIFU for the treatments of hemorrhage was proposed to save lives in emergency situations, such as in the prehospital phase, early in the hospital phase, or in the battlefield [23, 24]. In these applications, the applicators, which were comprised of ultrasound imaging probes and phase array transducers, were used for both detecting the hemorrhage location and cauterizing the blood vessels from

the skin surface. The imaging system was Doppler-based allowing real-time visualization of the internal bleeding situation during HIFU sonications. The specification of the HIFU transducer depends on the target site, size, and temperature required. In [24], it was demonstrated that the required acoustic intensity is in the order of $1500 \, W \, cm^{-2}$ to cauterize the blood vessels and stop the bleeding. In this case, the HIFU transducer with a crystal aperture of 74 mm and 23-annuli elements operating at 2.2 MHz was used.

4.3. Robots for HIFU Therapies. Sophisticated computer-controlled multichannel HIFU systems have a large number of degrees of freedom to steer the focal point in a wide region and often require a large number of individual power amplifiers and transducer elements making the system non-portable [95]. This also adds to the challenges to use these systems in battlefield or in other emergency cases.

Portable manipulator systems designed for HIFU hemorrhage detection and treatment incorporating with HIFU were proposed by Alvarado et al. and Seip et al. [17, 18]. As shown in Figure 8, both of the HIFU and ultrasound diagnostic probes were attached to the end effectors of the manipulators, which were mounted on a life support for trauma and transport (LSTAT). The focus of the ultrasound beams emitted by the HIFU transducer elements could steer along the ultrasound propagation direction and be controlled by the individual signal phases from the HIFU amplifiers.

In [18], the transducer employed has 20 concentric ring elements operating at 1 MHz. The manipulator system has 6 DOF in order to satisfy the minimum set of controllable force, and at the same time, keep the weight and complexity at a minimum. Operating this system verified that the targeted spots shown on the ultrasound image of biological tissues could be thermally ablated after 60 s of sonication

Advances in Haptics, Tactile Sensing, and Manipulation for Robot-Assisted Minimally Invasive Surgery, Noninvasive Surgery, and Diagnosis

85

with an acoustic power of 300 W. In the system developed in [17], the end-effector is detachable from the manipulator. The manipulator was designed to place, release, and later retrieve the end effector on the patient. The applicator was scanned, with fine mechanical motion, on the surface of the patient's skin during treatment. This design significantly improved the positioning accuracy and repeatability as the applicator registration can be better maintained with the low-profile and low-weight end effector when the injured soldier was moved over rough terrain. All of the non-invasive diagnoses and treatments were performed by the end effector under local or remote control during the transport of the patient. Another advantage of this detachable design is that the robot manipulator can be used for other applications when the treatment is in progress. For example, it can provide video feedback of the procedure or manipulate another medical device if necessary.

5. Conclusion

Modern robotics has been applied to facilitate the complex medical interventions including surgeries. Robot-assisted surgical platforms have evolved tremendously from the pre-electronics era to make the medical procedures safer, faster, more reliable, and comprehensive. The use of robotic manipulators in operating rooms is becoming even more justified due to the recent advancements in mechanical design, controls, and computer programming. Roboticists have also worked to improve robots' capabilities through adaptation, using sensory information to respond to changing conditions, and autonomy. Another advantage is that robots can be designed with infinite different morphologies to deal with any specific operational topology; the surgeons, however, are comparable to general-purpose machines applicable to many procedures but ideal to none. In addition to the historical and technological advancements in robotics, several surgical trends have also affected the focus of research. Primary factors include the increasing emphasis on minimally invasive and non-invasive techniques and the availability of 3D imaging, data. Despite the afore described positive aspects, there are still several challenges facing medical robotics. One of the primary antagonists in the application of any robotic system is the inability to actively process diverse sources of information, perform qualitative reasoning, and exercise meaningful judgments. This is a problem of artificial intelligence and is one of the main arguments against autonomous surgery at this time.

Within medicine, application of haptics is fairly limited; however, with the move towards less autonomous systems, the development of advanced haptic feedback for teleoperated medical robotics and haptic endoscopes and devices has become more prevalent. Most of the current research in this area focuses on bringing haptic feedback to medical tools to aid in operations, especially those in MIS. Haptic systems have been more widely used in surgical training programs to simulate tool behavior for medical trainees. Finally, the advance of multi-imaging modalities allows the treatments of NIS radiosurgery and NIS HIFU ablation with real-time image guidance.

Acknowledgment

This work was supported in part by the Japan Society for the Promotion of Science.

References

[1] C. R. Wagner, N. Stylopoulos, and R. D. Howe, "The role of force feedback in surgery: analysis of blunt dissection," in *Proceedings of the 10th Symposium on Haptic Interfaces for Virtual Environment and Teleoperator Systems*, pp. 1–7, Orlando, Fla, USA, 2002.

[2] M. J. Mack, "Minimally invasive and robotic surgery," *Journal of the American Medical Association*, vol. 285, no. 5, pp. 568–572, 2001.

[3] B. Davies, "A review of robotics in surgery," *Proceedings of the Institution of Mechanical Engineers, Part H*, vol. 214, no. 1, pp. 129–140, 2000.

[4] P. B. McBeth, D. F. Louw, P. R. Rizun, and G. R. Sutherland, "Robotics in neurosurgery," *American Journal of Surgery*, vol. 188, supplement 4, pp. 68–75, 2004.

[5] H. F. Fine, N. Simaan, and W. Wei, "A novel dual-arm dexterous ophthalmic microsurgical robot: applications for retinal vascular cannulation and stent deployment," in *Proceedings of the American Society of Retinal Specialists, Retina Congress*, New York, NY, USA, 2009.

[6] H. Su, A. Camilo, G. A. Cole et al., "High-field MRI-compatible needle placement robot for prostate interventions," *Studies in Health Technology and Informatics*, vol. 163, pp. 623–629, 2011.

[7] A. Üneri, M. A. Balicki, J. Handa, P. Gehlbach, R. H. Taylor, and I. Iordachita, "New steady-hand eye robot with microforce sensing for vitreoretinal surgery," in *Proceedings of the 3rd IEEE RAS and EMBS International Conference on Biomedical Robotics and Biomechatronics (BioRob '10)*, pp. 814–819, Tokyo, Japan, September 2010.

[8] M. Quirini, R. J. Webster, A. Menciassi, and P. Dario, "Design of a pill-sized 12-legged endoscopic capsule robot," in *Proceedings of the IEEE International Conference on Robotics and Automation (ICRA '07)*, pp. 1856–1862, April 2007.

[9] A. J. Madhani, G. Niemeyer, and J. K. Salisbury Jr., "Black Falcon: a teleoperated surgical instrument for minimally invasive surgery," in *Proceedings of the IEEE/RSJ International Conference on Intelligent Robots and Systems*, vol. 2, pp. 936–944, October 1998.

[10] M. C. Çavuşoğlu, W. Williams, F. Tendick, and S. S. Sastry, "Robotics for telesurgery: second generation Berkeley/UCSF laparoscopic telesurgical workstation and looking towards the future applications," *Industrial Robot*, vol. 30, no. 1, pp. 22–29, 2003.

[11] W. J. Peine, J. S. Son, and R. D. Howe, "A palpation system for artery localization in laparoscopic surgery," in *Proceedings of the 1st International Symposium on Medical Robotics and Computer-Assisted Surgery*, Pittsburgh, PA, USA, 1994.

[12] J. Dargahi, M. Parameswaran, and S. Payandeh, "Micromachined piezoelectric tactile sensor for an endoscopic grasper—theory, fabrication and experiments," *Journal of Microelectromechanical Systems*, vol. 9, no. 3, pp. 329–335, 2000.

[13] H. Fischer, R. Trapp, L. Schüle et al., "Actuator array for use in minimally invasive surgery," in *Proceedings of the 4th European Symposium on Martensitic Transformations*, 1997.

[14] D. G. Caldwell, N. Tsagarakis, and C. Giesler, "Integrated tactile/shear feedback array for stimulation of finger mechanoreceptor," in *Proceedings of the IEEE International Conference on Robotics and Automation (ICRA '99)*, pp. 287–292, May 1999.

[15] Cyberknife, "CyberKnife Radiosurgery: an emerging surgical revolution," 2011.

[16] M. S. Eljamel, "Validation of the PathFinder neurosurgical robot using a phantom," *International Journal of Medical Robotics and Computer Assisted Surgery*, vol. 3, no. 4, pp. 372–377, 2007.

[17] R. Seip, A. Katny, W. Chen, N. T. Sanghvi, K. A. Dines, and J. Wheeler, "Remotely operated robotic high intensity focused ultrasound (HIFU) manipulator system for critical systems for trauma and transport (CSTAT)," in *Proceedings of the IEEE International Ultrasonics Symposium (IUS '06)*, pp. 200–203, October 2006.

[18] P. V. Y. Alvarado, C. Y. Chang, and K. Hynynen, "Design of a manipulator system for hemorrhage detection and treatment using high intensity focused ultrasound," in *Proceedings of the IEEE/RSJ International Conference on Intelligent Robots and Systems (IROS '09)*, pp. 4529–4534, October 2009.

[19] L. Mettler, M. Ibrahim, and W. Jonat, "One year of experience working with the aid of a robotic assistant (the voice-controlled optic holder AESOP) in gynaecological endoscopic surgery," *Human Reproduction*, vol. 13, no. 10, pp. 2748–2750, 1998.

[20] R. H. Taylor, B. D. Mittelstadt, H. A. Paul et al., "Image-directed robotic system for precise orthopaedic surgery," *IEEE Transactions on Robotics and Automation*, vol. 10, no. 3, pp. 261–275, 1994.

[21] G. Madhavan, S. Thanikachalam, I. Krukenkamp et al., "Robotic surgeons," *Potentials, IEEE*, vol. 21, no. 3, pp. 4–7, 2002.

[22] J. Pransky, "ROBODOC—surgical robot success story," *Industrial Robot*, vol. 24, no. 3, pp. 231–233, 1997.

[23] P. Dario, E. Guglielmelli, and B. Allotta, "Robotics in medicine," in *Proceedings of the IEEE/RSJ/GI International Conference on Intelligent Robots and Systems*, pp. 739–752, September 1994.

[24] N. Abolhassani, R. Patel, and M. Moallem, "Needle insertion into soft tissue: a survey," *Medical Engineering and Physics*, vol. 29, no. 4, pp. 413–431, 2007.

[25] J. Kettenbach, D. F. Kacher, A. R. Kanan et al., "Intraoperative and interventional MRI: recommendations for a safe environment," *Minimally Invasive Therapy and Allied Technologies*, vol. 15, no. 2, pp. 53–64, 2006.

[26] Department of Defence, "Unmanned tele-operated robots as medical support on the battlefield," 2012, http://www.defence.gov.au/health/infocentre/journals/ADFHJ_apr05/ADFHealth_6_1_34-38.html.

[27] K. Cleary, A. Melzer, V. Watson, G. Kronreif, and D. Stoianovici, "Interventional robotic systems: applications and technology state-of-the-art," *Minimally Invasive Therapy and Allied Technologies*, vol. 15, no. 2, pp. 101–113, 2006.

[28] D. W. Borowski, V. Kanakala, A. K. Agarwal et al., "Single-port access laparoscopic reversal of hartmann operation," *Diseases of the Colon & Rectum*, vol. 54, no. 8, pp. 1053–1056, 2011.

[29] P. Bucher, F. Pugin, and P. Morel, "Single port access laparoscopic right hemicolectomy," *International Journal of Colorectal Disease*, vol. 23, no. 10, pp. 1013–1016, 2008.

[30] H. S. Park, T. J. Kim, T. Song et al., "Single-port access (SPA) laparoscopic surgery in gynecology: a surgeon's experience with an initial 200 cases," *European Journal of Obstetrics Gynecology and Reproductive Biology*, vol. 154, no. 1, pp. 81–84, 2011.

[31] D. Stewart, "A platform with six degrees of freedom," *Proceedings of the Institution of Mechanical Engineers*, vol. 180, no. 1, pp. 371–386, 1965.

[32] G. Brandt, A. Zimolong, L. Carrat et al., "CRIGOS: a compact robot for image-guided orthopedic surgery," *Transactions on Information Technology and Biomedicine*, vol. 3, no. 4, pp. 252–260, 1999.

[33] T. Tanikawa and T. Arai, "Development of a micro-manipulation system having a two-fingered micro-hand," *IEEE Transactions on Robotics and Automation*, vol. 15, no. 1, pp. 152–162, 1999.

[34] J. P. Merlet, "Optimal design for the micro parallel robot MIPS," in *Proceedings of the IEEE International Conference on Robotics and Automation*, pp. 1149–1154, May 2002.

[35] M. Shoham, M. Burman, E. Zehavi, L. Joskowicz, E. Batkilin, and Y. Kunicher, "Bone-mounted miniature robot for surgical procedures: concept and clinical applications," *IEEE Transactions on Robotics and Automation*, vol. 19, no. 5, pp. 893–901, 2003.

[36] B. Maurin, J. Gangloff, B. Bayle et al., "A parallel robotic system with force sensors for percutaneous procedures under CT-guidance," in *Proceedings of the 7th International Conference Medical Image Computing and Computer-Assisted Intervention (MICCAI '04)*, Lecture Notes in Computer Science, pp. 176–183, Springer, Berlin, Heidelberg, September 2004.

[37] T. C. Tsai and Y. L. Hsu, "Development of a parallel surgical robot with automatic bone drilling carriage for stereotactic neurosurgery," in *Proceedings of the IEEE International Conference on Systems, Man and Cybernetics (SMC '04)*, vol. 3, pp. 2156–2161, October 2004.

[38] J. Tokuda, S. E. Song, G. Fischer et al., "Preclinical evaluation of an MRI-compatible pneumatic robot for angulated needle placement in transperineal prostate interventions," *International Journal of Computer Assisted Radiology and Surgery*, vol. 7, no. 6, pp. 949–957, 2012.

[39] S. D'Angella, A. Khan, F. Cepolina et al., "Modeling and control of a parallel robot for needle surgery," in *Proceedings of the IEEE International Conference on Robotics and Automation*, pp. 3388–3393, 2011.

[40] A. Salimi, A. Ramezanifar, J. Mohammadpour et al., "ROBO-CATH: a patient-mounted parallel robot to position and orient surgical catheters," in *Proceedings of the ASME Conference on Dynamic Systems & Control*, pp. 1–9, Ft Lauderdale, Fla, USA, 2012.

[41] K. W. Grace, J. E. Colgate, M. R. Glucksberg, and J. H. Chun, "Six degree of freedom micromanipulator for ophthalmic surgery," in *Proceedings of the IEEE International Conference on Robotics and Automation*, pp. 630–635, May 1993.

[42] G. Brandt, K. Radermacher, S. Lavallée et al., "A compact robot for image guided orthopedic surgery: concept and preliminary results," in *Proceedings of the Computer Vision, Virtual Reality and Robotics in Medicine (CVRMed '97)*, pp. 767–776, Grenoble, France, 1997.

[43] H. Tang, Y. Li, and X. Zhao, "Cell immobilization and contour detection for high-throughput robotic micro-injection," in *Proceedings of the International Conference on Fluid Power and Mechatronics*, pp. 935–940, Beijing, China, 2011.

[44] Y. Li, J. Huang, and H. Tang, "A compliant parallel XY micromotion stage with complete kinematic decoupling," *IEEE Transactions on Automation Science and Engineering*, vol. 9, no. 3, pp. 538–553, 2012.

Advances in Haptics, Tactile Sensing, and Manipulation for Robot-Assisted Minimally Invasive Surgery, Noninvasive Surgery, and Diagnosis

87

[45] H. Tang, Y. Li, and J. Huang, "Design and analysis of a dual-mode driven parallel *XY* micromanipulator for micro/nano-manipulations," *Proceedings of the Institution of Mechanical Engineers C: Journal of Mechanical Engineering Science*, vol. 226, no. 12, pp. 3043–3057, 2012.

[46] P. S. Jensen, "Toward robot-assisted vascular microsurgery in the retina," *Graefe's Archive for Clinical and Experimental Ophthalmology*, vol. 235, no. 11, pp. 696–701, 1997.

[47] H. H. King, K. Tadano, R. Donlin et al., "Preliminary protocol for interoperable telesurgery," in *Proceedings of the International Conference on Advanced Robotics (ICAR '09)*, Munich, Germany, June 2009.

[48] G. Dogangil, O. Ergeneman, J. J. Abbott et al., "Toward targeted retinal drug delivery with wireless magnetic microrobots," in *Proceedings of the IEEE/RSJ International Conference on Intelligent Robots and Systems (IROS '08)*, pp. 1921–1926, Nice, France, September 2008.

[49] N. A. Patronik, M. A. Zenati, C. N. Riviere et al., "A study ex vivo of the effect of epicardial fat on the heart lander robotic crawler," in *Proceedings of the 5th European Conference of the International Federation for Medical and Biological Engineering*, pp. 227–230, Budapest, Hungary, 2011.

[50] M. Quirini, R. J. Webster, A. Menciassi, and P. Dario, "Design of a pill-sized 12-legged endoscopic capsule robot," in *Proceedings of the IEEE International Conference on Robotics and Automation (ICRA '07)*, pp. 1856–1862, Roma, Italy, April 2007.

[51] X. Wang and M. Q. H. Meng, "Perspective of active capsule endoscope: actuation and localisation," *International Journal of Mechatronics and Automation*, vol. 1, no. 1, pp. 38–45, 2011.

[52] G. Ciuti, P. Valdastri, A. Menciassi, and P. Dario, "Robotic magnetic steering and locomotion of capsule endoscope for diagnostic and surgical endoluminal procedures," *Robotica*, vol. 28, no. 2, pp. 199–207, 2010.

[53] G. S. Lien, C. W. Liu, M. T. Teng, and Y. M. Huang, "Integration of two optical image modules and locomotion functions in capsule endoscope applications," in *Proceedings of the 13th IEEE International Symposium on Consumer Electronics (ISCE '09)*, pp. 828–829, Kyoto, Japan, May 2009.

[54] M. Simi, P. Valdastri, C. Quaglia, A. Menciassi, and P. Dario, "Design, fabrication, and testing of a capsule with hybrid locomotion for gastrointestinal tract exploration," *IEEE/ASME Transactions on Mechatronics*, vol. 15, no. 2, pp. 170–180, 2010.

[55] M. H. Lee and H. R. Nicholls, "Tactile sensing for mechatronics—a state of the art survey," *Mechatronics*, vol. 9, no. 1, pp. 1–31, 1999.

[56] A. F. Rovers, *Design of a robust master-slave controller for surgery applications with haptic feedback [M.S. thesis]*, Technische Universiteit Eindhoven, 2003.

[57] A. Liu, F. Tendick, K. Cleary, and C. Kaufmann, "A survey of surgical simulation: applications, technology, and education," *Presence: Teleoperators and Virtual Environments*, vol. 12, no. 6, pp. 599–614, 2003.

[58] L. Margaret McLaughlin, P. João Hespanha, and G. S. Sukhatme, *Touch in Virtual Environments: Haptics and the Design of Interactive Systems*, Pearson Education, 2002.

[59] J. B. Cooper and V. R. Taqueti, "A brief history of the development of mannequin simulators for clinical education and training," *Postgraduate Medical Journal*, vol. 84, no. 997, pp. 563–570, 2008.

[60] G. C. Burdea, *Force and Touch Feedback for Virtual Reality*, John Wiley & Sons, 1996.

[61] M. C. Çavuşoğlu, I. Villanueva, and F. Tendick, "Workspace analysis of robotic manipulators for a teleoperated suturing task," in *Proceedings of the IEEE/RSJ International Conference on Intelligent Robots and Systems*, pp. 2234–2239, Maui, Hawaii, USA, November 2001.

[62] A. M. Okamura, "Methods for haptic feedback in teleoperated robot-assisted surgery," *Industrial Robot*, vol. 31, no. 6, pp. 499–508, 2004.

[63] C. R. Wagner, S. J. Lederman, and R. D. Howe, "A tactile shape display using RC servomotors," in *Proceedings of the 10th Symposium Haptic Interfaces for Virtual Environment and Teleoperator Systems Haptics*, pp. 354–355, 2002, Orlando, Fla, USA.

[64] S. Payandeh and Z. Stanisic, "On application of virtual fixtures as an aid for telemanipulation and training," in *Proceedings of the 10th Symposium on Haptic Interfaces for Virtual Environment and Teleoperator Systems*, pp. 18–23.

[65] B. Davies, M. Jakopec, S. J. Harris et al., "Active-constraint robotics for surgery," *Proceedings of the IEEE*, vol. 94, no. 9, pp. 1696–1704, 2006.

[66] B. Davies, "A review of robotics in surgery," *Proceedings of the Institution of Mechanical Engineers*, vol. 214, no. 1, pp. 129–140, 2000.

[67] A. J. Madhani, G. Niemeyer, and J. K. Salisbury, "Black Falcon: a teleoperated surgical instrument for minimally invasive surgery," in *Proceedings of the IEEE/RSJ International Conference on Intelligent Robots and Systems*, pp. 936–944, October 1998.

[68] S. Shimachi, S. Hirunyanitiwatna, Y. Fujiwara, A. Hashimoto, and Y. Hakozaki, "Adapter for contact force sensing of the da Vinci robot," *International Journal of Medical Robotics and Computer Assisted Surgery*, vol. 4, no. 2, pp. 121–130, 2008.

[69] M. Tavakoli and R. V. Patel, *Haptics for Teleoperated Surgical Robotic Systems*, World Scientific Publishing Co, 2008.

[70] M. E. H. Eltaib and J. R. Hewit, "Tactile sensing technology for minimal access surgery—a review," *Mechatronics*, vol. 13, no. 10, pp. 1163–1177, 2003.

[71] M. H. Lee and H. R. Nicholls, "Tactile sensing for mechatronics—a state of the art survey," *Mechatronics*, vol. 9, no. 1, pp. 1–31, 1999.

[72] A. Ataollahi, P. Polygerinos, P. Puangmali, L. D. Seneviratne, and K. Althoefer, "Tactile sensor array using prismatic-tip optical fibers for dexterous robotic hands," in *Proceedings of the 23rd IEEE/RSJ 2010 International Conference on Intelligent Robots and Systems (IROS '10)*, pp. 910–915, October 2010.

[73] O. S. Bholat, R. S. Haluck, W. B. Murray, P. J. Gorman, and T. M. Krummel, "Tactile feedback is present during minimally invasive surgery," *Journal of the American College of Surgeons*, vol. 189, no. 4, pp. 349–355, 1999.

[74] B. K. Mohamed, H. Moustapha, A. Jean-Marc et al., "Tactile interfaces: a state-of-the-art survey," in *International Symposium on Robotics*, pp. 1–9, Paris, France, 2004.

[75] T. H. Massie and K. J. Salisbury, "PHANToM haptic interface: a device for probing virtual objects," in *Proceedings of the International Mechanical Engineering Congress and Exposition*, pp. 295–299, Chicago, Ill, USA, November 1994.

[76] N. A. Langrana, G. Burdea, K. Lange, D. Gomez, and S. Deshpande, "Dynamic force feedback in a virtual knee palpation," *Artificial Intelligence in Medicine*, vol. 6, no. 4, pp. 321–333, 1994.

[77] W. J. Peine, J. S. Son, and R. D. Howe, "A palpation system for artery localization laparoscopic surgery," in *Proceedings of the 1st International Symposium on Medical Robotics and Computer-Assisted Surgery*, Pittsburgh, Pa, USA, 1994.

[78] M. V. Ottermo, O. Stavdahl, and T. A. Johansen, "Electromechanical design of a miniature tactile shape display for minimally invasive surgery," in *Proceedings of the 1st Joint Eurohaptics Conference and Symposium on Haptic Interfaces for Virtual Environment and Teleoperator Systems*, pp. 561–562, 2005.

[79] P. S. Wellman, E. P. Dalton, D. Krag, K. A. Kern, and R. D. Howe, "Tactile imaging of breast masses: first clinical report," *Archives of Surgery*, vol. 136, no. 2, pp. 204–208, 2001.

[80] M. I. Petra, D. J. Holding, and P. N. Brett, "Implementation of hardwired distributive tactile sensing for innovative flexible digit," in *Proceedings of the 1st International Conference on Bio-Medical Engineering and Informatics (BMEI '08)*, pp. 629–635, May 2008.

[81] M. Tavakoli, R. V. Patel, and M. Moallem, "Haptic interaction in robot-assisted endoscopic surgery: a sensorized end-effector," *The International Journal of Medical Robotics and Computer Assisted Surgery*, vol. 1, no. 2, pp. 53–63, 2005.

[82] S. Schostek, C. N. Ho, D. Kalanovic, and M. O. Schurr, "Artificial tactile sensing in minimally invasive surgery—a new technical approach," *Minimally Invasive Therapy and Allied Technologies*, vol. 15, no. 5, pp. 296–304, 2006.

[83] A. L. Trejos, J. Jayender, M. T. Perri, M. D. Naish, R. V. Patel, and R. A. Malthaner, "Robot-assisted tactile sensing for minimally invasive tumor localization," *International Journal of Robotics Research*, vol. 28, no. 9, pp. 1118–1133, 2009.

[84] H. Y. Yao, V. Hayward, and R. E. Ellis, "A tactile enhancement instrument for minimally invasive surgery," *Computer Aided Surgery*, vol. 10, no. 4, pp. 233–239, 2005.

[85] W. R. Provancher, M. R. Cutkosky, K. J. Kuchenbecker, and G. Niemeyer, "Contact location display for haptic perception of curvature and object motion," *International Journal of Robotics Research*, vol. 24, no. 9, pp. 691–702, 2005.

[86] D. A. Kontarinis, "Tactile display of vibratory information in teleoperation and virtual environments," *Presence-Teleoperators and Virtual Environments*, vol. 4, no. 4, pp. 387–402, 1995.

[87] L. B. Rosenberg and D. Stredney, "A haptic interface for virtual simulation of endoscopic surgery," *Studies in health technology and informatics*, vol. 29, pp. 371–387, 1996.

[88] A. Bicchi, G. Canepa, D. De Rossi, P. Iacconi, and E. P. Scilingo, "Sensorized minimally invasive surgery tool for detecting tissutal elastic properties," in *Proceedings of the 13th IEEE International Conference on Robotics and Automation*, pp. 884–888, April 1996.

[89] K. Cleary, "Medical robotics and the operating room of the future," in *Proceedings of the 27th Annual International Conference of the Engineering in Medicine and Biology Society (IEEE-EMBS '05)*, pp. 7250–7253, September 2005.

[90] S. Pather, B. L. Davies, and R. D. Hibberd, "The development of a robotic system for HIFU surgery applied to liver tumours," in *Proceedings of the 7th International Conference on Control, Automation, Robotics and Vision (ICARC '02)*, vol. 2, pp. 572–577, December 2002.

[91] K. Hynynen, V. Colucci, A. Chung, and F. Jolesz, "Noninvasive arterial occlusion using MRI-guided focused ultrasound," *Ultrasound in Medicine and Biology*, vol. 22, no. 8, pp. 1071–1077, 1996.

[92] K. Hynynen, O. Pomeroy, D. N. Smith et al., "MR imaging-guided focused ultrasound surgery of fibroadenomas in the breast: a feasibility study," *Radiology*, vol. 219, no. 1, pp. 176–185, 2001.

[93] N. McDannold, M. Moss, R. Killiany et al., "MRI-guided focused ultrasound surgery in the brain: tests in a primate model," *Magnetic Resonance in Medicine*, vol. 49, no. 6, pp. 1188–1191, 2003.

[94] M. Gautherie, *Methods of External Hyperthermic Heating*, Springer, New York, NY, USA, 1990.

[95] S. D. Sokka, J. Juste, and K. Hynynen, "Design and evaluation of broadband multi-channel ultrasound driving system for large scale therapeutic phased arrays," in *Proceedings of the IEEE Ultrasonics Symposium*, vol. 2, pp. 1638–1641, October 2003.

8

Task Allocation and Path Planning for Collaborative Autonomous Underwater Vehicles Operating through an Underwater Acoustic Network

Yueyue Deng, Pierre-Philippe J. Beaujean, Edgar An, and Edward Carlson

Department of Ocean and Mechanical Engineering, Florida Atlantic University, 777 Glades Road, Boca Raton, FL 33431, USA

Correspondence should be addressed to Pierre-Philippe J. Beaujean; pbeaujea@fau.edu

Academic Editor: Duško Katić

Dynamic and unstructured multiple cooperative autonomous underwater vehicle (AUV) missions are highly complex operations, and task allocation and path planning are made significantly more challenging under realistic underwater acoustic communication constraints. This paper presents a solution for the task allocation and path planning for multiple AUVs under marginal acoustic communication conditions: a location-aided task allocation framework (LAAF) algorithm for multitarget task assignment and the grid-based multiobjective optimal programming (GMOOP) mathematical model for finding an optimal vehicle command decision given a set of objectives and constraints. Both the LAAF and GMOOP algorithms are well suited in poor acoustic network condition and dynamic environment. Our research is based on an existing mobile ad hoc network underwater acoustic simulator and blind flooding routing protocol. Simulation results demonstrate that the location-aided auction strategy performs significantly better than the well-accepted auction algorithm developed by Bertsekas in terms of task-allocation time and network bandwidth consumption. We also demonstrate that the GMOOP path-planning technique provides an efficient method for executing multiobjective tasks by cooperative agents with limited communication capabilities. This is in contrast to existing multiobjective action selection methods that are limited to networks where constant, reliable communication is assumed to be available.

1. Introduction

Autonomous underwater vehicles (AUVs) represent one of the most challenging frontiers for robotics research. AUVs work in an unstructured environment and face unique perception, communication, and control difficulties. Currently, the state of the art in mission planning is dominated by single AUV operations using preplanned trajectories with offline postprocessing of the data collected during the mission. Multiple cooperative vehicle systems (MCVSs) hold great promise for use in large-scale oceanographic surveys, mine countermeasures (MCMs), and other underwater missions, due to better resource and task allocation [1–3].

Simultaneous use of multiple vehicles can improve performance, reduce mission time, and increase the likelihood of mission success. It is not necessary for all the vehicles in an operation to be the same. In fact, heterogeneity could become a powerful driver of multiple AUV (MAUV) operations.

Instead of using a single vehicle or a homogeneous fleet able to perform every possible mission, a fleet could comprise a variety of AUVs [4]. Different missions would be accomplished using different combinations of vehicles. The key to obtaining the greatest benefit from MAUV or combined operations is *cooperation*. Cooperative strategies among multiple underwater vehicles can be complex. In addition, communication is a critical aspect of vehicle cooperation and must not be trivialized, as underwater communication is notoriously difficult, slow, and limited in range [5].

The work presented here addresses the two closely coupled problems of multivehicle underwater operations, which are (1) how to assign tasks to each vehicle efficiently and robustly, and (2) how to ensure rapid and effective vehicle actions.

The main contributions of this article are the following: (1) first are the design and simulation of a path-planning controller that efficiently handles the cooperative operations of multiple underwater vehicles using underwater acoustic communication; this path-planning controller uses a grid-based multi-objective optimal programming (GMOOP) approach; GMOOP finds the optimal vehicle command decision given a set of objectives and constraints; (2) second are the design and simulation of a location-aided task allocation framework (LAAF) in the form of a task allocation algorithm specially designed for harsh underwater acoustic communication environments. The LAAF and GMOOP controllers are combined in a "task-plan-act" structure [6] to generate an optimized local system output in a timely manner to achieve fleetwide cooperation.

This paper is organized as follows. Section 2 briefly reviews the literature on multiple collaborative underwater vehicle research. The next two sections present the LAAF task allocation protocol and the GMOOP multiple-objective control method. Section 5 presents the mission scenarios that employ these methods. Section 6 shows the simulation results. The final section presents a summary of our research and concluding remarks.

2. Literature Review on Collaborative AUVs

This section briefly states the research problem of interest and reviews existing research work on underwater acoustic communication, task allocation strategies, path planning, and control for multiple collaborative AUVs.

2.1. Problem Statement. Our mission interest is to search and classify underwater targets. In this mission, two classes of vehicles are involved. Every search vehicle (SV) is equipped with wide field of view target detection sonar. These sensors cover a large volume of water but do not have sufficient resolution to discern the specificity of the targets. In contrast, every classify vehicle (CV) is equipped with a narrow field-of-view, high-resolution sonar. These sonar systems cover a small volume of water with sufficient resolution for target classification. In this work, the main focus is the task acquisition and execution of the CVs' mission. The target distribution is assumed to be known a priori to SVs. Similar approach of using heterogeneous robots in a cooperative mission has been demonstrated [4, 7, 8] and viewed as a feasible and cost-effective means to carry out search and classify missions.

2.2. Underwater Acoustic Communication. Cooperative control and optimization for multiple collaborative autonomous underwater vehicles has become a vital area of research [1, 2, 4, 6]. Although some of the prior work are closely related to this research, few address the same problems specifically related to underwater environments with realistic acoustic communication constraints [9–11]. Underwater acoustic networking is an enabling technology for various collaborative missions involving multiple autonomous underwater vehicles [5]. Unfortunately, underwater acoustic communication is slow and limited in range, which impacts MCVS controllers.

The proposed research problem is significantly different from those reported using unmanned aerial vehicles (UAVs)

as the intervehicle communication can only be achieved via sound propagation [12, 13]. In addition, all UAVs are mostly equipped with high precision navigation sensors onboard (with GPS) whereas the navigation of most underwater vehicles is still based on the traditional dead-reckoning techniques or motion sensors' fusion.

2.3. Dynamic Task Allocation. In dynamic task allocation, every robot assignment is dynamically adjusted with changes in the environment or group performance. While coordination algorithms for task allocation use only local sensing [14], these algorithms do not take advantage of the underwater acoustic network capability and formal analysis tools are lacking.

Task allocation models can be classified as either centralized or distributed. In centralized models, a central agent exists and plays the role of arbitrator. This arbitrator aggregates information from their team members, plans optimally for the entire group, and finally propagates the task assignments to other team members. This model has the advantage of finding the optimal solution, but it generally falls short in harsh underwater communication and partially observable environment as these controllers cannot handle tight MCVS coordination due to limited flexibility and long response time [6, 15].

The distributed task allocation approach handles this shortcoming [16, 17]. In a distributed multirobot system, each robot operates independently under local sensing, with high-level system coordination arising from interactions with other robots as well as the task environment [18]. In this approach, agents rely on a predefined negotiation framework that allows them to decide what activity to do next, what information to communicate, and to whom. A difficulty with this approach is that it requires agents to possess accurate knowledge about their environment and assumes that consistent communication among agents is available throughout the mission. Both conditions are difficult to maintain in heterogeneous underwater MCVS.

Task allocation among multiple robots is commonly accomplished by market-based methods, which are tested as one of the most successful solutions [12, 19–21]. In traditional task assignment problems, market-based approaches are widely used due to low computational complexity. The original form of auction algorithm was first proposed by Bertsekas [19, 22]. This algorithm operates whereby unassigned agents bid simultaneously for objects thereby raising their prices. The auction algorithm is widely used in task allocation for cooperating UAVs. In [23], Nygard et al. present a network flow optimization model for allocating UAVs to targets. This problem is formulated as a linear programming problem to obtain decisions for allocating the UAVs. The network model has been further studied by Schumacher et al. [12] and Mitchell. In [24], the authors present a team theoretic approach that allows UAVs to perform decision-making independently whenever UAVs cannot exchange information.

Task Allocation and Path Planning for Collaborative Autonomous Underwater Vehicles Operating through
an Underwater Acoustic Network

91

Typically, the task allocation problem divides a mission into individual subtasks and assigns robots to each subtask. Auctions are powerful tools for allocating resources effectively, especially in situations where optimal allocation is expensive to achieve or where the environment is not completely known.

Another well-accepted auction algorithm, effective for general multiagent task allocation, is Challenger [18]. The difficulty with this auction algorithm is that it becomes stagnant when cluster-typed targets exist in the search space and it easily fails in harsh underwater environment. It is further challenged in an underwater acoustic mobile ad hoc network (MANET) due to limited network capacity and high latency [25].

As the size of system grows or the network condition deteriorates, market-based approaches are not applicable as the increased demands in communication bandwidth. This leads to emergent coordination through negotiation or self-organized mechanisms. In such methods, individual robots coordinate their actions based solely on local sensing or local interactions. The advantages and disadvantages of the three task allocation methods are listed in Table 1.

2.4. Path-Planning and Multiple-Objective Optimization. Path planning typically involves defining vehicle waypoints based on the minimization of a positive cost metric (typically a function of mission time or energy consumption), under certain constraints and objectives. In [27], a motion control algorithm is proposed, which constructs adjoint equations based on the vehicle velocity, the time to reach the target, and the optimal usage of the battery life.

Cooperative strategies among heterogeneous vehicles are difficult to devise. MCVSs with heterogeneous agents need an advanced formation structure, a dynamic task, and resource allocation algorithm as well as powerful techniques to construct and optimally solve a path-planning problem. Additional complexity arises from the challenges of limited navigation and communication capabilities in the underwater environment.

For a network of vehicles to communicate, additional constraints, such as transmission delay and throughput, would need to be considered. In this case, queue length and bandwidth efficiency must be accounted for. With the objective to maximize throughput and queue length, a receding horizon control algorithm is proposed in [28], which yields unique, piecewise continuous optimal controllers that limit and route traffic.

There has been significant work reported in MCVS architecture applicable to multiple AUVs as well as MCVS simulation platform [13, 29–32]. However, the research findings and conclusions from these works are based on the assumption of idealized underwater communication conditions. In [32], a simulation environment for the coordinated operation of multiple autonomous underwater vehicles is presented with five clearly defined control architecture layers: the physical layer, the abstraction layer, the functional layer, the coordination layer, and the organization layer. One limitation of this simulator is its model of the acoustic environment: simple time delays are used and no message routing is considered in this simulation platform.

Extensive research work has been done on cooperation and coordination of mobile agents. Traditional methods for cooperative underwater vehicles include swarmed cooperative schemes for homogeneous vehicles and low-level cooperative or merely coordinated techniques for simple task accomplishment [27, 29, 33]. For the coordinated maneuver of multiple vehicles, [27] presents a control maneuver-integrated acoustic navigation system for a formation of three AUVs and one surface craft to gradient searching and following missions. In [30], the authors present a theoretical study of the coordination of the geometrical movements of one flotilla of autonomous underwater vehicles. However, these techniques are insufficient for mission planning when a fleet of vehicles operate in a MANET with a very limited communication bandwidth.

To better address the dynamic system with behavior-based strategies, an interval programming (IvP) method is described in [34, 35] to solve multi-objective optimal problems: a mathematical programming model finds an optimal decision given a set of competing objective functions.

3. Location-Aided Task Allocation Framework Protocol

We propose a location-aided task allocation framework (LAAF) that addresses these challenges by extending the radio network auction algorithm developed by Chavez [18] and Bertsekas [19] to the AUV network. LAAF uses three types of strategy: centralized, negotiation, and self-organized [26]. This approach does not guarantee an optimal allocation, but it is especially suited to dynamic environments, where execution time might deviate significantly from estimates and where the ability to adapt to changing conditions is the key to success. In this framework, each robot considers its local plans when bidding and multiple targets can be allocated to a single robot during the negotiation process. The rest of this section provides a description, within the context of a search-classify AUV mission, of the LAAF auction algorithms.

3.1. Cost Function. The mission of interest is to search and classify targets with a prior-known distribution. Each target is modeled as a step utility function and considered to be classified when it is within the sensor range of a CV. Assume that t_n is the time at which vehicle n reaches its final waypoint. A cost function that penalizes the maximum completion time (to cover all targets) is traditionally assigned to each vehicle. Unfortunately, such cost function may lead to exceedingly long trajectories for most of the vehicles. To avoid this problem, the authors minimize a *cost-to-the-fleet* (CTF) function assigned to the SVs and CVs working cooperatively to identify the targets:

$$ \text{CTF} = \sum_{\tau \in \text{targets}} \left(P_\tau + \sum_{n \in \text{classify vehicles}} c_{n\tau} \right). \tag{1} $$

TABLE 1: Comparison of centralized, negotiation, and self-organized task allocation [26].

Task allocation	Advantages	Disadvantages
Centralized	Close to global optimum	Limited to small number of robots; not applicable in worst-case network; computation intensity
Negotiation	Only require short-range communication	Not scalable, solution not global optimum
Self-organized	Robust, scalable	Very difficult to achieve global optimum

Here $c_{n\tau}$ is the positively defined cost incurred by CV n to service target τ and is defined as the time needed by this CV to reach the target (including foreseeable vehicle collision avoidance). P_τ is the positively defined cost of allocating the optimal CV to classify the target τ. P_τ is a function of two variables computed in the auction algorithm (Section 3.2): the Task Allocation Round Time (TART) and Effective Task Allocation Time (ETAT). Both TART and ETAT depend on the network conditions.

TART and ETAT are computed by the simulation tool using the simulator clock. TART is calculated from the time a new classify task is announced to the time that the winner acknowledge message reaches the auctioneer. The role of the auctioneer is assigned to SVs as they carry long-range search sensors. Compared to TART, ETAT ends when the self-determination process is complete. If the task allocation keeps failing due to exceedingly unreliable communication, the auctioneer will stop the task allocation process past a predefined timeout. If the task allocation timeout is reached, TART is set to a very large value (equivalent to infinity) [26]. The relationship between P_τ, TART, and ETAT is

$$P_\tau = \begin{cases} \text{TART} & \text{if all bids are successfully received,} \\ \text{ETAT} & \begin{cases} \text{if the auctioneer times out} \\ \text{and at least one bid is received.} \end{cases} \end{cases} \quad (2)$$

The output action commands generated using the GMOOP model produces the local optimal value of $c_{n\tau}$. The corresponding messages are designed to use as few data bytes as possible [36]. The message size is labeled Bytes for Task Allocation (BTA).

3.2. Generic Auction Algorithm. The well-accepted auction algorithm developed by Bertsekas [19, 22] is used as a reference in our analysis. We will refer to this algorithm as the *generic* auction algorithm from now on. This algorithm determines which robot can best complete a given task, based on their proposed bids. The protocol requires a single controller and an auctioneer. For each auction, the auctioneer needs prior knowledge of the number of bidders and the maximum expected round-trip time of a message in the network. This protocol is characterized by the following assumptions.

(1) A single vehicle can be an auctioneer, a bidder, or neither.

(2) Only one type of tasks can be auctioned among a given set of auctioneers and bidders, and all auctioneers and bidders have a prior knowledge of all task types.

(3) All bidders have a cost function which accepts the auction data and returns a bid. The lowest bid wins the auction.

(4) Network broadcast is available, so that a single transmission may reach all the nodes in the network.

Each bidder replies to each auction announcement with a bid. The bid is an estimate of the time required for the vehicle to accomplish the task, which consists of the time needed to complete all equivalent or higher-priority tasks plus the transit time to the Area of Interest (AOI). Upon winning an auction, a bidder adds the AOI to its task queue. When an AOI task becomes active, the bidder proceeds directly to the AOI. Once in the AOI, if a target is successfully identified, the task is marked as completed. If a search over the entire AOI produces no positively identified targets, the task is marked as incomplete.

3.3. Location-Aided Auction Framework. The reference auction algorithm may fail occasionally due to the unreliable underwater acoustic network. In this paper, the simulated acoustic modem (FAU DPAM) performs poorly beyond 4000 m. In the simulator, the medium model will consistently return a Frame Error Rate (FER, or message error rate) close to 100% beyond 4 km [37]. As shown in Figure 1, even under best-case conditions, the maximum useful range of the modem is not expected to exceed 4096 meters. Therefore, a moving classify vehicle may not be able to return a bid during an auction due to poor communication. Note that the performance curves shown in Figure 1 have been derived by fitting a Nakagami-m model with actual field data [37]. These data were collected under environmental conditions closely resembling those that used the simulation.

3.3.1. Overview. A solution to this problem is to add a self-decision capability to each classify vehicle, based on its knowledge about the overall fleet topology. Each CV accepts or denies the target assignment according to its location relative to the target, any known locations of other vehicles, and any neighbor decision received via the acoustic modem. Once decisions have been made, the same CV will disseminate the results to the entire network.

The benefits of self-determination are that a CV can (1) always attempt to classify a target if needed, (2) utilize the

Task Allocation and Path Planning for Collaborative Autonomous Underwater Vehicles Operating through an Underwater Acoustic Network

93

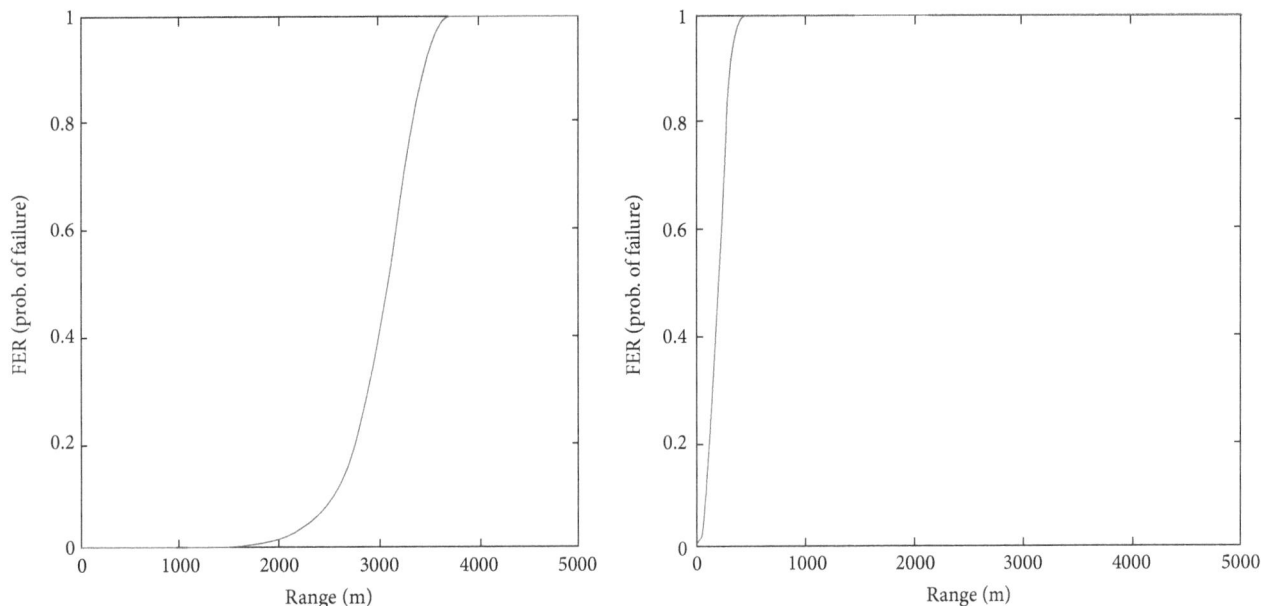

(a) Best-case environmental conditions: Nakagami-m = 2.0 and noise PSD = 55.0

(b) Worst-case environmental condition: Nakagami-m = 1.5 and noise PSD = 85.0

FIGURE 1: Medium model estimation of FER versus range.

available topology information periodically updated as the mission processes, and (3) reduce the need for underwater communication and, in turn, the response time between detection and classification. None of these features is possible in a generic auction cycle.

The shortcoming of this self-determination mechanism is that multiple vehicles may attempt to classify the same target, due to the limited accuracy of the topology information. This can be alleviated by the fact that as the classify vehicles are approaching the same target, the acoustic communication improves. Therefore, the classify vehicles have a more accurate knowledge of the topology and decide whether to abort the target classification.

Based on these observations, a location-aided auction framework (LAAF) has been developed. LAAF is an extension of the generic auction algorithm that incorporates a negotiation among all bidders and utilizes the available topology information of the local vehicle. It is designed to meet the challenges of high latency and limited bandwidth of the acoustic network linking the underwater agents. LAAF uses a master-slave architecture which handles most of the allocation work through the acoustic network. If the agents are too far apart to communicate, they identify their individual tasks by reasoning on their available world information. Figure 2 gives an overview of LAAF: when a search vehicle discovers any new target(s), the search vehicle starts an auction for the task(s). It broadcasts an auction announcement message that has a header field to distinguish between single and multiple items.

In a scenario of a single-item auction, once the auctioneer has received bids from all bidders, it chooses the best bid and sends out a winner notification message to all the bidders. The auctioneer then waits for the winner-acknowledge message

from the winner and closes the current auction. If the auction times out before the auctioneer receives the winner acknowledge message, the auction will be reinitiated.

In the multi-item scenario, a bid message received by the auctioneer contains the optimal configuration for classifying the targets by a set of vehicles, and the auctioneer simply broadcasts the winners to the whole network. In this case, the winner notification is not necessary in LAAF as ETAT is used to evaluate the task assign performance. There are two possibilities that an auctioneer transitions to an *auction close* state: (1) the auctioneer successfully received a winner acknowledge message, so the target(s) are successfully assigned through auction process; (2) due to poor network conditions, the auction times out and the classify vehicles are left to determine their own tasks based on their knowledge of the fleet topology.

3.3.2. LAAF Bidder Agent's Policy. Algorithm 1 shows the LAAF algorithm used by a bidder to respond to bid requests from an auctioneer. When a bidder is in the bidding state, it keeps track of the current cost to classifying all the known unclassified targets (base cost), which is added to any further bid calculations. The policy of the bidder in the case of single-item auction is intuitive; it adds the cost to classify the new task to its base cost and sends it back to the auctioneer.

Since the time slot allocated for each node by the Medium Access Control (MAC) is fairly long (typically 5 seconds), LAAF processes multi-items within an auction cycle. In comparison, single-item task allocation generally requires more auction cycles to complete the task allocation. When the bidder agent receives a multiple-item auction request, it selects the target(s) to bid for. An intuitive solution is to

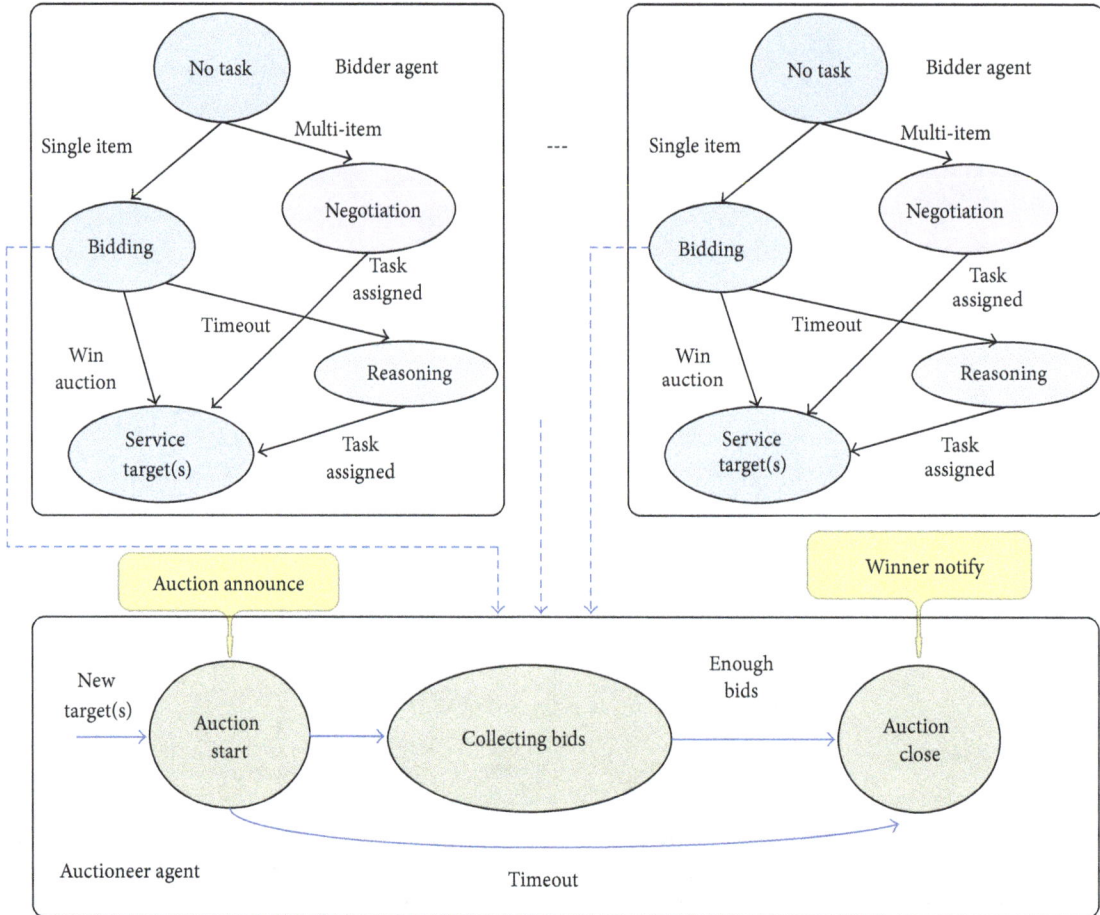

FIGURE 2: Location-Aided Auction Framework.

compute all the possible combinational bids and send the bids information back to the auctioneer to select the winner(s). The problem with this method is that for n concurrent targets, each bid message contains $P(n, 1) + P(n, 2) + \cdots + P(n, n)$ combinational bids, where $P(n, i)$ is the permutation function [36]. This approach requires a large amount of data to be passed between the bidder and the auctioneer in a slow underwater acoustic network.

To overcome this problem, each bidder initiates a negotiation process with other bidders. When a bidder receives a multiple-item auction, it calculates a bid for each target in the task announcement and disseminates the bid information through the network. In this case, only n bids are disseminated by each classify vehicle. Once a bidder has collected enough negotiation messages from other bidders, it determines the optimal combination of targets' assignment. By finding the minimum value of all the combinational bids, this bidder finds the optimal list of vehicles to accomplish the multiple-target auction.

At the end of this auction, the auctioneer acknowledges the winner. In case of an acknowledgment timeout, the auctioneer reannounces the auction (targets) and starts a new auction process with the same target(s).

Simulation results show that when multiple classify tasks are acquired, a CV should reschedule the task list to achieve

optimality [36]. For example, if a CV is on its way to classify a remote target and receives a new task near the current route, this vehicle should adjust the priorities in its task list.

4. Path-Planning and Multiple-Objective Optimization

The GMOOP model is designed to find the optimal solutions for a search-classify mission using an action determination map subject to static and dynamic constraints and objectives [36]. Derived from the IvP algorithm [35], the GMOOP model has a similar solution strategy of representing and optimizing over multiple competing objective functions. The main differences are that GMOOP model uses grid maps and addresses the impact of unreliable underwater acoustic communications. In particular, the output of a GMOOP model is the optimized values for command variables.

4.1. Potential Field. The main function of the GMOOP model is to dynamically create paths followed by each vehicle. To do so, it uses a path-planning controller to move the vehicle from one location in the mission field to the next, such that it minimizes a positive cost metric (here the traverse

```
Algorithm Bidder

(1) loop
(2)     WAIT-FOR-MESSAGE
(3)     msg ← GET-MESSAGE
              • Messages of unknown type are silently ignored.
(4)     timer ← SET-TIMER
(5)     while NOT-EXPIRED  (timer)
(6)       if MESSAGE-TYPE (msg) = "AUCTION-ANNOUNCEMENT"  then
(7)          auction-id, auction-type, auction-data ←
                          PARSE-ANNOUNCEMENT-MESSAGE (timer)
(8)          if MESSAGE-TYPE (msg) = "SINGLE"
(9)             bid ← CREATE-BID (auction-id, auction-type, auction-data)
(10)            SEND-BID (bid)
(11)         else if MESSAGE-TYPE (msg) = "MULTIPLE"
(12)            NEGOTIATE-BID (bid)
(13)         end if
(14)       else if MESSAGE-TYPE (msg) = "AUCTION-WINNER"  then
(15)          auctioneer ← GET-SENDER (msg)
(16)          auction-id, auction-data ← PARSE-WON-MESSAGE (msg)
                  • Note that we acknowledge regardless of whether or not we find the bid.
(17)          SEND-WIN-ACKNOWLEDGEMENT (auctioneer, auction-id)
(18)          bid ← FIND-BID (msg)
(19)          if  bid ≠ nil then
(20)             ADD-TASK (auction-data)
(21)             CLOSE-BID (bid)
(22)          end if
(23)       end if
(24)       FREE-MESSAGE (msg)
(25)     end while
(26)     START-SELF-DETERMINATION
(27) end loop
```

ALGORITHM 1: Bidder agent response to auctioneer in LAAF.

length), while subjecting certain constraints and fulfilling certain objectives.

From (1), we know that the general expression of the search-classify mission overall cost to the fleet is CTF, where P_τ is the price paid to find the optimal classify vehicle for target τ. Thus, in the path-planning problem only, the objective is to minimize the cost metric $c_{n\tau}$, which is the cost incurred by classify vehicle n to service target τ.

The problem space can be formulated as a set of states denoting vehicle locations connected by directional arcs, each of which has an associated cost. The vehicle starts from an initial state (at the start location) and moves across arcs to other states until it reaches the goal G_t. A potential field is defined between every combination of neighboring states X and Y to calculate the cost $c(X, Y)$ of traversing an arc from state X to state Y. The optimal neighbor state Y^* is found by minimizing this cost function. The minimum cost function is the variable $c_{n\tau}$ used in (1).

4.2. Model Elements. The GMOOP algorithm (Figure 3) uses a set of functions that address specific behavior patterns. These functions are defined over an action variable space or action determination grid map. Grid maps are widely used in vehicle localization and environment estimation [38–41]. The GMOOP algorithm produces an action determination grid map which results from the weighted combination of a series of objective functions. These behavior functions are constrained by the objectives of every vehicle. Collision avoidance from nearby vehicles and fast target engagement are two typical behavior patterns.

Each behavior pattern uses a mathematical function of the relative distance between two vehicles or between a vehicle and a target [36]. The *action determination grid map* contains the value of a specific objective function given a vehicle velocity vector. The variation in time of such a map can be used to predict the trajectory of the vehicle. Each vehicle also carries a *vehicle distribution grid map*, which contains the location of other vehicles in the search space and areas already covered by a vehicle. Due to uncertainty in the navigation information and acoustic communication, the position information for each vehicle is the combination of a mean value and a probabilistic Rayleigh model.

The *objective function*, defined by the command variables, is applied to the action determination grid map and produces

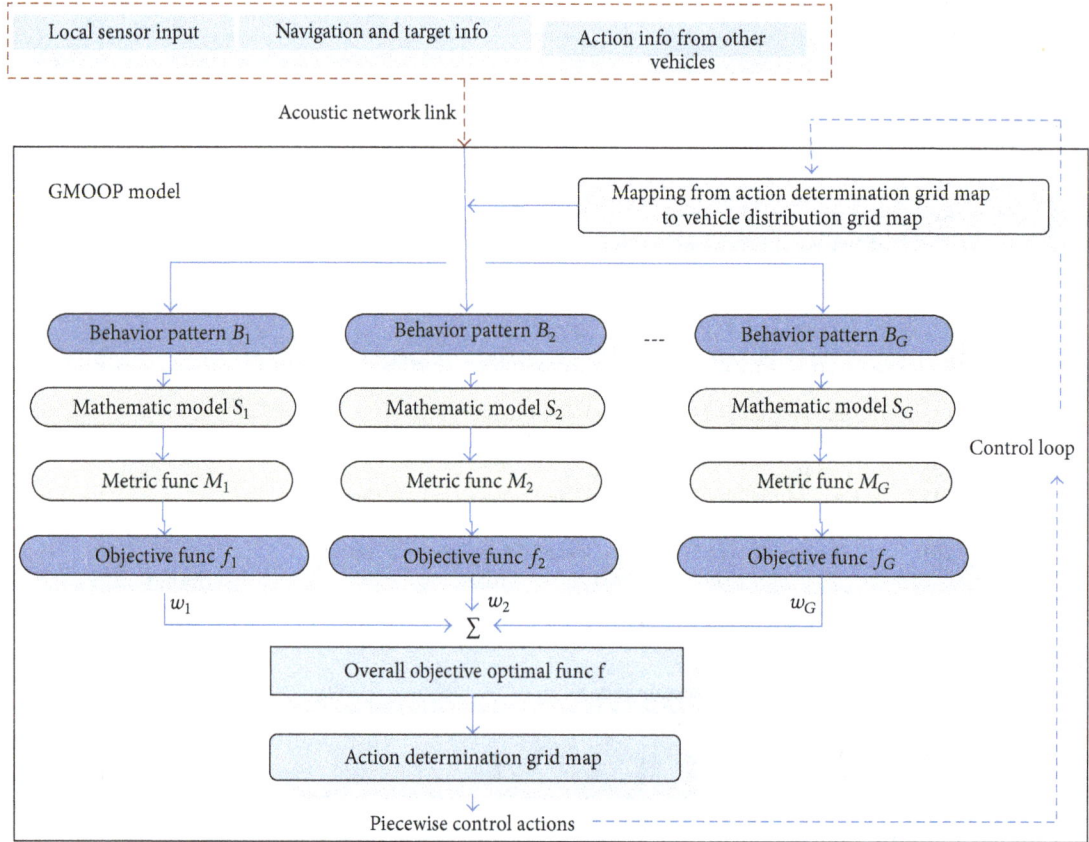

FIGURE 3: Components and building processes of the GMOOP model.

a Control Action Output (CAO). The planar form of the overall objective function is given by

$$\text{CAO} = \max \left\{ \sum_{g=1}^{G} \frac{w_g f_g(u,v)}{\text{mid}} \sum_{g=1}^{G} w_g = 1 \right\}. \qquad (3)$$

f_g is a behavior function, designed following the potential field approach [42]: neighbor vehicles exert repulsive forces, while the target applies attractive force to the robot. w_g is the weight for this behavior. (u,v) represents the north and east component of the vehicle velocity in geographical coordinate.

5. Multiple Autonomous Underwater Vehicle Control Model Used by GMOOP

The mission assigned to an underwater vehicle impacts the underwater navigation and underwater communication performance. The simulated missions emphasize rapid target classification using cooperating AUVs which are operating in a congested coastal area. Therefore, collision between vehicles is a real concern. The scenarios considered in this section emphasize the continuous trade-off between collision avoidance, remaining within communication range, and target engagement.

5.1. GMOOP Control Loop. Here we consider the case of a single vehicle controlled by the GMOOP controller. The

required operation involves a vehicle moving through time and space, where periodically, at fixed time intervals, a decision is made as to how to control the next move of the vehicle. As depicted in Figure 4, the next decision occurs at time t_m, while the output decision c_m is computed by building and solving the GMOOP problem in the time interval $[t_{m-1}, t_m]$. The control loop builds and solves the GMOOP model iteratively (Figure 5). Each objective function is defined over a common decision space where each decision precisely spells out the next action for the vehicle to implement starting at time t_m.

5.2. Vehicles and Environment. The mission of interest is to search and classify underwater targets. Search vehicles act as auctioneers while classify vehicles act as bidders in the task planning process. Both the search and classify vehicles use the GMOOP controller with a group of different predefined behaviors $B_{Tg}^i(t)$ (Figure 3) defined as

$$B_{Tg}^i(t) = B_{Tg}^i\left(\vec{p}_d, \{\vec{u_{Sm}}, \vec{u_{Cn}}\}, \{\vec{p_{Sm}}, \vec{p_{Cn}}\}, t\right),$$

$$m \in [1, 2, \ldots, M], \text{M is the total number of SV,} \qquad (4)$$

$$n \in [1, 2, \ldots, N], \text{N is the total number of CV.}$$

Here $g \in [1, 2, \ldots, G]$ is the number of behavior types, $i \in [1, 2, \ldots, I]$ is the vehicle number, and $T \in \{S(earch), C(lassify)\}$ is the vehicle type. The vehicle position

Task Allocation and Path Planning for Collaborative Autonomous Underwater Vehicles Operating through
an Underwater Acoustic Network

97

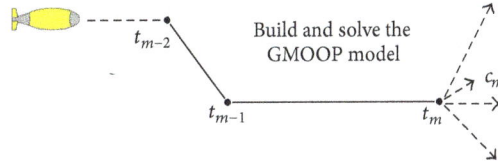

FIGURE 4: The time interval of a decision-making cycle.

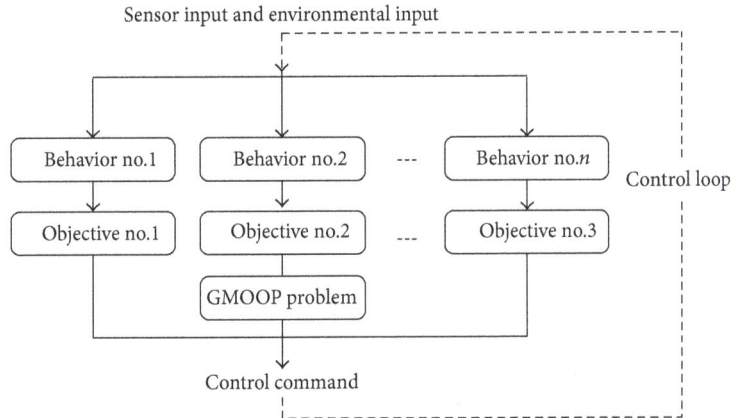

FIGURE 5: The decision control loop structure.

vector is $\vec{p}_d$, the velocity vector is $\{\vec{u_{Sm}}, \vec{u_{Cn}}\}$, and the other vehicles' position vector is $\{\vec{p_{Sm}}, \vec{p_{Cn}}\}$.

The output is a set of predefined behavior patterns function of the target, vehicle location, and vehicle task information:

$$CV \begin{cases} \text{Fastest to engage target,} \\ \text{Repulsion from spot by CV,} \\ \text{Repulsion from spot by SV,} \end{cases} \quad (5)$$

$$SV \begin{cases} \text{Follow a predefined trajectory,} \\ \text{Repulsion from spot by CV,} \\ \text{Repulsion from spot by SV,} \\ \text{Remain within communication range.} \end{cases} \quad (6)$$

The vehicle task information includes (a) search or classification information from onboard sensors, (b) destination information assigned by the task auctioneer, (c) other vehicle's actions and position information obtained through the acoustic network, and (d) the vehicle position.

6. Simulation Results

6.1. Simulator. The simulation tool is the MANET simulator presented in [25, 26, 37], upgraded to support MCVS operations. Several routing protocols can be used in the MANET [25], if necessary. The simulator requires a breadth of inputs, including the number of vehicles, vehicle types, vehicle identification and initial position, target distribution, communication performance, environmental conditions, and priority of various behavior patterns.

The event-driven simulator uses a dedicated process, labeled world, to maintain the clock and environment objects, the true vehicle models, target models, and the sensor (perception) models. Each vehicle object represents an autonomous entity in the environment. The vehicles are created by the user, separate from the simulated environment, and are added to the world before the simulation begins. The world makes the necessary connections between the vehicle systems and the environment. A vehicle requires a helm to maneuver, a protocol stack to communicate, and a sensor to detect objects in the environment.

Figure 6 shows a sequence diagram between mission- and path-planning controllers and other vehicle modules. It depicts a sequence of messages between multiple modules. The new task has a forward flow from vehicle sensor model to path-planning controller via the mission-planning controller. This new task decision is made by the mission-planning controller based on local vehicle sensor detections as well as by group state from network channels.

6.2. Simulation Using Reliable Acoustic Communications. The target search space is a 1000-by-1000-meter box. The simulation results presented in this paper use one SV and two CVs with a predefined number of targets. The acoustic modem MAC uses time division multiple access (TDMA): every vehicle has a preassigned 5-second transmission slot. Since three vehicles are present in this network, each vehicle transmits information every 15 seconds. We assume that the acoustic modems are operated under mild conditions, that is, at full power with background noise PSD of 55 dB re $1\,\mu\text{Pa}/\sqrt{\text{Hz}}$ and some fading (represented by the Nakagami coefficient $m = 2$) [37]. This would correspond to operating

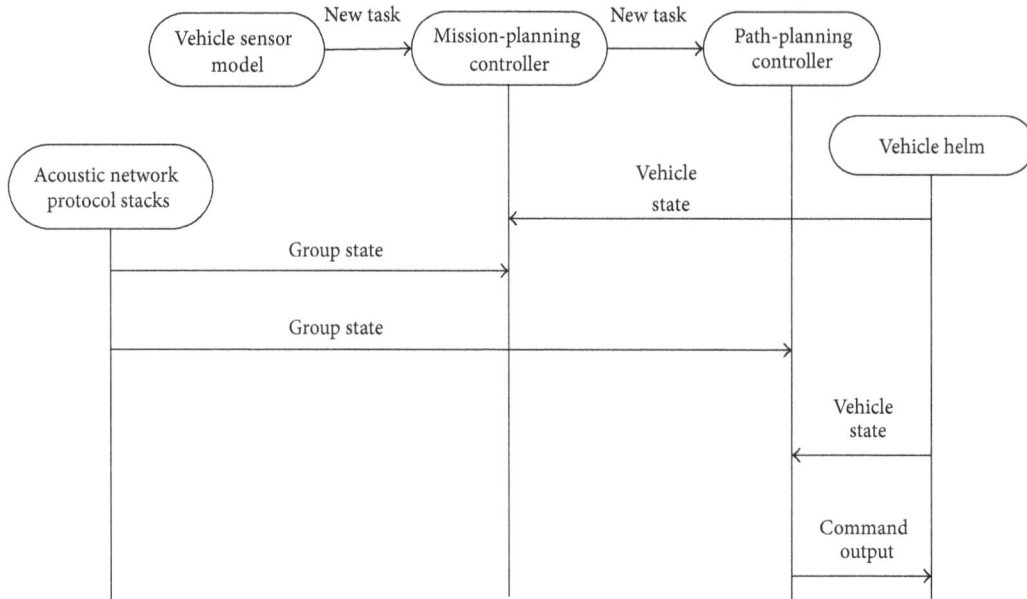

FIGURE 6: A sequence diagram of mission and path-planning controllers.

the vehicles in shallow waters over sandy bottom in calm seas. We also keep the relative distance between any two vehicles at any given time to be less than 300 m. As a result, messages are almost guaranteed to reach their destination (the simulated FER versus range is shown in Figure 1(a)). The results presented here do not require any data routing, as the vehicles always remain within reliable communication range of one another.

The SV is configured to broadcast target announcements to the CVs at specific times, simulating the detection of new targets. Once a CV identifies an unaccomplished target mission in its list, the GMOOP path-planning controller generates an optimal control action every second, with the objective of approaching the target as quickly as possible under the constraint of collision avoidance.

6.2.1. Performance Comparison between Auction Algorithms. Figure 7 shows the specific trajectories of a search-classify mission when the two different auction algorithms are used. The CVs have preassigned classification tasks (T1 and T2 for CV no. 1 and CV no. 2, resp.). At the mission time of 8 seconds, the search vehicle (SV) identifies two new targets T3 and T4 simultaneously. The two-target classify tasks are announced sequentially in the case of a generic auction algorithm and are announced simultaneously in the case of LAAF. In the case of a generic auction algorithm, CV no. 1 wins both classify task bids. Both new tasks are added to CV no. 1's task list and CV no. 2 stops as it accomplishes the classify task of T2. In the case of the LAAF, the negotiation result is that the concurrent new tasks of T3 and T4 are assigned to both classify vehicles. Thus, CV no. 1 continues to approach to T4 after finishing servicing T1 and CV no. 2 heads for T3 after servicing T2.

The differences in using the two auction algorithms are further illustrated in Figure 8. If the generic auction

TABLE 2: Comparison between generic auction and LAAF, best-case scenario.

Auction Method	TART (s)	BTA (bytes)	ETAT (s)
Generic auction algorithm	52.7	384	52.7
Location-aided auction	42.7	320	12.6

algorithm is used, the two successive auction processes are shown using black and red arrows, respectively. If LAAF is used, only one cycle of task allocation is needed, and the bidders negotiate to select the targets to service (the negotiation processes are highlighted in red in this case). The simulation results are given in Table 2. All three metrics clearly show that LAAF beats the generic auction algorithm and leads to a quicker completion of the mission.

6.2.2. Path-Planning Results. In search and classify missions, there are occasions where the AUVs are so close to one another that collision becomes likely. The scenario considered in this section centers around the need to keep a minimum standoff distance while simultaneously transiting to a destination as quickly and directly as possible. A GMOOP problem is created and solved every second through the control loop. The output of each action determination is a commanded vehicle speed vector, with components in the north and east directions.

Figure 9 shows a best-case scenario in which a four-target mission is to be completed by two classify vehicles and one search vehicle. The search vehicle (SV) is assumed to remain at the same location throughout the mission and broadcasts the four targets' location information at certain predefined times. The first and second targets (labeled Tar no. 1 and Tar no. 2) are announced simultaneously by the search vehicle at $t = 8$ seconds. Subsequently, the third

Task Allocation and Path Planning for Collaborative Autonomous Underwater Vehicles Operating through
an Underwater Acoustic Network

99

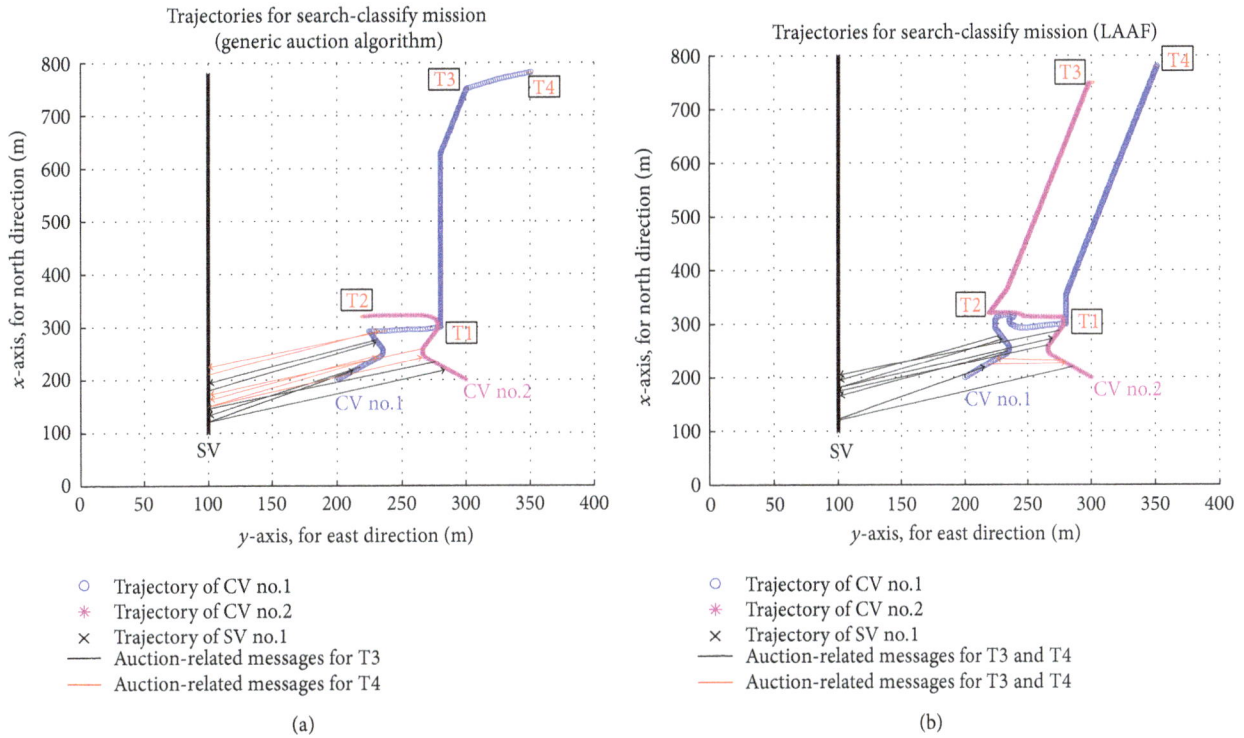

FIGURE 7: Trajectories for search-classify mission using a generic auction algorithm (a) and LAAF (b).

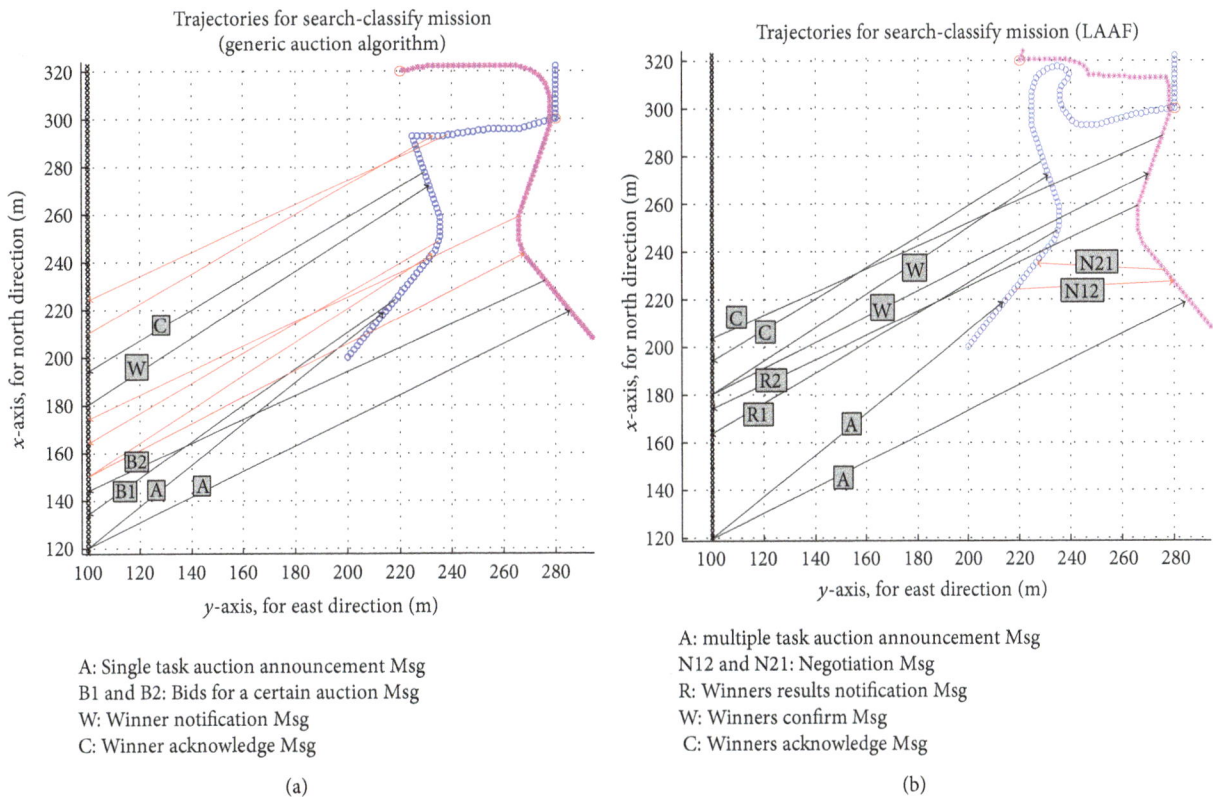

FIGURE 8: Communication messages for task allocation.

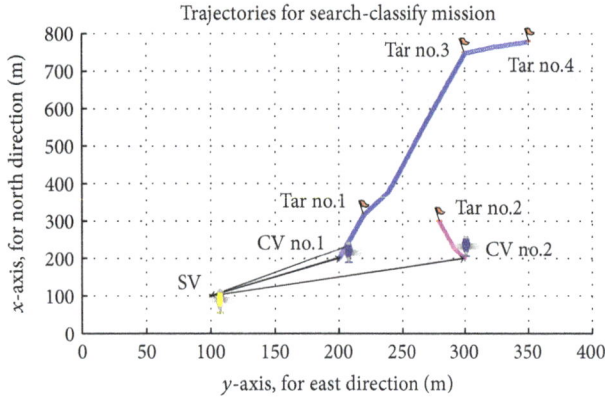

FIGURE 9: Four targets mission with 2 classify vehicles and 1 search vehicle.

TABLE 3: Task allocation metrics evaluated for each target, best-case scenario.

Target no.	Actual time Announced (s)	Winner	TART (s)	ETART (s)	BTA (bytes)
1	12.7	CV no.1	29.7	24.7	160
2	42.7	CV no.2	34.7	29.7	160
3	132.7	CV no.1	29.7	24.7	160
4	162.7	CV no.1	29.7	24.7	160

and fourth targets (labeled Tar no. 3 and Tar no. 4) are announced at $t = 128$ seconds. The mission is complete once each of the four targets has been classified; otherwise the mission times out and is considered a failure. Again, each vehicle has a single 5-second time window to transmit either LAAF task allocation or GMOOP communication messages. The task allocation message has priority over the GMOOP communication message.

As shown in Figure 9, all four targets are allocated to the "best" candidates and are classified under the GMOOP controller. The metrics for each classified target are listed in Table 3. Although two targets are scheduled by the search vehicle at the same time in our scenario, they are announced as independent tasks and broadcasted separately to the bidders. In this scenario, there is a time lag of 30 seconds (due to the acoustic modem MAC) between two successive targets announcement. The winning bidder of Tar no. 2 is CV no. 2, indicating that when the two classify vehicles send their bids, CV no. 2 is closer to the bidding target. The total mission time is 214 seconds.

The GMOOP-controlled trajectories turn out to be simple in this scenario, because the two classify vehicles keep a relatively large distance from each other. Thus the only objective considered in the GMOOP model is to transit to the target as quickly as possible.

6.3. Simulated Impact of Using Unreliable Acoustic Communications on GMOOP and LAAF. We now assume that the acoustic modems are operated under more severe conditions,

TABLE 4: Allocation metrics evaluated for each target, worst-case scenario amongst successful missions.

Target no.	Actual time announced (s)	Winner	TART (s)	ETART (s)	BTA (bytes)
1	12.7	CV no.1	29.7	24.7	160
2	42.7	CV no.2	154.7	149.7	128
3	132.7	CV no.1	29.7	24.7	160
4	162.7	CV no.1	154.7	149.7	128

that is at full power with background noise PSD of 85 dB re $1\,\mu\text{Pa}/\sqrt{\text{Hz}}$ and severe fading (represented by the Nakagami coefficient $m = 1.5$) [37]. Operating the vehicles over a shallow water reef in rough seas is a good example. We still keep the relative distance between any two vehicles at any given time to be less than 300 m. As a result, messages are much less likely to reach their destination (the simulated FER versus range is shown in Figure 1(b)).

Here, we only use LAAF and GMOOP in the simulations. Figure 10 shows scenarios where two classify vehicles are tasked to classify four targets. Eight simulation runs were carried out using the same initial scenario state; four runs are shown. Once again, each bid is sent once (no retry). The initial task allocation messages between SV and CVs are sketched using black lines. We find that the entire mission is completed so long as the four classify tasks reach the CVs. From Figure 10(b), we note that the trajectories of both classify vehicles are the same as shown in Figure 9 (with the best environmental conditions). Figures 10(b) and 10(d) show two successful mission, while Figures 10(a) and 10(c) show two failed missions due to timeout. In Figure 10(a), the search vehicle is able to transmit the Tar no. 1 location information to CV no. 1 at $t = 12.7$ seconds, but this bidding announcement does not reach CV no. 2. Thus the winning bidder for Tar no. 1 is CV no. 1. At $t = 42.7$ seconds, the bidding announcement successfully reaches both bidders, and CV no. 1 wins again in this bidding due to a closer range to Tar no. 2. The bidding announcements for Tar no. 3 and Tar no. 4 do not reach any of the CVs, causing the mission to fail.

The performance metrics for each target are listed in Table 4. Compared to Table 2, we note that, for Tar no. 2 and Tar no. 4, there is a great increase in terms of TART and ETART but a decrease in BTA. This is because the winning bidder cannot receive the winner notification message due to the poor network condition. After the predefined timer expires, the classify vehicles switch to the self-determination mode defined in the LAAF auction architecture and finally pick targets on its own.

7. Conclusion

In this paper, the authors introduced GMOOP and LAAF. The GMOOP model is intended to balance the competing objectives and constraints for a vehicle in the mission. The LAAF task allocation algorithm is specially designed for the harsh underwater network.

Task Allocation and Path Planning for Collaborative Autonomous Underwater Vehicles Operating through
an Underwater Acoustic Network

101

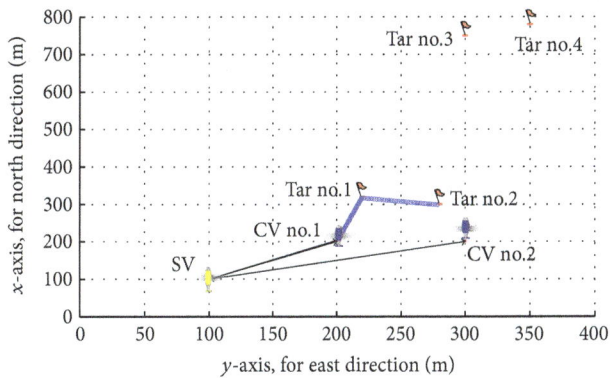

(a) Trajectories for search-classify mission

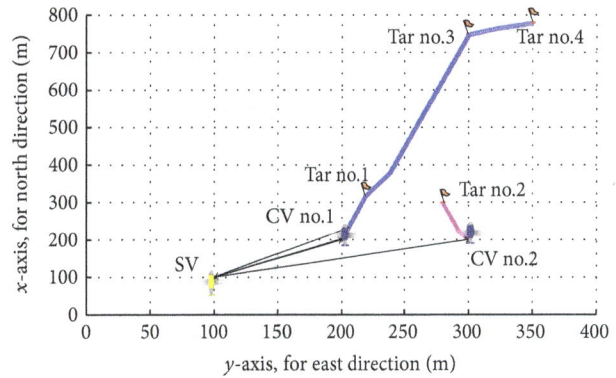

(b) Trajectories for search-classify mission

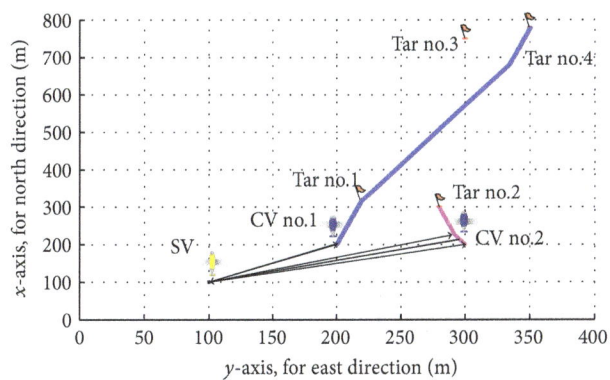

(c) Trajectories for search-classify mission

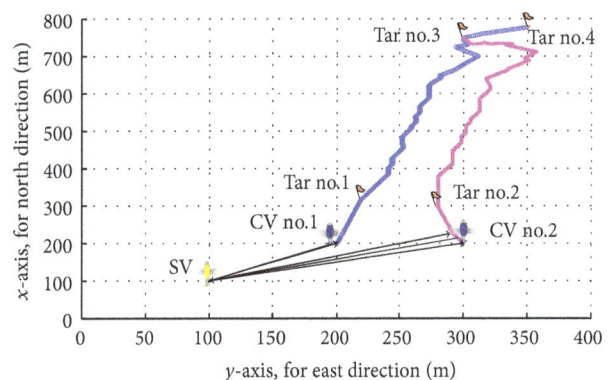

(d) Trajectories for search-classify mission

FIGURE 10: Four executions of the same mission using 4 targets, 2 classify vehicles, and LAAF when acoustic communication is unreliable.

The GMOOP and LAAF algorithms are implemented in the mission- and path-planning module of a simulator. The simulation results show that the location-aided auction strategies performed significantly better than the generic auction algorithm in terms of effective task allocation time and bytes usage. In a simplified task allocation case, the EFAT was reduced by 76.1% and the bytes used for task allocation were reduced by 16.7% due to the self-determination mechanism and multiple-target handling technique. These improvements are essential for communication-based underwater operations in which mission time and bandwidth are critical criteria. In addition, LAAF could still be used with some level of success if underwater acoustic communications become unreliable.

References

[1] R. L. Wernli, "AUVs-a technology whose time has come," in *Proceedings of the International Symposium on Underwater Technology*, pp. 309–314, April 2002.

[2] M. Herman and J. S. Albus, "Overview of the multiple autonomous underwater vehicles (MAUV) project," in *Proceedings of the IEEE International Conference on Robotics and Automation*, vol. 1, pp. 618–620, April 1988.

[3] T. R. Cuff and R. W. Wall, "Support platform and communications to manage cooperative AUV operations (OCEANS '06)," in *Proceedings of the Oceans Conference in Asia Pacific*, pp. 1–8, May 2007.

[4] D. Spenneberg, C. Waldmann, and R. Babb, "Exploration of underwater structures with cooperative heterogeneous robots," in *Proceedings of the Oceans '05—Europe*, pp. 782–786, June 2005.

[5] E. M. Sozer, M. Stojanovic, and J. G. Proakis, "Underwater acoustic networks," *IEEE Journal of Oceanic Engineering*, vol. 25, no. 1, pp. 72–83, 2000.

[6] E. H. Turner and R. M. Turner, "A schema-based approach to cooperative problem solving with autonomous underwater vehicles," in *Proceedings of the Ocean Technologies and Opportunities in the Pacific for the 90's (Oceans'91)*, vol. 2, pp. 1067–1073, October 1991.

[7] S. M. Smith, K. Ganesan, P. E. An, and S. E. Dunn, "Strategies for simultaneous multiple autonomous underwater vehicle operation and control," *International Journal of Systems Science*, vol. 29, no. 10, pp. 1045–1063, 1998.

[8] D. Stilwell and B. Bishop, "Platoons of underwater vehicles," *IEEE Control Systems Magazine*, vol. 20, no. 6, pp. 45–52, 2000.

[9] F. Arrichiello, D. N. Liu, S. Yerramalli et al., "Effects of underwater communication constraints on the control of marine robot

teams," in *Proceedings of the 2nd International Conference on Robot Communication and Coordination*, April 2009.

[10] M. Vajapeyam, S. Vedantam, U. Mitra, J. C. Preisig, and M. Stojanovic, "Distributed space-time cooperative schemes for underwater acoustic communications," *IEEE Journal of Oceanic Engineering*, vol. 33, no. 4, pp. 489–501, 2008.

[11] G. Hollinger, S. Yerramalli, S. Singh, U. Mitra, and G. Sukhatme, "Distributed coordination and data fusion for underwater search," in *Proceedings of the IEEE International Conference on Robotics and Automation (ICRA '11)*, pp. 349–355, Shanghai, China, May 2011.

[12] C. Schumacher, P. Chandler, M. Pachter, and L. S. Pachter, "Constrained optimization for UAV task assignment," in *Proceedings of the AIAA Guidance, Navigation, and Control Conference and Exhibit*, pp. 3152–3165, Austin, Tex, USA, 2003.

[13] R. M. Turner and E. H. Turner, "Simulating an autonomous oceanographic sampling network: a multi-fidelity approach to simulating systems of systems," in *Proceedings of the Oceans '00*, pp. 905–911, September 2000.

[14] H. Bojinov, A. Casal, and T. Hogg, "Emergent structures in modular self-reconfigurable robots," in *Proceedings of the IEEE International Conference on Robotics and Automation*, pp. 1734–1741, San Francisco, Calif, USA, April 2000.

[15] B. Jouvencel, V. Creuze, and P. Baccou, "A new method for multiple AUV coordination a reactive approach," in *Proceedings of the 8th International Conference on Emerging Technologies and Factory Automation (ETFA '01)*, vol. 1, pp. 51–55, October 2001.

[16] B. P. Gerkey and M. J. Matarić, "A formal analysis and taxonomy of task allocation in multi-robot systems," *International Journal of Robotics Research*, vol. 23, no. 9, pp. 939–954, 2004.

[17] L. Liu and D. A. Shell, "Multi-level partitioning and distribution of the assignment problem for large-scale multi-robot task allocation," in *Proceedings of the Robotics: Science and Systems Conference (RSS '11)*, Los Angeles, Calif, USA, June 2011.

[18] A. Chavez, A. Moukas, and P. Maes, "Challenger: a multi-agent system for distributed resource allocation," in *Proceedings of the 1st International Conference on Autonomous Agents*, pp. 323–331, Marina del Rey, Calif, USA, February 1997.

[19] D. P. Bertsekas, "The auction algorithm: a distributed relaxation method for the assignment problem," *Annals of Operations Research*, vol. 14, no. 1, pp. 105–123, 1988.

[20] S. Sariel and T. R. Balch, "Efficient bids on task allocation for multi-robot exploration," in *Proceedings of the 19th International Florida Artificial Intelligence Research Society Conference (FLAIRS '06)*, pp. 116–121, May 2006.

[21] M. J. Matarić, G. S. Sukhatme, and E. H. Ostergaard, "Multi-robot task allocation in uncertain environments," *Autonomous Robots*, vol. 14, no. 2-3, pp. 255–263, 2003.

[22] D. P. Bertsekas, "An Auction Algorithm for Shortest Paths," webpage: PSU-auction-28, 1991.

[23] K. E. Nygard, P. R. Chandler, and M. Pachter, "Dynamic network flow optimization models for air vehicle resource allocation," in *Proceedings of the American Control Conference*, pp. 1853–1858, Arlington, Va, USA, June 2001.

[24] J. W. Mitchell, P. Chandler, M. Pachter, and S. J. Rasmussen, "Communication delays in the cooperative control of wide area search munitions via iterative network flow," in *Proceedings of the AIAA Guidance, Navigation, and Control Conference and Exhibit*, pp. 2003–5665, Austin, Tex, USA, August 2003.

[25] E. Carlson, P. Beaujean, and E. An, "Location-aware routing protocol for underwater acoustic networks," in *Proceedings of the MTS/IEEE Oceans '06*, Boston, Mass, USA, September 2006.

[26] Y. Deng, *Task allocation and path planning for acoustic networks of AUVs [Ph.D. thesis]*, Florida Atlantic University, Boca Raton, Fla, USA, 2010.

[27] B. Jouvencel, V. Creuze, and P. Baccou, "A new method for multiple AUV coordination a reactive approach," in *Proceedings of the 8th International Conference on Emerging Technologies and Factory Automation (ETFA '01)*, vol. 1, pp. 51–55, October 2001.

[28] J. T. Napoli, T. J. Tarn, J. R. Morrow Jr., and E. An, "Optimal communication control for cooperative autonomous underwater vehicle networks," in *Proceedings of the IEEE International Conference on Robotics and Automation*, pp. 1624–1631, Barcelona, Spain, April 2005.

[29] A. Martins, J. M. Almeida, and E. Silva, "Coordinated maneuver for gradient search using multiple AUVs," in *Proceedings of the Oceans '03*, vol. 1, pp. 347–352, September 2003.

[30] R. Kumar and J. A. Stover, "A behavior-based intelligent control architecture with application to coordination of multiple underwater vehicles," *IEEE Transactions on Systems, Man, and Cybernetics A*, vol. 30, no. 6, pp. 767–784, 2000.

[31] J. Albus and D. Blidberg, "A control system architecture for multiple autonomous undersea vehicles (MAUV)," in *Proceedings of the 5th International Symposium on Unmanned Untethered Submersible Technology*, vol. 5, pp. 444–466, June 1987.

[32] J. B. de Sousa and A. Gollu, "Simulation environment for the coordinated operation of multiple autonomous underwater vehicles," in *Proceedings of the Winter Simulation Conference*, pp. 1169–1175, December 1997.

[33] Y. C. Sun and C. C. Cheah, "Coordinated control of multiple cooperative underwater vehicle-manipulator systems holding a common load," in *Proceedings of the MTTS/IEEE TECHNO-Ocean '04*, vol. 3, pp. 1542–1547, November 2004.

[34] M. R. Benjamin, "Multi-objective autonomous vehicle navigation in the presence of cooperative and adversarial moving contacts," in *Proceedings of the Oceans '02 MTS/IEEE*, vol. 3, pp. 1878–1885, October 2002.

[35] M. R. Benjamin, *The interval programming: a multi-objective optimization model for autonomous vehicle control [Ph.D. thesis]*, Brown University, Providence, RI, USA, 2002.

[36] Y. Deng, P. P. J. Beaujean, E. An, and E. A. Carlson, "A path planning control strategy for search-classify task using multiple cooperative underwater vehicles," in *Proceedings of the MTS/IEEE Oceans '08*, Quebec, Canada, September 2008.

[37] E. Carlson, P. Beaujean, and E. An, "An Ad Hoc wireless acoustic network simulator applied to multiple underwater vehicle operations in shallow waters using high-frequency acoustic modems," *Journal of Underwater Acoustics*, vol. 56, pp. 113–139, 2006.

[38] H. Kenn and A. Pfeil, "A sound source localization sensor using probabilistic occupancy grid maps," in *Proceedings of the Mechatronics and Robotics Conference*, pp. 802–807, 2004.

[39] B. Yamauchi, A. Schultz, and W. Adams, "Mobile robot exploration and map-building with continuous localization," in *Proceedings of the IEEE International Conference on Robotics and Automation*, pp. 3715–3720, Leuven, Belgium, May 1998.

[40] H. Moravec and A. Elfes, "High resolution maps from wide angle sonar," in *Proceedings of the IEEE International Conference on Robotics and Automation*, pp. 116–121, St. Louis, Mo, USA, 1985.

[41] A. M. Law and W. D. Kelton, *Simulation Modeling & Analysis*,
McGraw-Hill, 2nd edition, 1991.

[42] O. Khatib, "Real-time obstacle avoidance for manipulators
and mobile robots," in *Proceedings of the IEEE International
Conference on Robotics and Automation*, pp. 500–505, St. Louis,
Mo, USA, March 1985.

9

Robots Learn Writing

Huan Tan,[1] Qian Du,[2] and Na Wu[3]

[1] *Department of Electrical Engineering and Computer Science, Vanderbilt University, Nashville, TN 37240, USA*
[2] *Institute of Robotics and Automatic Information System, Nankai University, Tianjin 300071, China*
[3] *Graduate School of Decision and Technology, Tokyo Institute of Technology, Tokyo 152-8552, Japan*

Correspondence should be addressed to Huan Tan, huantan@ieee.org

Academic Editor: Huosheng Hu

This paper proposes a general method for robots to learn motions and corresponding semantic knowledge simultaneously. A modified ISOMAP algorithm is used to convert the sampled 6D vectors of joint angles into 2D trajectories, and the required movements for writing numbers are learned from this modified ISOMAP-based model. Using this algorithm, the knowledge models are established. Learned motion and knowledge models are stored in a 2D latent space. Gaussian Process (GP) method is used to model and represent these models. Practical experiments are carried out on a humanoid robot, named ISAC, to learn the semantic representations of numbers and the movements of writing numbers through imitation and to verify the effectiveness of this framework. This framework is applied into training a humanoid robot, named ISAC. At the learning stage, ISAC not only learns the dynamics of the movement required to write the numbers, but also learns the semantic meaning of the numbers which are related to the writing movements from the same data set. Given speech commands, ISAC recognizes the words and generated corresponding motion trajectories to write the numbers. This imitation learning method is implemented on a cognitive architecture to provide robust cognitive information processing.

1. Introduction

Robots are expected to generate human-like behaviors in dynamic environments [1, 2]. However, it is very difficult for robots to develop skills or behaviors totally from scratch without any initial knowledge. As stated in Sloman's paper, robots should learn both altricial and precocial behaviors after their "birth" [3]. Therefore, it is reasonable that robots have some basic and simple initial knowledge with motion primitives [4], or some basic and simple initial skills to explore the world to develop new knowledge and skills to survive or complete tasks. Upon these initial knowledge and skills, humans can teach robots more complex behaviors or skills to complete much more complex tasks.

Imitation learning (also called learning from demonstration, programming by demonstration) is now considered as a powerful tool for transferring skills between robots (especially humanoid robots) [5]. Unlike the traditional teaching-executing mode, where robots simply record the trajectory programmed by human operators and move the angles and end effector along the trajectory, since 1970s, the researchers had tried to train robots to learn simple motion patterns [6]. In 1980s, Atkeson trained a robot to learn how to balance an inverted pendulum in an upright position through practice [7]. From then, many methods in imitation learning have been proposed in various areas [8]. In 2000s, researchers found biological evidence and models of imitation learning in animals [9]. Gradually, imitation learning has been divided into two parts [10]. one is to train robots to learn the dynamics of movements [11] and the other is to train robots to learn the primitives in a behavior sequence [12].

The motivation of this paper is to find a method that robots can learn motion models and semantic knowledge simultaneously in the current popular imitation learning framework. In the experiment part, a humanoid robot, named ISAC, is trained to learn writing numbers from a human teacher.

The rest of this paper is organized as follows. Section 2 introduces the current related work; Section 3 explains the system framework and the algorithms used in this framework; Section 4 explains the implementation on a cognitive architecture; Section 5 explains the experiment setup and the experimental results; Section 6 discusses the experimental results and the future work; Section 7 concludes the work in this paper.

2. Related Work and Motivation

2.1. Motion Learning. Demonstrations of motions are given by human teachers or other robots and a robot student tries to record the demonstrations. There are many different kinds of methods for demonstrating the motions: learning from observation [13], from joystick operation [14], by manually moving the arm of a robot [10], and from the sensor on the human body [15, 16].

Sometimes the dimension of the recorded data is reduced by projecting the data from a high-dimensional data space to a low-dimensional data space, named latent space. Correspondingly, the data needs to be reconstructed from the low-dimensional data space to the high-dimensional data space. The "dimension reduction" and the "reconstruction" are not always required in current imitation learning research. In some situations where the dynamics of the demonstrations need to be analyzed or several inner correlations need to be analyzed, the "Dimension Reduction" and the "Reconstruction" can be applied. Many dimension reduction methods are proposed to extract the features of the data such as principal component analysis [17], factor analysis [18], ISOMAP [19], local linear embedding [20], and MDS [21]. A typical example of using dimension reduction technology is [10], in which Calinon and Billard proposed a method to utilize the dimension reduction methods to establish a strong coupling relationship between the data in the latent space and the data space, and use the data distribution in the latent space to ensure the generated behavior has similar inner dynamics and constraints to the demonstrations.

The learned motion models are stored in the memory (database) of robots, where robots store the learned knowledge or skills. Linear Global Model (LGM) [22], Gaussian Process (GP) [23], Locally Weighted Regression (LWR) [24], Locally Weighted Projection Regression (LWPR) [25], Principal Curve (PC), Gaussian mixture model [26], and Artificial Neural Network (ANN) [27] are used for representing the models in the memory.

2.2. Semantic Knowledge Learning. Robots need to understand the learned motions, and it means that robots need to relate these motion models to corresponding semantic knowledge. This is normally done by labeling the motion models with a semantic name or with a semantic description of related tasks.

2.3. Generation. Given a similar and slightly different situation (where robots need to complete a same type of task with different parameters), a command, or an outside trigger (signal, image, etc.), required actions are planned and required motion models are retrieved by searching their corresponding behavior names in the "labeled behavior models."

If needed, the parameters of the motions are modified to adapt to the similar but slightly different situation. The generated behaviors are described as actions with specified parameters. Dynamic Movement Primitives (DMPs) [16] are widely for generating motions which have similar dynamics to the demonstrations and can achieve various targets. Calinon et al. proposed a method to minimize the weighted distance between the generated motions and the learned motions both in the latent space and in the original data space [10]. Peters used reinforcement learning [28] methods for robots to adapt the parameters of motion models to generate similar motions in similar but slightly different situations [29]. Theodorou applied optimal control [30] in reinforcement learning environment for robots to learn the motion models of demonstrations and generate similar motions using DMP [31, 32].

If the data is stored in a latent space, the generated trajectories of motions need to be projected from the latent space to the original data space, for example, joint space.

2.4. Motivation. Robots need to learn the motions of the behaviors and the semantic meaning of the behaviors at two stages as described in former sections. One of the learning stages is still like a programming process, in which the behavior names are assigned to the motion models manually by human teachers.

An important problem is that robots can learn the motions of the demonstrations and use the learned motions in a similar but slightly different situation, but how can robots use such learned knowledge in other area besides executing movements, for example, recognition, semantic understanding, reasoning, and planning, Especially in writing, the learned movements should relate to semantic meanings of the letters, numbers or symbols, and robots may use the learned movements to find their higher level semantic meanings. When we see someone demonstrate how to a character, we can have direct thought about the meaning of that character in our brain, and when we have the method of writing a character in our brain, we can evaluate the results with a real character. A game which may be familiar with most of us is that someone writes letters on our back using fingers and we try to guess what he/she is writing. Obviously, we, humans, can use the sensing information on the back to construct the trajectory of the movement of the finger in our brain, and try to compare it with our learned knowledge about the letters. In this paper, the robot uses encoders to sense the movement of the joints in the joint space and tries to match the sensed movement of the joints with the learned knowledge about the numbers. Then, human teachers do not need to teach the robot that what the number is and the robot can automatically relate the learned movements of writing letters to the corresponding letters.

3. System Design

In this paper, we proposed a general framework, using which robots can complete two learning stages mentioned above simultaneously, which is shown in Figure 1.

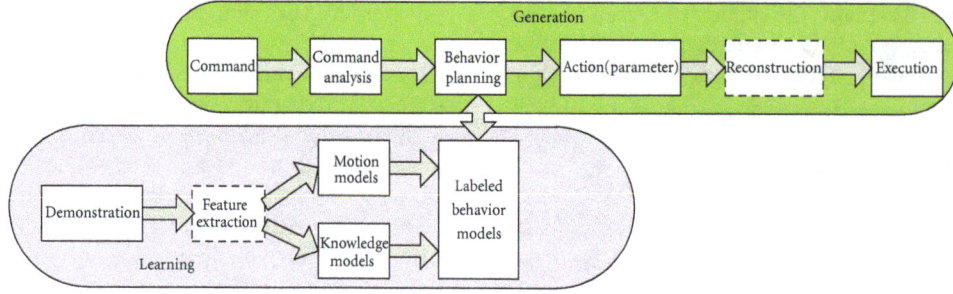

FIGURE 1: Proposed framework for imitation learning.

From Figure 1, the contribution of this proposed framework is that the information for both motion models and the knowledge models is both from one single source. The robot uses the information of the demonstrations to learn motion models and the knowledge models. In this paper, ISAC learn how to write numbers and to automatically relate the motions of writing numbers to their semantic knowledge models.

3.1. Demonstration. Demonstrations are given by human teachers. In this paper, demonstrations are shown by manually moving the right arm of ISAC.

The recorded data is $\theta_s = \{\theta_v, t\}$, which is an $N \times 7$ matrix. θ_v records the angles of six joints on the right arm of ISAC, and t is the temporal information.

3.2. Feature Extraction. For most situations, robots need to learn two features: motions and semantic meaning. The extracted information is stored in corresponding models.

3.2.1. Motion Models Learning. As mentioned in the introduction section, there are many methods to represent the data of motions. In this paper, we use a modified ISOMAP algorithm [33] to project the 6-dimensional data space to a 2-dimensional space. The movements of the writings are in a 3-dimensional Cartesian space which is driven by 6 joints. However, the features of the writings are 2-dimensional because the characters are written on a 2-dimensional plane. So it is reasonable to use the features on a 2-dimensional plane for other use. The additional motivations of using this modified ISOMAP algorithm is to visualize the sampled data on 2-dimensional plane for researchers to find the features of motions easily, and to make the trajectory on the 2-dimensional plane does not have overlapped parts inside itself or have intersections inside itself. Using this algorithm, the spatial and temporal characteristic of the sample data can be visualized on a 2-dimensional plane.

We want to mention that dimension reduction is not necessary for all applications. In this paper, the dimension reduction is convenient for extracting the features and using the features in recognition.

The algorithm of the original ISOMAP algorithm is specified as follows.

(1) Sample the points on the demonstration trajectory:

$$\theta_s = \{\theta_v, t\}. \tag{1}$$

(2) Compute the geodesic distance matrix D_{Ms}:

$$D_{Gs}(i, j) = \begin{cases} \left\| \theta_{vi} - \theta_{vj} \right\|, & \text{if } \left\| \theta_{vi} - \theta_{vj} \right\| \leq d \\ 0, & \text{otherwise} \end{cases}$$

$$D_{Ms}(i, j) = \min \left(D_{Gs}(i, j), D_{Gs}(i, k) + D_{Gs}(k, j) \right),$$

$$k = 1, 2, \ldots, N, \tag{2}$$

D_{Ms} is iteratively calculated until the values of the elements are converged.

In the original ISOMAP algorithm, $\|\theta_{vi} - \theta_{vj}\|$ is defined as an Euclidean spatial distance between two points: θ_{vi} and θ_{vj}. In our modified ISOMAP algorithm, $\|\theta_{vi} - \theta_{vj}\|$ is defined as a temporal distance between two points: θ_{vi} and θ_{vj}. θ_v is used to record the angles of the six joints: $\theta_v = \theta_{v1}, \theta_{v2}, \ldots, \theta_{vN}$.

(3) Compute the inner products,

$$\tau(D_{Ms}) = -\frac{1}{2} HSH, \tag{3}$$

where $S_{ij} = D_{Msij}{}^2$, and $H = \delta_{ij} - (1/N) (\delta_{ij} = 1$, when $i = j$, $\delta_{ij} = 0$, when $i \neq j$).

(4) Compute the new coordinates of the sampled points in the latent space X:

$$x_i = \left(\sqrt{\lambda_{s1}} \alpha_{s1}, \sqrt{\lambda_{s2}} \alpha_{s2} \right)^T, \tag{4}$$

where λ_1 and λ_2 are the two largest eigenvalues of $\tau(D_M)$ with two corresponding eigenvectors: α_1 and α_2.

The modified ISOMAP method, which reflects both the temporal and spatial relationship between sampled data points on a two-dimensional plane, is used to train robots to learn the motions of writing the letters on a two-Dimensional plane [33].

The original ISOMAP is an extension of the MDS, which constructs the distance matrix by connecting the sampled points through the neighbors. The original ISOMAP is used to describe the distance between the sampled neighbor points. In order to find the temporal information of the sampled trajectory, in our algorithm the neighbors are strictly defined as temporal neighbors. The spatial relationships are not defined but calculated by this modified ISOMAP algorithm. The modification is designed in

$$D_{\text{Gt}}(i,j) = \begin{cases} \left\| \theta_{vi} - \theta_{vj} \right\|, & \text{if } \left| t_i - t_j \right| \le s, \\ 0, & \text{otherwise,} \end{cases} \quad (5)$$

where s is the temporal threshold value. In (1), d is the spatial threshold value.

Using this method, D_{Gt} and the corresponding $\tau(D_{\text{Mt}})$ are calculated. The sampled points in the latent space are represented as $y_i = (\sqrt{\lambda_{t1}}\alpha_{t1}, \sqrt{\lambda_{t2}}\alpha_{t2})^T$, where λ_{t1} and λ_{t2} are the two largest eigenvalues of $\tau(D_{\text{Mt}})$ with two corresponding eigenvectors: α_{t1} and α_{t2}.

Jenkins and Matarić proposed spatial-temporal ISOMAP algorithm in 2004 [34]. Their method is comprehensive and defines the types of neighbors in detail. For different neighbors, the construction of the distance matrix is different in their method. In our method, we simply add temporal constraints on the construction of the distance matrix and strictly assume all the distances should be temporal related. This kind of method is simple but convenient for computation. Both Jenkins's method and our method are effective for describing the spatial-temporal characteristics of the sampled data points.

In current imitation learning, behaviors are special robotic movements in certain task-related situations. This means we can assume that the sampled data from demonstrations of one behavior always lie on the same manifold in the data space. The results of projecting data from the latent space to the data space must be on the same manifold as the demonstration. Therefore, it is reasonable to assume that there exists a relationship between the data in the data space and the latent space and it can be described as a function:

$$\theta_{vi} = f(x_i, W), \quad (6)$$

where x_i is a data point in the latent space and θ_{vi} is a corresponding data point in the original data space.

Therefore, $f(X, W)$ is designed as a generalized linear regression model:

$$\theta_{vi} = W\Phi(x_i). \quad (7)$$

$\Phi(x)$ is composed of R basis functions:

$$\Phi_i(x) = \exp\left(-\frac{(x_i - c_i)^2}{\Sigma_i}\right), \quad i = 1, 2, \dots, R, \quad (8)$$

where c_i is the center of the ith basis function and Σ_i is the bandwidth. The centers are uniformly distributed in the latent space and the bandwidth is designed for the basis functions to cover the latent space.

W is a $(D - 1) * R$ matrix which projects the data from the latent space to the data space. However, Bishop has verified that the number of basis functions must typically grow exponentially with the dimensionality of the input space [35]. This means the advantage of dimension reduction in calculation and storage eventually arrives at a certain value as the number of the dimensionality increases. In Section 4, comparisons of results using different number of basis functions are given.

Assuming the projection matrix W is known, the probabilities of the distributions of the points in data space are

$$p(\theta_i \mid x_i, W, \beta) = \left(\frac{\beta}{2}\right)^D \exp\left(-\frac{\beta}{2}\|f(x, W) - \theta\|^2\right). \quad (9)$$

The log likelihood of the probability of the distribution of points in data space is the multiplication of the distribution probability of each point:

$$L(W, \beta) = \sum_{i=1}^{N} \ln p(t_i \mid x_i, W, \beta). \quad (10)$$

Maximizing the log likelihood function can be achieved by differentiating the log likelihood function with respect to W:

$$\sum_{i=1}^{N} (W\Phi(x_i) - \theta_i)\Phi(x_i)^T = 0. \quad (11)$$

Rewrite (11):

$$W\sum_{i=1}^{N} \Phi(x_i)\Phi(x_i)^T = \sum_{i=1}^{N} \theta_i\Phi(x_i)^T. \quad (12)$$

Projection matrix W can be calculated from (11),

$$W = \sum_{i=1}^{N} \theta_i\Phi(x_i)^T * \left(\sum_{i=1}^{N} \Phi(x_i)\Phi(x_i)^T\right)^\dagger, \quad (13)$$

where $\left(\sum_{i=1}^{N} \Phi(x_i)\Phi(x_i)^T\right)^\dagger$ is the Moore-Penrose pseudo inverse matrix of $\sum_{i=1}^{N} \Phi(x_i)\Phi(x_i)^T$.

The sampled trajectory is projected from a 6-dimensional data space to a 2-dimensional space using the original ISOMAP algorithm and a modified ISOMAP algorithm in "feature extraction" block.

In the latent space, we have data points set X, which is a two-dimensional space. As stated in previous sections, in "dimension reduction," the temporal information is only used to calculate the neighborhood graph. But in the "behavior planning" stage, the temporal information should be combined into the model and set as the enquiry point.

Data points in the latent space follow (14):

$$x_i = f(t_i), \quad i = 1, 2, \dots, N. \quad (14)$$

Using Gaussian process [22], we can get a kernel method-based model of the demonstration in the latent space. The points on the two-dimensional plane are described as $x_i = \{x'_i, x''_i\}$, and one GP model is used in one dimension in the latent space. GP has been widely used [36–39] for representing the sampled data points because its robustness and nonparametric characteristics.

Assume the N two-dimensional data points in the two-dimensional latent space has the following probabilistic distribution:

$$p\left(\vec{z'} | \vec{x'}\right) = \mathcal{N}\left(\vec{z'} | \vec{x'}, \beta^{-1}I\right),$$
$$p\left(\vec{z''} | \vec{x''}\right) = \mathcal{N}\left(\vec{z''} | \vec{x''}, \beta^{-1}I\right). \tag{15}$$

Take the calculation in the first dimension as an example: $p(\vec{z'}) = \mathcal{N}(\vec{z'} | 0, C_N)$, where the covariance matrix $C(n, m) = k(t_n, t_m) + \beta^{-1}\delta_{nm}$, and $\vec{z}$ is a vector of target values.

$k(t_n, t_m)$ is the kernel function. Normally,

$$k(t_n, t_m) = \theta_0 \exp\left\{-\frac{\theta_1}{2}\|t_n - t_m\|^2\right\} + \theta_2 + \theta_3 t_n^T t_m, \tag{16}$$

and x_n is considered as the timing step in the demonstration.

In the "Generation" stage, a new time step $t_{enquiry}$ is given as an enquiry point and GP is used to calculate the corresponding data value $z'_{enquiry}$.

$$p\left(\left\{\vec{z'}, z'_{enquiry}\right\}\right) = \mathcal{N}\left(\left\{\vec{z'}, z'_{enquiry}\right\} | 0, C_{N+1}\right). \tag{17}$$

The covariance matrix is:

$$C_{N+1} = \begin{pmatrix} C_N & \vec{k} \\ \vec{k}^T & c \end{pmatrix}, \tag{18}$$

where $k = k(t_n, t_{enquiry})$ for $n = 1, 2, \ldots, N$.

Using Bayesian method, $z'_{enquiry}$ is calculated using (19)

$$z'_{enquiry} = \vec{k}^T C_N^{-1} \vec{z'}. \tag{19}$$

Using the same method, $z''_{enquiry}$ can be calculated using (20):

$$z''_{enquiry} = \vec{k}^T C_N^{-1} \vec{z''}. \tag{20}$$

In this part, the input is a data set including the sampled data points in a 6-dimensional joint space, and the output is a GP model for the trajectory in the latent space. Given an enquiry point (normally a timing point), the output of this model, which is (19) and (20), is a corresponding data point on the trajectory.

Using the following equation, we can project the data from the low-dimensional latent space to the original data space:

$$\theta_{vi} = W\Phi(x_i). \tag{21}$$

These data points will be used for robot to learn how to generate the required movement trajectories. In robotic imitation learning, robots need to fit the recorded movement trajectory with the model used in a generator [16, 40]. The fitting process is considered as learning a pattern in a generator.

In our system, we used the Dynamic Movement Primitives (DMP) [11], proposed by Ijspeert, as the pattern generator.

DMP is configured as

$$\tau\dot{z} = \alpha_z(\beta_z(g - y) - z),$$
$$\tau\dot{y} = z + f. \tag{22}$$

g is the goal state, z is the internal state, f, an RFWR model, is calculated to record the dynamic of the demonstration and to guarantee convergence of the new generated trajectories, y is the position generated by the DMP differential equations, and $\dot{y}$ is the generated velocity correspondingly. α_z, β_z, and τ are the constants in this equation.

The fitting (or learning) is to train robots to learn the model:

$$f = \frac{\sum_{i=1}^{N} \Psi_i w_i v}{\sum_{i=1}^{N} \Psi_i}. \tag{23}$$

and v satisfies the following equation:

$$\tau\dot{v} = \alpha_z(\beta_z(g - x) - v),$$
$$\dot{x} = v. \tag{24}$$

Ψ_i is a receptive basis function, which is distributed in the space:

$$\Psi_i = \exp\left(\frac{1}{-2h_i^2}(x - c_i)^2\right). \tag{25}$$

c_i is the center of the basis function, which is distributed in the space, and h_i is the bandwidth.

The target is to use the sampled points as x and use iterative learning method for robots to adapt the parameters w_i. After the learning, the parameters are fixed and do not need to change at the generation stage:

$$\Delta_i^{(n+1)} = \exp\left(-\frac{1}{2}\left(x^{(n+1)} - c_i\right)^T D\left(x^{(n+1)} - c_i\right)\right),$$
$$w_i^{(n+1)} = \lambda w_i^{(n)} + \Delta_i^{(n+1)}, \tag{26}$$

The subscript $(n+1)$ denotes that this is the $(n+1)$ iteration, $x^{(n+1)}$ is the data point used to update the model at the $(n+1)$ iteration, and $\Delta_i^{(n+1)}$ is computed as the weighted distance between the data point $x^{(n+1)}$ and the center of the basis function, which is used to update the weight w_i.

3.2.2. Semantic Knowledge Learning. Recording and writing the characters are not enough for robots to interact with humans and robots should understand the semantic meaning of the motions and correlate the motion models and the semantic knowledge models automatically.

The general algorithm is shown in Figure 2.

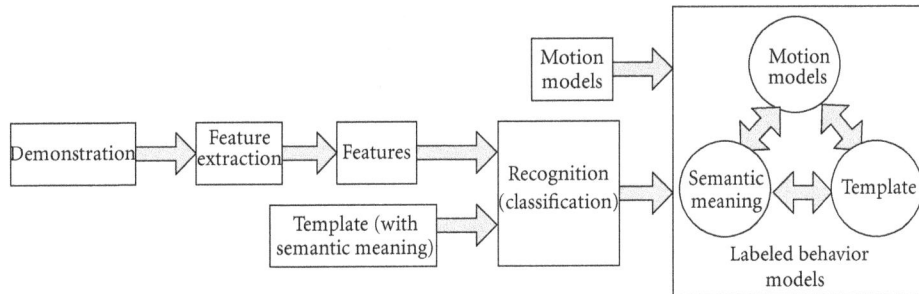

FIGURE 2: Learning of motion models and semantic knowledge.

The extracted feature of the demonstration is used to compare with the templates. The classification results automatically assign the semantic meaning of the template to the corresponding models. The learned motions should have relationship with its semantic-related templates. For example, the motions of writing characters should have relationships to the shapes or topologies of characters, and so on.

In this paper, ISAC is trained to write numbers and to learn the semantic meaning of the numbers automatically. As stated below, the demonstration is writing, the templates should be the shapes or topologies of the numbers. Correspondingly, in this part, the original ISOMAP method, which reflects the spatial topology of sampled data points, is used to train robots to learn semantic meaning of the motions. In order to reflect the overall spatial topology of the sampled data points, all of the neighbor points are considered as temporal neighbors. For simplicity, the modified ISOMAP algorithm can also be applied in this modeling part while the threshold value of temporal distance s is set as the size of the sampled data points.

Using the original ISOMAP algorithm, the demonstration of the motions of writing letters is projected onto a two-dimensional plane with corresponding projection matrices. The recorded trajectories in the latent space are normalized in the same scale. In this paper, the range of the x-axis is $[0, 1]$ and the range of the y-axis is also $[0, 1]$. The reason for normalization is obvious because the given demonstration by human teachers could have different scaling. In order to compare the demonstrations with the commands shown at the generation stage, these processed demonstrations should be normalized.

The technology of establishing models of the recorded trajectories writing numbers and recognizing the numbers based on the templates are not the concentrations of this paper; readers may be interested in other literatures to find many advanced word segmentation and recognition methods.

In this paper, an Optical Character Recognition (OCR) software tool, TesseractOCR (developed by Hewlett-Packard and currently maintained by Google), is used for ISAC to recognize comparing the knowledge model with the characters in the database. In practical application, the results of recognizing a single number are not good by using Tesseract-OCR. Therefore, the picture of a knowledge model

is resized to normal size of a letter and placed behind a sentence "This is." Then the recognition is "This is **." By abandoning "This is," ISAC obtains the semantic knowledge of this recognized picture.

After recognition, the motion models are automated-assigned a semantic meaning and a corresponding template based on the recognition results.

The labeled motion models are stored in the "labeled behavior model" block.

In "Behavior Modeling" block, the projection matrices are calculated using a typical learning algorithm. The trajectories in the latent space and their corresponding projection matrices are stored in "Behavior Models" block. Given a command, ISAC parses the commands or recognizes the commands and converts them into actions in the "command analysis" block and retrieves the behavior models from the "Behavior Models" block. The "reconstruction" block projects the trajectory from the latent space to the joint space to generate new behaviors.

3.3. Generation. In the generation stage, the command is sent to robots and robots need to analyze the command and convert the command into actions with specified parameters. Then the required motion model is obtained by searching the semantic names of the actions with specified parameters. If the motion models are stored in the latent space, a reconstruction is needed to project the motion model from the latent space to the original data space; otherwise, the motion models will be used directly. At last, robots need to execute the motions to complete tasks.

3.3.1. Command Analysis. For most situations of imitation learning, robots are given a task-related situation. The starting states and the goal states are given in this situation and robots need to use learned behaviors to complete the task (achieving the goal state). In this paper, the speech command is used for robots to understand the task.

Using speech, robots needs to listen to the command from a human operator, recognize the required information, and convert the information into actions with specified parameters [41]:

$$\text{Action(parameter)} + \text{Action(parameter)}$$
$$+ \cdots + \text{Action(parameter)}. \tag{27}$$

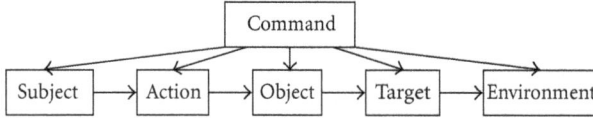

FIGURE 3: Command analysis.

Figure 3 displays the general method of analyzing the command.

Robots break down the commands into different parts by finding matching words in the lexicons using certain grammar rules. Subjects, actions, objects, targets, and environments are predefined in the lexicons. Using certain rules, the commands are converted into actions as shown in (7). This is a typical natural language processing method. Readers may refer to the book written by [42].

For the example of writing numbers, the lexicon design is

Subject: ISAC

Action: Write

Object: Zero, One, Two, Three, Four, Five, Six, Seven, Eight, Nine, and Ten.

The grammar design is Action+Object.

Receiving a command from a human teacher, ISAC extracts the "object" information from the commands and retrieves the corresponding behavior model from the "Labeled Behavior Models" block by searching the required "Object" in the behavior names. The implementation is using the Microsoft speech recognition library.

3.3.2. Reconstruction and Execution. If the motion models are stored in the latent space, using (6), the models can be projected from the latent space to the original data space.

In this paper, a GP-based model is used to describe the motion models in the latent space. Therefore, the (6) is rewritten as:

$$\theta = W\Phi(z), \tag{28}$$

where z is the data point obtained from the GP model given an enquiry point.

Then the required data in the joint space is obtained and robots move the actuator following the generated trajectory.

Using (28), required trajectories could be computed as θ_d.

Using forward kinematics, positions and orientations can be computed as

$$X_d = \text{Forward Kinematics } (\theta_d). \tag{29}$$

The target of generating a similar motion trajectory to complete a task is to minimize the error between the demonstrated trajectory and the generated trajectory.

We define the quadratic cost at each timing step is:

$$L_k = \left(X_{dk} - X_{gk}\right)^T W_k \left(X_{dk} - X_{gk}\right) \tag{30}$$

X_d is a desired trajectory (a demonstrated trajectory in this paper), X_g is a generated trajectory, and k is the timing step.

L represents the weighted error between the demonstrated trajectory and the generated trajectory at timing step k. The target is to minimize the overall cost:

$$\Phi(N) + \sum_{k=1}^{N-1} L_k. \tag{31}$$

while $\Phi(N)$ is the terminal cost, which is normally defined as:

$$\Phi(N) = \left(X_{dN} - X_{gN}\right)^T W_N \left(X_{dN} - X_{gN}\right). \tag{32}$$

For simplicity, in our algorithm, W_k and W_N are defined as a unity diagonal matrix.

The control process is an integration of sensing and planning. In this paper, we do not focus on the low level actuator control. Because the regulators on ISAC are commercial devices just like "black-boxes", we assume that a regulator can automatically adjust control output to achieve the control goals when a required reference input is given.

The initial position and orientation are computed by:

$$X_0 = \text{Forward Kinematics } (\theta_0). \tag{33}$$

At timing step k, θ_{sk-1} is the sensed joint angles at timing step $k-1$ and assume the θ_{gk} is planned based on the current sensing information:

$$X_{gk} - X_{gk-1} = J * \left(\theta_{gk} - \theta_{sk-1}\right), \tag{34}$$

where X_{gk-1} is computed by:

$$X_{gk-1} = \text{Forward Kinematics } (\theta_{sk-1}), \tag{35}$$

and J is the Jacobian matrix.

The target is to minimize

$$L_k = \left(X_{dk} - X_{gk}\right)^T W_k \left(X_{dk} - X_{gk}\right). \tag{36}$$

Rewrite (36):

$$L_k = \left(X_{dk} - J\left(\theta_{gk} - \theta_{sk-1}\right) + X_{gk-1}\right)^T \\ \times \left(X_{dk} - J\left(\theta_{gk} - \theta_{sk-1}\right) + X_{gk-1}\right). \tag{37}$$

Minimize this cost function by differentiating L_k with respect to θ_{gk} and set the derivative to zero, we get

$$\theta_{gk} = \frac{1}{2}\left(X_{dk} - X_{gk-1} + J\theta_{sk-1}\right)^T \\ - \frac{1}{2}J^T\left(X_{dk} - X_{gk-1} + J\theta_{sk-1}\right). \tag{38}$$

At each timing step, θ_{gk} is given to the regulator as reference input for low-level actuator control.

Deliberation and commitment

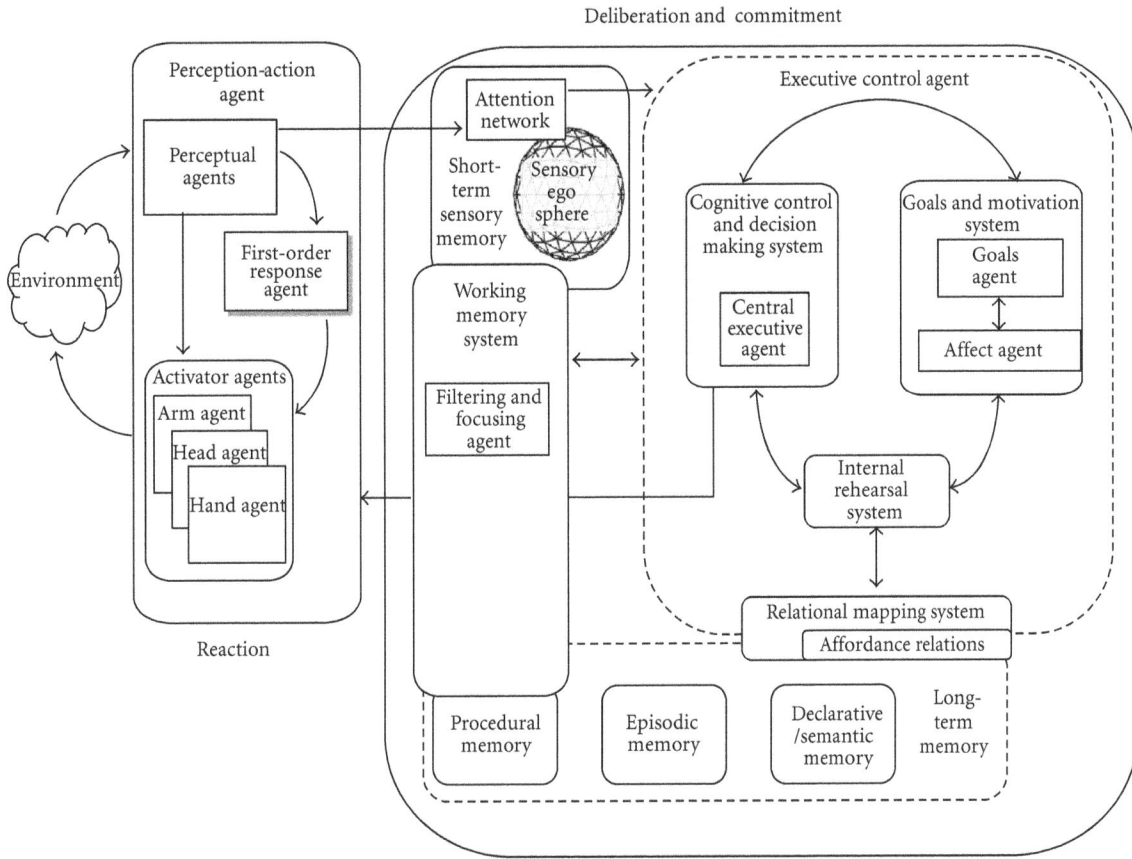

FIGURE 4: ISAC cognitive architecture.

3.4. Implementation. This framework is implemented on a cognitive architecture, named ISAC cognitive architecture, developed by the Center for Intelligent Systems of Vanderbilt University [43–45].

Figure 4 displays the system design of the ISAC cognitive Architecture.

3.4.1. Perceptual Agents (PA). The PA obtains the sensory information from environment. Normally, encoders on the joints of the robot, cameras on the head of the robot, and the force feedback sensor on the wrist of the robot are implemented in this agent.

3.4.2. Short-Term Memory (STM). The obtained information is sent to and stored in the STM. The Sensory Ego Sphere (SES) is implemented in the STM, which performs spatio-termporal coincidence detection, mediates the salience of each percept, and facilitates perceptual binding.

3.4.3. Working Memory System (WMS). The WMS stores the task-related information in chunks. This component is especially important in the generation stage.

3.4.4. Central Executive Agent (CEA). The CEA provides central processing, decision making, and control policy generating for different task goals which is stored in the

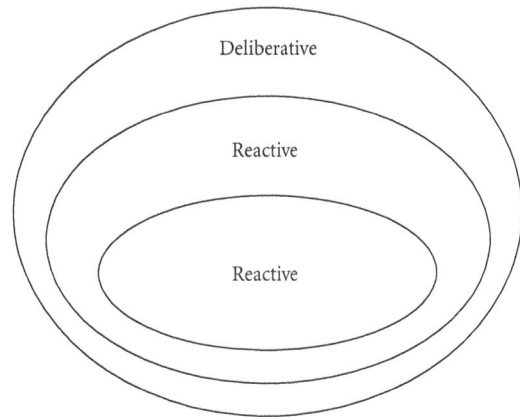

FIGURE 5: Work loops.

Goals Agent (GA). In hierarchy architecture, this component accesses all of the sensed information and makes decision for tasks.

3.4.5. Goal Agent (GA). Correspondingly, the GA stores the motivations or goals of tasks in situations.

3.4.6. Long-Term Memory (LTM). The LTM stores the memory especially the knowledge for long term use. Procedural,

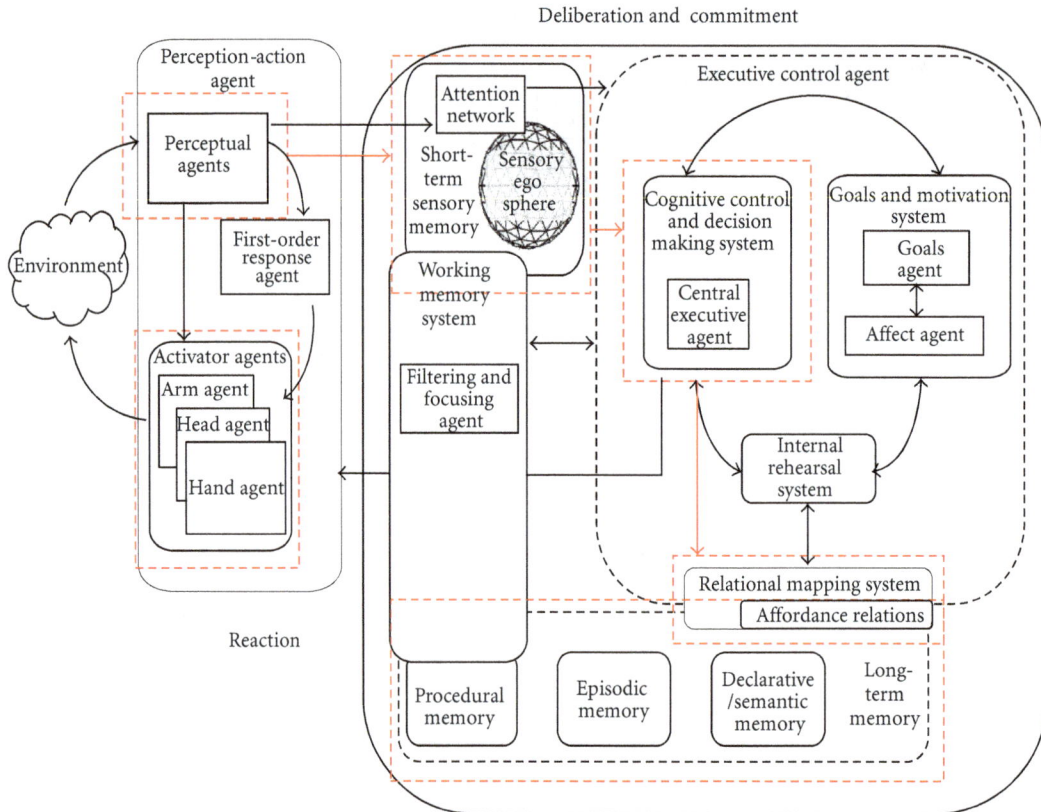

FIGURE 6: Learning stage in ISAC cognitive architecture.

FIGURE 7: Work loop of the learning stage.

semantic, and episodic knowledge are stored in this component. In imitation learning, the learned skill or knowledge is stored as procedural and episodic knowledge using a mathematical model.

3.4.7. Internal Rehearsal System (IRS). The IRS evaluates the results of the decisions made from the CEA through internal rehearsal.

3.4.8. Work Loops. Using this architecture, we can develop three work loops: reactive, routine, and deliberative.

Reactive Loop is inside the perception-action agent. The perceptual agent collects the sensory information from the environment. Using the first-order agent, necessary actions are taken by the actuator agent to affect the environment and the robotic body. This control loop is used for robots to process the emergency or unexpected change in the environment.

The Routine Loop is within the the Perception-Action Agent, filtering and focusing Agent, the STM, and the WMS. This loop completes routine tasks which are well defined in

the WMS. The robot obtains the task-related information from the WMS and sends this information to the actuators through Filtering and Focusing Agent. Actuators are driven by the received information to complete tasks. The Routine Loop also involves the Reactive Loop to avoid the unexpected changes in the environment.

The Deliberative Loop is used for robots to learn new behaviors or skills through modeling, knowledge coupling, and so forth, and to complete new tasks or select behaviors to complete tasks using reasoning, decision making, and so forth. The CEA is the central component in this loop. It retrieves the stored knowledge from the LTM, receives the environmental information from the STM and the WMS, and uses the IRS to evaluate current situation to make decisions or establish models for the sensed information. When the decision is made, the task-related information is sent to the WMS and the system will use the Routine Loop to complete the task. The Deliberative Loop involves the Reactive Loop and the Routine Loop. Our system is largely based on the Deliberative Loop.

Figure 5 displays the relationships among the tree work loops.

Deliberation and commitment

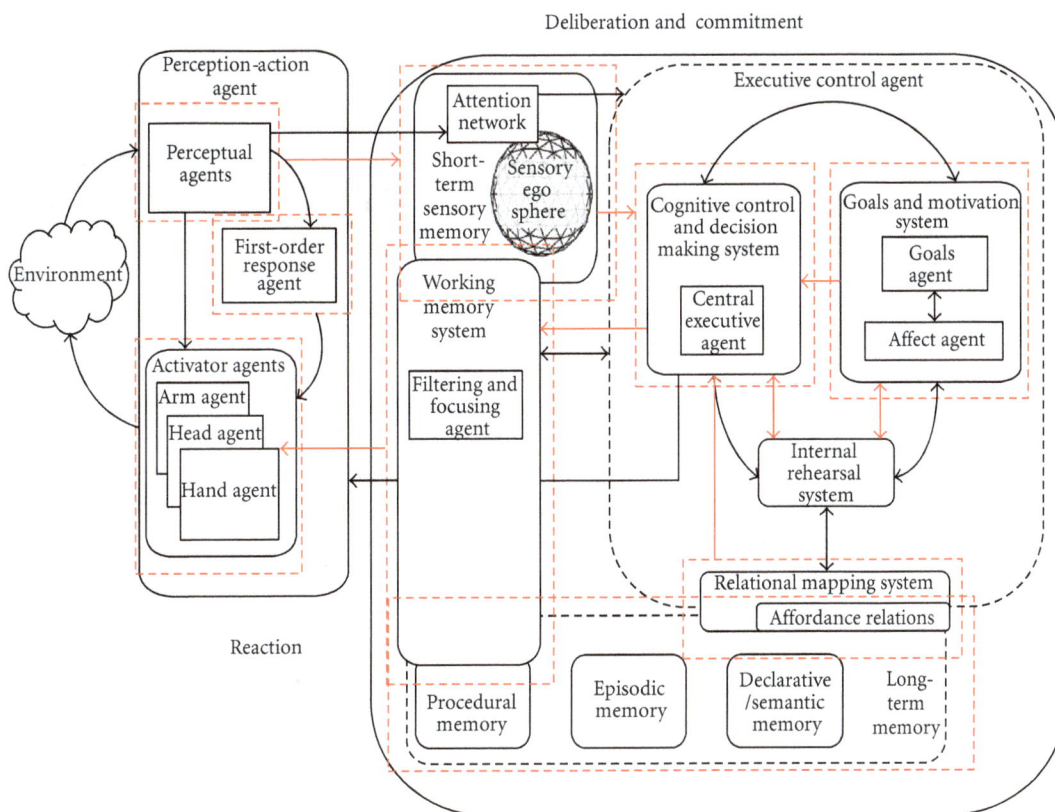

FIGURE 8: Generation stage in ISAC cognitive architecture.

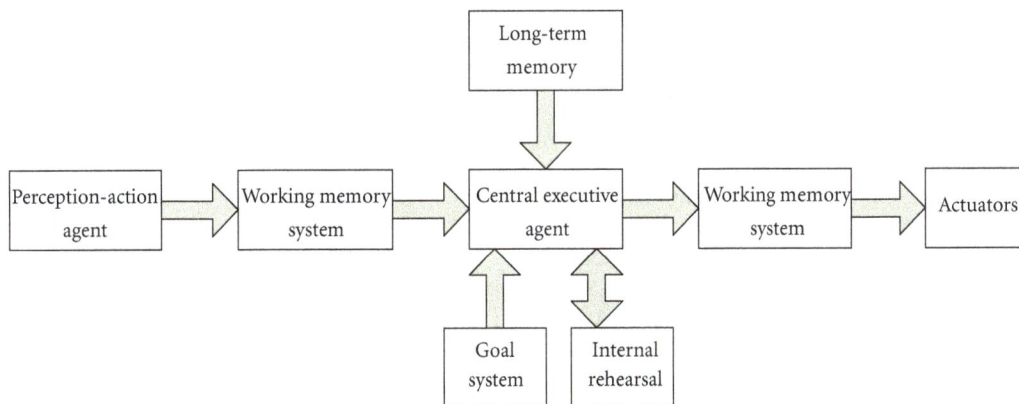

FIGURE 9: Work loop of the generation stage.

3.4.9. Learning Stage. At the learning stage as shown in Figure 6, ISAC collects the information from the encoders using the PA and sends the sensory information to the CEA. The CEA obtains the original ISOMAP algorithm and the modified ISOMAP algorithm from the LTM and calculates the motion models and the knowledge models. Using the RMS, ISAC establishes the labeled behavior models upon the prelearned semantic knowledge and stores the models in the LTM.

The work loop could be displayed as in Figure 7.

3.4.10. Generation Stage. At the Generation Stage as shown in Figure 8, given a speech command, ISAC collects the speech information using the PA, and sends the speech information to the CEA through the STM. By analyzing the speech command, the CEA generates the corresponding actions with specified parameters. The required behavior model is obtained from the LTM through searching the RMS. Upon the obtained behavior model, the CEA plans the motions according to the goal and sends the motion information to the WMS. The WMS stores the task-related

FIGURE 10: ISAC robot.

FIGURE 11: Letters for demonstrations.

FIGURE 12: Demonstrations.

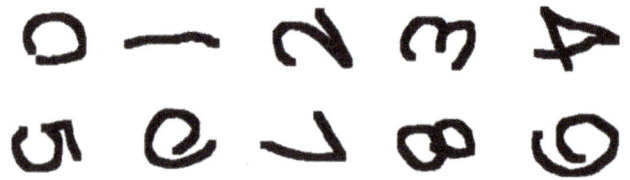

FIGURE 13: Dimension reduction results of the original ISOMAP algorithm.

information and sends the control commands to the AA to execute the motions.

The work loop could be displayed as in Figure 9.

4. Experimental Results

A humanoid robot, named ISAC, is used to validate out proposed system (as shown in Figure 10). ISAC is a stationary pneumatic driven humanoid robot, which has seven Degrees-of-Freedom (DOFs) on each of its arm (including a freedom (open and close) for the end effector). In this system, we only used the right of ISAC to demonstrate the movement trajectories of writing and to write the required numbers. The pen is always grasped using the end-effector, and we only use six DOFs of the right arm of ISAC. Two cameras mounted on the robot are used for ISAC to observe the environment, and we developed an OpenCV-based program to capture and process the images obtained from the cameras. A personal computer, with a 1 GHz CPU, is used to control the arm of ISAC, a personal computer, with a 2.4 GHz CPU, is used to process the images, and a laptop, with a 2.4 GHz CPU, is used to store the semantic knowledge models and the movement trajectory models.

ISAC is shown how to write letters by manually moving its right arm as shown in Figure 12. Figure 11 displays the letters used in the demonstrations. The topologies of the movements of writing letters in the Cartesian space are also the same as shapes of the letters.

The collected data is projected onto a 2-dimensional plane using the original ISOMAP algorithm and the modified ISOMAP algorithm. Figure 13 displays the obtained model using the original ISOMAP algorithm, and Figure 14

displays the obtained model using the modified ISOMAP algorithm.

In practical application, in order to use the dimension reduction results in the recognition part, the image on the 2-dimensional plane has been dilated.

In Figure 13 the dimension reduction results display the shapes and topologies of the distributions of the sampled joint angles from the demonstrations in the latent space. From this figure, the shapes and the topologies of the data distributions are similar to the real letters on the paper and the movement of the end-effector for writing letters in the Cartesian space. It is necessary to emphasize here that without using the kinematics model which calculates the position of the end-effector in the Cartesian space using the joint angles, the dimension reduction results can still get approximate descriptions of the letters on the paper.

In Figure 14, the motion models are represented in the latent space. These models are obtained using the modified ISOMAP algorithm. From the experimental results, we can find two features of the models using this algorithm. (1) Each trajectory does not overlap with itself; (2) each trajectory, which is generated automatically using the modified ISOMAP, always starts from one side and ends at another side. The second feature is guaranteed by the definition of the neighbors in the algorithm. Intuitively, because neighbors are defined as temporal neighbors, the temporal distance of the first point and the last point is the largest in the distance matrix. Therefore, the algorithm always put the starting point and the ending point at two opposite sides of the graph.

Through recognition, the knowledge models in Figure 13 are recognized with the pre-learned numbers using the Tesseract-OCR.

In practical application, the pictures in Figure 13 should be preprocessed in order to be compatible with the Tesseract-OCR. There are several steps of preprocessing.

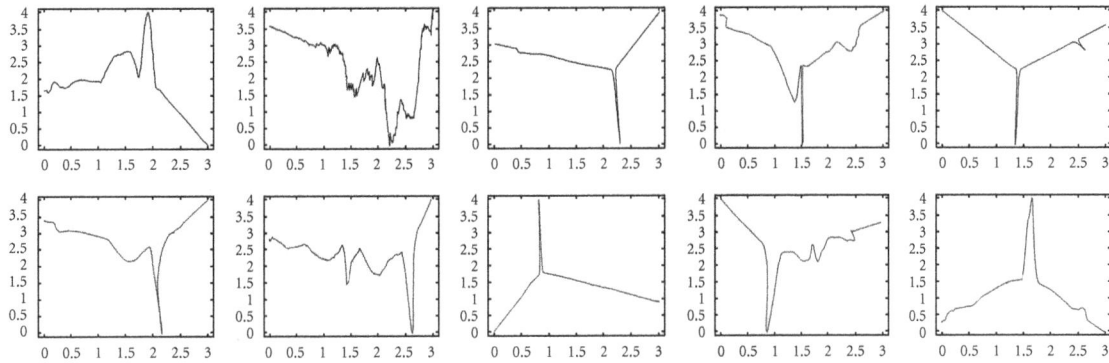

FIGURE 14: Motion models in the latent space using the modified ISOMAP algorithm.

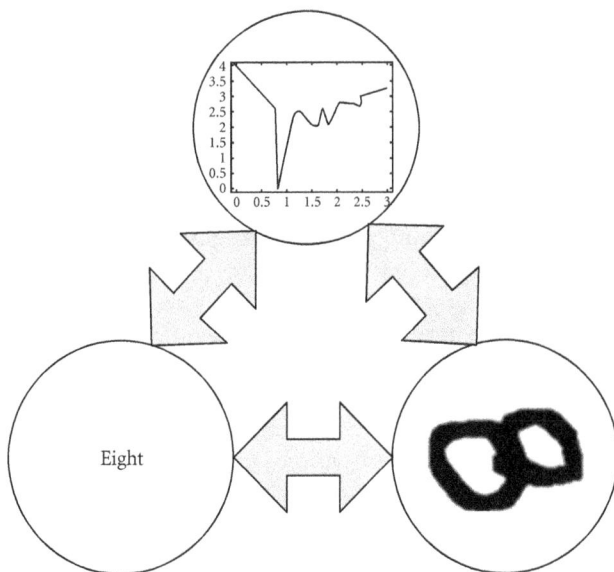

FIGURE 15: Labeled Behavior Models.

(1) Picture i is rotated 0 degree, the result is picture i_1.

(2) Picture i is rotated 90 degree, the result is picture i_2.

(3) Picture i is rotated 180 degree, the result is picture i_3.

(4) Picture i is rotated 270 degree, the result is picture i_4.

(5) Picture i is flipped horizontally, the result is picture i_5.

(6) Picture i_5 is rotated 90 degree, the result is picture i_6.

(7) Picture i_5 is rotated 180 degree, the result is picture i_7.

(8) Picture i_5 is rotated 270 degree, the result is picture i_8.

The obtained eight pictures are all recognized using the Tesseract-OCR. If the recognized result is included in the predefined white list: $\{0, 1, \ldots, 9\}$, it is accepted.

If the recognized result is (6) or (9), it needs to be further processed. Our method is to determine the starting point of writing (6) or (9). If the starting point is near the edge of the image, it is (6); otherwise, it is (9).

Upon the recognition results, the labeled behavior models are established. Figure 15 is an example of the models.

Upon receiving a command from humans, ISAC analyzes the command and converts the command into the actions with specified parameters: Write (Six) as shown in Figure 16. The required motion model is obtained from the labeled behavior models and ISAC executes the motions.

Figure 17 displays the numbers written by ISAC on the papers.

5. Discussion

In this paper, we proposed a framework for robots to learn the motion models and semantic knowledge models simultaneously and only one dataset from the demonstrations is used for both learning stages.

In the current imitation learning frameworks, the motion learning has been highlighted for robots to learn to complete some interesting tasks.

Some researchers are working on incorporating concepts and ideas from cognitive science into robotics research. A typical application using cognitive architectures to implement cognitive processes or cognitive control loops for robotic control and learning [46].

If we consider the whole robotic learning frameworks as hierarchy architectures, the emergent problem is that there seems a gap exists between the utilization of cognitive architecture and the motion learning. As we know, the reasoning and the planning in the cognitive architecture are often implemented in a symbolic way and traditional Artificial Intelligence (AI) methods are used. Therefore, how to connect the symbolic representation and the mathematical models-based motion models?

In our paper, a framework is proposed, in which the connection is based on the natural language processing, which labels the motion models with suitable behavior names and analyzes the command in the cognitive architecture with the behavior names. In this paper, we further propose to train robots to learn the semantic (or symbolic) knowledge by using the same dataset from the demonstrations automatically, which enhance our proposed framework.

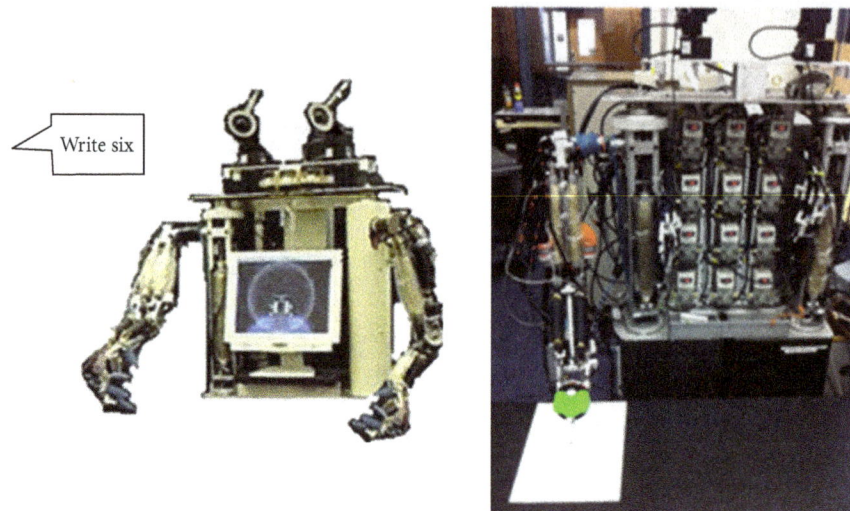

FIGURE 16: Command analysis and generation.

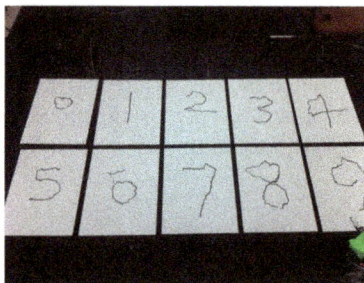

FIGURE 17: Letters written by ISAC.

As we know, when humans see, listen, and feel the behaviors from other humans, we can relate what we see, listen, and feel to our learned procedural, episodic, and semantic knowledge. The framework proposed in this paper is inspired from the daily cognitive brain work of humans.

6. Future Study and Conclusion

For the application of "writing," using the same methods in this paper (dimension reduction using the original and modified ISOMAP algorithm and a letter recognition technology) robots can learn how to write letters and relate the motion models and knowledge models to their corresponding semantic knowledge models.

In other areas, for example, "music playing," robots can learn how to play music (hitting drums, playing guitar, and playing piano) and relate the motions required to play music to their corresponding semantic knowledge models. The ISOMAP algorithms may not complete such learning in these areas. However, readers can simply find that the tempos of hitting drums are in correspondence to the tempos of moving hands up and down. If the tempos of the sound of hitting drums can be extracted as the templates and the tempos of moving hands can be extracted as the knowledge

model, this framework can also be used for robots to learn the motions of play music and the semantic meaning of these motions simultaneously and automatically.

The crucial point of the application of this framework is to find the features of the motions which are in correspondence with the inner features of behavior which is strongly related to the semantic models.

This paper proposes a framework for robots to learn the motion models and semantic knowledge models simultaneously using one data set from the demonstrations. A modified ISOMAP algorithm is used for robots to extract semantic information from the demonstrations. The implementation is on a cognitive architecture with several extensions of current algorithms. Semantic analysis of the command is also implemented in this framework. The experiments are carried out on a humanoid, and the experimental results demonstrate the effectiveness of this framework.

References

[1] R. A. Brooks, "A robust layered control system for a mobile robot," *IEEE journal of robotics and automation*, vol. 2, no. 1, pp. 14–23, 1986.

[2] R. Brooks, "How to build complete creatures rather than isolated cognitive simulators," in *Architectures for Intelligence*, K. VanLehn, Ed., pp. 225–239, Lawrence Erlbaum Associates, New York, NY, USA, 1991.

[3] A. Sloman and J. Chappell, "The altricial-precocial spectrum for robots," in *Proceedings of the International Joint Conferences on Artificial Intelligence (IJCA '05)*, pp. 1187–1192, Edinburgh, UK, 2005.

[4] A. Stoytchev, "Toward learning the binding affordances of objects: a behavior-grounded approach," in *Proceedings of the AAAI Symposium on Developmental Robotics*, pp. 21–23, 2005.

[5] S. Schaal, "Learning from demonstration," in *Advances in Neural Information Processing Systems*, M. J. M. C. Mozer and T. Petsche, Eds., pp. 1040–1046, The MIT Press, Cambridge, Mass, USA, 1997.

[6] M. Uchiyama, "Formation of high speed motion pattern of mechanical arm by trial," *Transactions, Society of Instrument and Control Engineers*, vol. 19, pp. 706–712, 1978.

[7] C. Atkeson and J. McIntyre, "Robot trajectory learning through practice," in *Proceedings of the IEEE Conference on Robotics and Automation*, pp. 1737–1742, San Francisco, Calif, USA, 1986.

[8] B. D. Argall, S. Chernova, M. Veloso, and B. Browning, "A survey of robot learning from demonstration," *Robotics and Autonomous Systems*, vol. 57, no. 5, pp. 469–483, 2009.

[9] A. Billard, "Learning motor skills by imitation: a biologically inspired robotic model," *Cybernetics and Systems*, vol. 32, no. 1-2, pp. 155–193, 2001.

[10] S. Calinon, F. Guenter, and A. Billard, "On learning, representing, and generalizing a task in a humanoid robot," *IEEE Transactions on Systems, Man, and Cybernetics*, vol. 37, no. 2, pp. 286–298, 2007.

[11] A. Ijspeert, J. Nakanishi, and S. Schaal, "Learning attractor landscapes for learning motor primitives," in *Advances in Neural Information Processing Systems*, S. Becker, S. Thrun, and K. Obermayer, Eds., vol. 15, pp. 1547–1554, The MIT Press, 2003.

[12] R. Dillmann, O. Rogalla, M. Ehrenmann, R. Zollner, and M. Bordegoni, "Learning robot behaviour and skills based on human demonstration and advice: the machine learning paradigm," in *Proceedings of the 9th International Symposium of Robotics Research (ISRR '99)*, pp. 229–238, Snowbird, Utah, USA, October 1999.

[13] Y. Kuniyoshi, M. Inaba, and H. Inoue, "Learning by watching: extracting reusable task knowledge from visual observation of human performance," *IEEE Transactions on Robotics and Automation*, vol. 10, no. 6, pp. 799–822, 1994.

[14] T. Inamura, M. Inaba, and H. Inoue, "Acquisition of probabilistic behavior decision model based on the interactive teaching method," in *Proceedings of the 9th International Conference on Advanced Robotics,*, pp. 523–528, 1999.

[15] R. M. Voyles and P. K. Khosla, "A Multi-agent system for programming robots by human demonstration," *Integrated Computer-Aided Engineering*, vol. 8, no. 1, pp. 59–67, 2001.

[16] A. J. Ijspeert, J. Nakanishi, and S. Schaal, "Movement imitation with nonlinear dynamical systems in humanoid robots," in *IEEE International Conference on Robotics and Automation*, pp. 1398–1403, Washington, DC, USA, May 2002.

[17] I. Jolliffe, *Principal Component Analysis*, Springer, New York, NY, USA, 1986.

[18] D. J. Bartholomew, "The foundations of factor analysis," *Biometrika*, vol. 71, no. 2, pp. 221–232, 1984.

[19] J. B. Tenenbaum, V. de Silva, and J. C. Langford, "A global geometric framework for nonlinear dimensionality reduction," *Science*, vol. 290, no. 5500, pp. 2319–2323, 2000.

[20] S. T. Roweis and L. K. Saul, "Nonlinear dimensionality reduction by locally linear embedding," *Science*, vol. 290, no. 5500, pp. 2323–2326, 2000.

[21] C. K. I. Williams, "On a connection between kernel PCA and metric multidimensional scaling," *Machine Learning*, vol. 46, no. 1–3, pp. 11–19, 2002.

[22] C. Bishop, *Pattern Recognition and Machine Learning*, Springer, New York, NY, USA, 2006.

[23] C. Rasmussen, "Gaussian processes in machine learning," in *Advanced Lectures on Machine Learning*, pp. 63–71, The MIT Press, Cambridge, Mass, USA, 2004.

[24] C. G. Atkeson, A. W. Moore, and S. Schaal, "Locally weighted learning," *Artificial Intelligence Review*, vol. 11, no. 1–5, pp. 11–73, 1997.

[25] S. Vijayakumar and S. Schaal, "Locally weighted projection regression: an O (n) algorithm for incremental real time learning in high dimensional space," in *Proceedings of The 17th International Conference on Machine Learning*, pp. 288–293, Stanford, Calif, USA, 2000.

[26] S. Chernova and M. Veloso, "Confidence-based policy learning from demonstration using Gaussian mixture models," in *Proceedings of the 6th International Joint Conference on Autonomous Agents and Multiagent Systems*, p. 233, 2007.

[27] H. Abdi, "A neural network primer," *Journal of Biological Systems*, vol. 2, no. 3, pp. 247–283, 1994.

[28] R. Sutton and A. Barto, *Reinforcement Learning: An Introduction*, The MIT press, 1998.

[29] J. Peters, S. Vijayakumar, and S. Schaal, "Reinforcement learning for humanoid robotics," in *Proceedings of the IEEE-RAS International Conference on Humanoid Robotis*, pp. 1–20, Karlsruhe, Germany, 2003.

[30] P. Dyer and S. R. McReynolds, *The Computation and Theory of Optimal Control*, Academic Press, 1970.

[31] E. Theodorou, J. Buchli, and S. Schaal, "Reinforcement learning of motor skills in high dimensions: a path integral approach," in *IEEE International Conference on Robotics and Automation (ICRA '10)*, pp. 2397–2403, 2010.

[32] E. A. Theodorou, J. Buchli, and S. Schaal, "A generalized path integral control approach to reinforcement learning," *The Journal of Machine Learning Research*, vol. 11, pp. 3137–3181, 2010.

[33] H. Tan and K. Kawamura, "A computational framework for integrating robotic exploration and human demonstration in imitation learning," in *Proceedings of the IEEE International Conference on System, Man and Cybernetics*, pp. 2501–2506, Anchorage, Alaska, USA, 2011.

[34] O. C. Jenkins and M. J. Matarić, "A spatio-temporal extension to isomap nonlinear dimension reduction," in *Proceedings of the 21st International Conference on Machine Learning (ICML '04)*, p. 56, July 2004.

[35] C. M. Bishop, M. Svensén, and C. K. I. Williams, "GTM: the generative topographic mapping," *Neural Computation*, vol. 10, no. 1, pp. 215–234, 1998.

[36] S. Calinon and A. Billard, "A probabilistic programming by demonstration framework handling constraints in joint space and task space," in *Proceedings of the IEEE International Conference on Intelligent Robots and Systems*, pp. 367–372, September 2008.

[37] D. Grimes, R. Chalodhorn, and R. Rao, "Dynamic imitation in a humanoid robot through nonparametric probabilistic inference," in *Proceedings of the Robotics: Science and Systems (RSS '06)*, The MIT Press, 2006.

[38] A. Shon, K. Grochow, A. Hertzmann, and R. Rao, "Gaussian process Cca for image synthesis and robotic imitation," Tech. Rep. UW-CSE-TR-2005-06-02, University of Washington CSE Department, 2005.

[39] M. Schneider and W. Ertel, "Robot learning by demonstration with local gaussian process regression," in *Proceedings of the IEEE International Conference on Intelligent Robots and Systems (IROS '10)*, pp. 255–260, October 2010.

[40] H. Tan, E. Erdemir, K. Kawamura, and Q. Du, "A potential field method-based extension of the dynamic movement primitive algorithm for imitation learning with obstacle avoidance," in *Proceedings of the IEEE International Conference on Mechatronics and Automation*, pp. 525–530, Beijing, China, 2011.

[41] H. Tan, Q. Du, and N. Wu, "A Framework for cognitive robots to learn behaviors through imitation and interaction

with humans," in *Proceedings of the IEEE International Multi-Disciplinary Conference on Cognitive Methods in Situation Awareness and Decision Support*, pp. 235–238, New Orleans, La, USA, 2012.

[42] S. J. Russell and P. Norvig, *Artificial Intelligence : A Modern Approach*, Prentice Hall, Upper Saddle River, NJ, USA, 3rd edition, 2010.

[43] K. Kawamura, R. Peters II, R. Bodenheimer et al., "Multiagent-based cognitive robot architecture and its realization," *International Journal of Humanoid Robotics*, vol. 1, pp. 65–93, 2004.

[44] K. Kawamura, S. M. Gordon, P. Ratanaswasd, E. Erdemir, and J. F. Hall, "Implementation of cognitive control for a humanoid robot," *International Journal of Humanoid Robotics*, vol. 5, no. 4, pp. 547–586, 2008.

[45] H. Tan, "Implementation of a framework for imitation learning on a humanoid robot using a cognitive architecture," in *The Future of Humanoid Robots: Research and Applications*, R. Zaier, Ed., pp. 189–210, InTech Open Access Publishing, 2012.

[46] H. Tan and C. Liang, "A conceptual cognitive architecture for robots to learn behaviors from demonstrations in robotic aid area," in *Proceedings of 33rd Annual International Conference of the IEEE Engineering in Medicine and Biology Society*, pp. 1248–1262, Boston, Mass, USA, 2011.

10

Robotics for Natural Orifice Transluminal Endoscopic Surgery: A Review

Xiaona Wang and Max Q.-H. Meng

Department of Electronic Engineering, The Chinese University of Hong Kong, Shatin, NT, Hong Kong

Correspondence should be addressed to Xiaona Wang, xnwang@ee.cuhk.edu.hk

Academic Editor: Yangmin Li

Natural Orifice Transluminal Endoscopic Surgery (NOTES) involves accessing the abdominal cavity via one of the bodies' natural orifices, for example, mouth, anus, or vagina. This new surgical procedure is very appealing from patients' perspectives because it eliminates completely abdominal wall aggression and promises to reduce postoperative pain, in addition to all other advantages brought by laparoscopic surgery. However, the constraints imposed by both the mode of access and the limited technology currently available make NOTES very challenging for the surgeons. Redesign of the instruments is imperative in order to make this emerging operative access safe and reproducible. In this paper, we survey on the state-of-the-art devices used in NOTES and introduce both the flexible instruments based on improvement of current endoscopic platforms and the revolutionary concept of robotic platforms based on the convergence of communication and micromechatronics technologies. The advantages and limitations of each category are addressed. Potential solutions are proposed to improve the existing designs and develop robust and stable robotic platforms for NOTES.

1. Introduction

Past decades have evidenced a steady decrease in the invasiveness of surgical interventions. The first laparoscopic cholecystectomy was performed in the mid 1980s. The laparoscopic surgery, or minimally invasive surgery, was welcomed by the patients undergoing this alternative for the rapid recovery. The new procedures perform operations in the abdomen through small incisions, usually 0.5–1.5 cm, which are much smaller than that needed by the traditional technique (over 10 cm). Since the 1990s, the laparoscopic surgery gradually prevails over the traditional open surgery. It has been proved to decrease postoperative morbidity, shorten hospitalization and convalescence, and improve cosmesis while matching the outcomes of equivalent open procedures. Today, the majority of open surgical procedures have been replicated or even replaced by laparoscopic techniques.

At the same time, the endoscopy technique has been evolving from pure diagnostic devices to therapeutic devices thanks to the development of the microelectronic techniques.

In addition to the lighting and imaging parts, the endoscopes are designed to possess multiple working channels, enabling equipment and insertion of various instruments, such as the ultrasonic device, the laser cauter, the biopsy instruments, and polyp removal tools. It is not surprising that endoscopy and surgery would eventually work together, and this theoretic point of fusion is being turned into reality. With the objective of preventing port-site complications associated with laparoscopy, further decreasing discomfort and removing the scar on the body surface, Natural Orifice Transluminal Endoscopic Surgery (NOTES) has been proposed (Figure 1) [1].

NOTES involves accessing the abdominal cavity via one of the bodies' natural orifices, for example, mouth, anus, vagina, or urethra. A flexible endoscope is advanced into the peritoneal cavity after puncturing one of the viscera such as stomach, colon, vagina, or bladder. Conventional endoscopic instruments are introduced through the working channels of the endoscope in order to perform the operation. NOTES not only provides all the advantages of laparoscopic surgery but also offers several potential benefits, including cosmetic result, lower anesthesia requirements, less pain, even faster recovery, and a decreased incidence of wound-related complications.

It is believed that NOTES may become the next major paradigm shift in surgery following laparoscopy [2]. Nevertheless, as an emerging surgical procedure, the safety issue of NOTES becomes a major concern. It requires continuous clinical practice and assessment and depends largely on the development of appropriate surgical tools. Since in literature there is a lack of a thorough survey on the instrumentation for NOTES, in this paper we hope to make up this gap by investigating the state-of-the-art instruments developed for NOTES, including both the prototype flexible endoscopes and the new concept robotic platforms. The paper is organized as follows: Section 2 introduces the background and status quo of NOTES; Section 3 and Section 4 present, respectively, the flexible endoscopes and the robotic platforms. The advantages and limitations are discussed, followed by potential solutions to the problems; Section 5 concludes the paper.

2. Background of NOTES

The first attempt of NOTES was reported in 2004 by Kalloo et al. in a porcine model [3]. They penetrated the gastric wall and operated in the abdominal cavity using an orally introduced flexible endoscope via a sterile overtube. The pioneer work demonstrated the feasibility and safety of an oral transgastric peritoneoscopy. The idea of "no-scar" abdominal surgery immediately captured the medical community and the general public. Since then, many researchers have used transluminal flexible endoscopy in animal models to perform various intraperitoneal procedures, ranging from tubal ligation to splenectomy [4].

In 2006, the first transgastric NOTES appendectomy performed in humans was reported by Rao and Reddy [5]. Swanström described the first case of human transgastric cholecystectomy in 2007 [6]. More and more clinical trials of NOTES commenced thereafter. To date, thousands of NOTES procedures have been carried out all around the world. Of all the natural orifices accessed by NOTES, the transvaginal route seems to be the most safe and feasible for clinical applications, because it minimizes the concerns about a secure enteral closure. The disadvantage is that it is possible only in the female. Sánchez-Margallo et al. reported transvaginal cholecystectomy performed in pig models without using laparoscopic assistance [7]. However, true NOTES, as an "incisionless" operation, has not been described in human clinical trials, because all cases reported were assisted by laparoscopic or percutaneous methods.

Before NOTES can be successfully and responsibly used in clinical care, several critical issues must be resolved, among which an appropriated instrumentation is of most importance [8–11]. Conventional flexible endoscopes are inadequate for performing complex transluminal surgical procedures. The limitations include the lack of a multitasking platform, the number and size of access channels, the inability to position and fix the instruments to allow robust

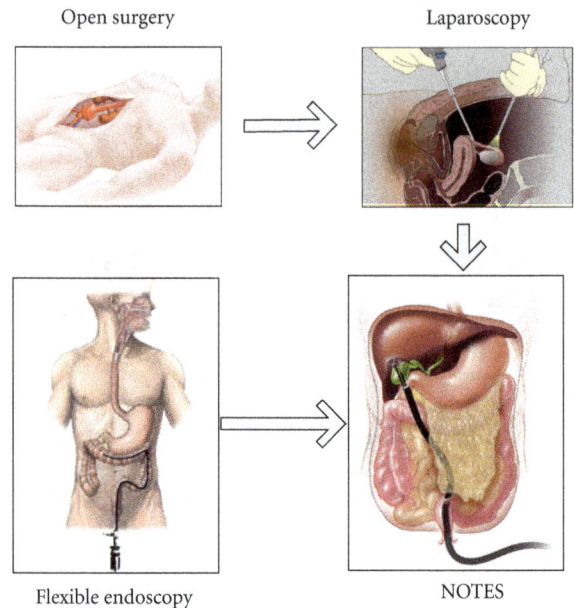

FIGURE 1: Natural orifice transluminal endoscopic surgery is a fusion of the therapeutic endoscopy and the laparoscopy surgery, which is a "scarless" surgery further decreasing the invasiveness.

retraction and exposure, and the unavailability of adequate triangulation, NOTES has encouraged a lot of research and development in the medical device industry. Many surgical devices are being developed, including advanced forms of endoscopes and robotic platforms. The details will be introduced in the following sections.

3. Flexible Endoscopic Platforms

3.1. Prototype of Endoscope. Flexible endoscopes are the main tool used in minimally invasive surgery. A typical endoscope is 10 mm in diameter and 70–180 mm in length. They are inserted directly into the hollow organ/cavity for examination or therapeutic treatment. Degani et al. [12] developed a 12 mm (diameter) × 300 mm (length) snake-like robot named "HARP". It consists of an inner and an outer snake made of rigid cylindrical links connected by spherical joint. By pulling/relaxing the cables stringing the joints, the probe can be made rigid/flexible, which is similar to the ShapeLock technology. Preliminary experiments in pigs with the prototype can reach the target, lock the outer snake, and replace the inner snake with colonoscope forceps. Abbott et al. [13] developed two generations of ViaCath systems for teleoperated endoluminal surgery. The endoluminal instruments consist of a connector, a flexible shaft, and an actuated joint with end-effector. The second generation instrument is 120 cm in length, 7.2 mm in diameter with a single lumen inside the shaft for the actuation cables, providing 9 degrees of freedom. The mechanical properties of the prototype have been evaluated. But it still needs validations with phantoms and animal models.

In comparison, the flexible endoscopic platforms for NOTES need to be stronger and more rigid to perform surgical procedures. Most prototypes are modified based on the therapeutic endoscopes by augmenting the number and the diameter of the channels. Swanström et al. partnered with USGI Medical developed Transport multilumen operating platform for NOTES [14, 15]. As shown in Figure 2, Transport is a 16 mm access device, with four large working channels, one for a standard 6 mm endoscope, and three others for large diameter instruments. It adopts the ShapeLock design, which allows independent steering of the tip and then lock into position once it is maneuvered to the operative site. The large 4 mm and 6 mm channels allow passage of stronger and flexible surgical tools. The force delivery at the tip can reach 0.89 kg and the instrument application force 0.1 kg [16].

Triangulation is one of the most essential concepts in laparoscopy, which means separation of the working hands from each other and to have the "eye" in-between the two "hands", as shown in Figure 3. USGI Medical developed a prototype called "Cobra" [15], which adds three independent arms controlled by a robotic interface to the Transport to provide some degree of triangulation, as shown in Figure 4.

Olympus modified a standard dual-channel therapeutic scope named the "R" scope for NOTES [17]. The multibending section of the scope allows it to be positioned near the target area and then locked. The device has two movable 3.8 mm instrument channels: one moves vertically and the other swings horizontally, which offer a reasonable simulation of triangulation, as shown in Figure 5. The two instruments can be manipulated during the operation with a knob and a lever that surround the angulation control knobs of the R-scope. Once the operator has decided on the knob or lever positions, these can be locked into position.

3.2. Limitations of Flexible Endoscope. The improved flexible endoscopes are useful for some antegrade intra-abdominal procedures. However, there still exist deficiencies due to their inherent features. One major point is that the flexibility makes it difficult to aggressively retract tissues. Due to the small channel size of the endoscopes, the end effectors of most instruments are small and feeble. A larger size and more ports permit some degree of triangulation but are still inadequate. Some instruments are fixed and require the device be removed to exchange tools and then reintroduced. In addition, the complexity of the devices does not allow a smooth and controlled movement of the tip and therefore the instrument, which challenges the precise maneuvers. The main limitations of the present flexible endoscope and the requirements for instruments used for NOTES are listed as follows [18–20].

   (i) Platform stability: The inherent flexibility of current endoscopes impedes achieving a stable operation field for NOTES. The ideal instrument would be capable of atraumatic insertion and positioning but then be able to fix in position to free the surgeon's hands to manipulate multiple instruments.

   (ii) Retraction: The lack of rigidity also limits the counter forces down the endoscope which can be applied to adequately retract tissues and apply strong sutures

or clips. New methods need to be developed to allow vigorous traction and large organ retraction for exposure.

   (iii) Triangulation: A critical concept in laparoscopy, which means separation of the working hands from each other and to have the "eye" in-between the two "hands". This is not the case for the flexible endoscope with inline instrumentation and optics. Independent movement of multichannel therapeutic instruments is desirable.

   (iv) Size: The flexible endoscopes usually have a diameter of 5–15 mm with only one or two working channels. Each channel is 2-3 mm in diameter, which limits the size of surgical instruments and abilities to triangulate and maneuver tissue properly. The device for NOTES should contain at least 2 instrument channels in addition to the imaging part, to enlarge the range of motion and increase the degrees of freedom.

   (v) Image: The image quality of current flexible endoscopes is comparable to laparoscopes for the most parts. However, there is an orientation problem when working in a retroflexed position, that is, the image might be inverted or reversed. An orthophoric imaging system with adequate lighting intensity is necessary to distinguish different anatomical structures in the intraperitoneal space.

Currently, the flexible endoscopes are not applicable for fine surgeries. There is a need to make significant improvements and design more aggressive instruments for NOTES. The problems mentioned above can be resolved to some extent with scope-handling expertise. However, a better long-term solution will be to redesign the endoscopic access devices and carry out the procedures in a completely new manner.

4. Robotic Platforms

At present, development of instrumentation to facilitate NOTES techniques is still in its infancy, but is critical for broadly applicable NOTES. Robotic technology is likely to allow us to make more agile and precise instruments than currently available tools, and perform procedures that cannot be done by conventional minimally invasive techniques. To address the limitations of the existing medical robots, such as bulky and expensive system, limited view field, and inconvenient operation, in recent years novel concepts of miniature/modular robots are being developed, which provide potential solutions to the laparoscopic and NOTES procedures. This section introduces the attempts on these robotic platforms.

4.1. Imaging Robot. Imaging robot is useful for providing visual feedback during the medical procedure. In the past decade, the wireless capsule endoscopy has been proved an established procedure for examination of the gastrointestinal tract [21]. The imaging robot encloses lighting, imaging,

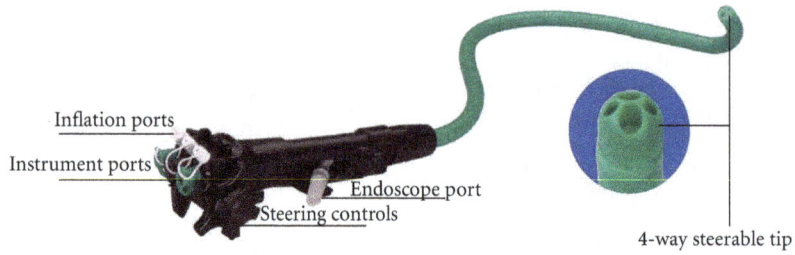

FIGURE 2: The shape-lock Transport endoscope with four operating channels (USGI Medical).

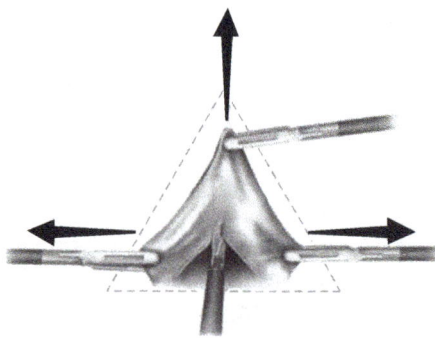

FIGURE 3: Demonstration of triangulation, a critical technique during laparoscopic surgery.

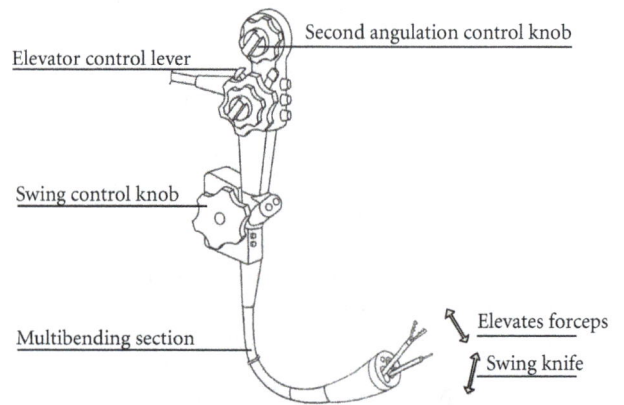

FIGURE 5: The R-scope (Olympus Medical Systems).

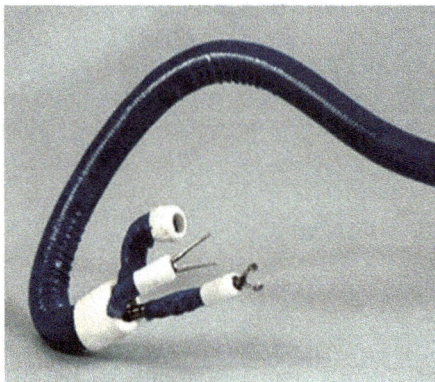

FIGURE 4: The Cobra triangulating scope (USGI Medical).

wireless transmission units, and button batteries within a capsule-sized of 11 mm × 26 mm. After swallowed, the capsule can take pictures and send them to the external data recorder wirelessly while travelling through the digestive tract by natural peristalsis. And at the completion of the examination, the capsule is excreted naturally. The wireless capsule endoscope enables completely noninvasive and painless examination. The swallowable feature and the technical breakthrough give inspirations for the design and development of medical microrobotics. Equipped with a guiding or an actuating mechanism, the capsule has great

potential to be applied for assisting the surgical operations [22, 23].

Rentschler et al. developed a mobile camera robot with a diameter of 12 mm and a length of 75 mm [24], as shown in Figure 6. It consists of two wheels driven independently by 6 mm DC motors to make forward, reverse, and turning motions. A tail is set in the middle of the wheels to prevent counter-rotation. An adjustable-focus image sensor is carried between the wheels to provide visual feedback during the movement of the robot. This robot has been tested in porcine model experiment, during which it was inserted into the peritoneal cavity through the transgastric incision, and retracted back through the esophagus by a standard upper endoscope after exploration of the abdominal cavity.

Rentschler and Oleynikov also introduced a fixed-base imaging robot with a body of 15 mm diameter. It is mounted on a spring-loaded foldable-tripod platform that allows a 45-degree angle forward tilting and 360-degree panning [25]. The objective is to enhance visualization and provide indepth perception of the abdominal cavity. The three legs can be retracted during insertion and abducted by the torsion springs after entry. Light-emitting diodes (LEDs) are equipped to provide illumination. The robot has been evaluated in canine and porcine model experiments to provide augmented visual feedback and enhance the field

and a battery. The structure module has a motor actuated joint, which can make ±90 degree bending with a torque of 6.5 mN·m and 0 degree to 180 degree rotation with a torque of 2.2 mN·m. The biopsy module has a foldable grasping mechanism, driven by the motor, to miniaturize the size during insertion and generate the grasping force for tissue sampling.

The modular robots are easy to be assembled into different topologies by magnetic attraction, while the disassembly method is a much more complex problem. Diller et al. [40] proposed to control the assembly and disassembly using an electrostatic anchoring surface, which can selectively keep specific modules from moving. The work is inspiring although the dimensions of the modules in the proposed work are in 1 mm and the system works in 2D surface, which is not appropriate for surgical applications.

4.4. Limitations of Robotic Platform. Comparing with the flexible endoscopes, the robotic approach facilitates the operations for minimally invasive surgery such as NOTES by allowing the use of multiple surgical instruments, improving triangulation and ergonomics, and providing relatively stable and rigid platform. The modular and cooperative design simplifies the structure of each individual functional robot and miniaturizes the size to enable the passage through a single port. Experimental evaluations confirm the effectiveness of the microrobots in providing extra visualization and task assistance [24, 26]. Nevertheless, the in vivo microrobots are all in nonsurvival animal evaluation stage and not mature enough for clinical use. While this approach is promising, the technology needs to address current limitations and be further developed to ease the operation and ensure safety in the future. Some limitations and potential solution are listed as follows.

(i) Platform stability: Most imaging and operative microrobots are anchored on the abdominal wall by magnetic coupling to avoid extra incisions. However, the magnetic attraction force diminishes exponentially with respect to the distance between the internal magnetic joint and the external handle. Experimental studies show that the force will be inadequate to retract the tissue at a distance larger than 15 mm, resulting in unstable platform in humans with thicker abdominal walls. An alternative way is to exchange the magnetic anchor by needles after the microrobots are guided and deployed magnetically inside the abdominal cavity. Obviously this will lead to a longer learning curve and less dexterity of the system. Stronger magnetic field is preferable to provide secure anchoring of the microrobot. For the reconfigurable microrobot, except the robust assembly by the magnetic joints, the undocking method is also critical for breaking up the topology and releasing the modules. Properly designed electromagnetic field may be feasible to generate powerful while controllable attraction force. Other than the magnetic anchoring problem, there are also reports about failure of the mechanical parts during experiments. Besides using

light but strong materials to handle this problem, parallel mechanical design may carry more force than the serial one.

(ii) Wireless control: The imaging and motorized microrobots introduced in literature are all tethered to the external power and control system. Future platform may be optimized by incorporating an on-board power supply and wireless controller, which enables completely independent deployment of each individual microrobot and reduces the possible confliction among different microrobots and other instruments. The technology of wireless capsule endoscope offers successful experiences that can be adopted for next-generation self-contained microrobot.

(iii) Versatile and robust operation tools: The operative microrobots reported mainly consist of some basic operation tools such as the grasper and the cautery, so that most microrobots can only work with the flexible endoscope as an assistant. Equipment of more end effectors such as scissors, needle driver, and dissector will broaden the range of the operations. Besides, robust graspers with increased rigidity, bigger jaws, and better control over positioning are desired especially for the clinical settings of thickened or diseased tissues. The image quality of the camera robot needs to be improved to match that of the conventional laparoscopes and flexible endoscopes. More integration of robotic control systems will facilitate additional functionality of the surgical system. Ultimately, it is expected that completely independent operations be performed by the versatile microrobots.

(iv) Articulated instruments: In the operative microrobots, tools are attached directly to the magnetic joint which is coupled to the external magnetic handle. The advantage of the design lies in simple structure and compact size. However, one degree of freedom confines the operation in a small working space and thus restricts performance of some complicated operations. Multiple micro DC motors may be equipped to actuate the articulated instrument, providing extended workspace and better dexterity.

In summary, each system has its advantages and limitations. It is not obvious which system is uniquely superior to the others in terms of the evaluation based on size, image quality, maneuverability, stability, and ability to provide triangulation. Doctors and engineers are still exerting their efforts on improvement and novel design of the robotic systems.

5. Conclusions

Like the introduction of laparoscopic procedures, which have great impact on the surgical treatment in the past 30 years, NOTES may become another paradigm shift of the surgery. However, before the wide adoption of the novel procedure, clinical and engineering limitations must be addressed. One critical demand is the development of new operation platforms. At present, most clinical trials have been performed with conventional laparoscopic instruments or flexible endoscopes revised based on the gastrointestinal endoscopes. Nevertheless, the former can only be employed in specific transvaginal operations and the latter platforms lack rigidity and cannot fulfill the requirements of triangulation and retraction. New-concept instruments need to be developed for NOTES, among which robotics provide a promising way. Pilot study has validated the feasibility of using operative and imaging microrobots for task assistance. Additional investigations are to be carried out to evaluate the outcomes in survival models in animals and ultimately safety and efficacy in humans. With more advanced robotics proposed and developed, we can expect that NOTES procedures become more mature and more widely accepted in the future.

References

[1] S. S. Garud and F. F. Willingham, "Natural orifice transluminal endoscopic surgery," *Endoscopy*, vol. 44, no. 9, pp. 865–868, 2012.

[2] S. V. Kantsevoy, B. Hu, S. B. Jagannath et al., "Transgastric endoscopic splenectomy: is it possible?" *Surgical Endoscopy*, vol. 20, no. 3, pp. 522–525, 2006.

[3] A. N. Kalloo, V. K. Singh, S. B. Jagannath et al., "Flexible transgastric peritoneoscopy: a novel approach to diagnostic and therapeutic interventions in the peritoneal cavity," *Gastrointestinal Endoscopy*, vol. 60, no. 1, pp. 114–117, 2004.

[4] L. L. Swanstrom, Y. Khajanchee, and M. A. Abbas, "Natural orifice transluminal endoscopic surgery: the future of gastrointestinal surgery," *The Permanente Journal*, vol. 12, no. 2, pp. 42–47, 2008.

[5] G. V. Rao and N. Reddy, "Transgastric appendectomy in humans," in *Proceedings of the Society of American Gastrointestinal and Endoscopic Surgeons (SAGES) Annual Conference*, Dallas, Tex, USA, 2006.

[6] L. L. Swanström, "Natural orifice transluminal endoscopic surgery," *Endoscopy*, vol. 41, no. 1, pp. 82–85, 2009.

[7] F. M. Sánchez-Margallo, J. M. Asencio, M. C. Tejonero et al., "Technical feasibility of totally natural orifice cholecystectomy in a swine model," *Minimally Invasive Therapy and Allied Technologies*, vol. 17, no. 6, pp. 361–364, 2008.

[8] J. Pearl and J. Ponsky, "Natural orifice transluminal endoscopic surgery: past, present and future," *Journal of Minimal Access Surgery*, vol. 3, no. 2, pp. 43–46, 2007.

[9] L. L. Swanstrom, M. Whiteford, and Y. Khajanchee, "Developing essential tools to enable transgastric surgery (NOTES)," *Surgical Endoscopy*, vol. 22, no. 3, pp. 16–20, 2008.

[10] S. J. Bardaro and L. Swanström, "Development of advanced endoscopes for natural orifice transluminal endoscopic surgery (NOTES)," *Minimally Invasive Therapy and Allied Technologies*, vol. 15, no. 6, pp. 378–383, 2006.

[11] D. Rattner and A. Kalloo, "White paper—ASGE/SAGES working group on natural orifice translumenal endoscopic surgery," *Surgical Endoscopy*, vol. 20, no. 2, pp. 329–333, 2006.

[12] A. Degani, H. Choset, A. Wolf, and M. A. Zenati, "Highly articulated robotic probe for minimally invasive surgery," in *Proceedings of the IEEE International Conference on Robotics and Automation (ICRA '06)*, pp. 4167–4172, Orlando, Fla, USA, May 2006.

[13] D. J. Abbott, C. Becke, R. I. Rothstein, and W. J. Peine, "Design of an endoluminal NOTES robotic system," in *Proceedings of the IEEE/RSJ International Conference on Intelligent Robots and Systems (IROS '07)*, pp. 410–416, San Diego, Calif, USA, October 2007.

[14] J. P. Pearl and J. L. Ponsky, "Natural orifice translumenal endoscopic surgery: a critical review," *Journal of Gastrointestinal Surgery*, vol. 12, no. 7, pp. 1293–1300, 2008.

[15] http://www.usgimedical.com/eos/index.htm .

[16] L. Swanström, P. Swain, and P. Denk, "Development and validation of a new generation of flexible endoscope for NOTES," *Surgical Innovation*, vol. 16, no. 2, pp. 104–110, 2009.

[17] J. Yonezawa, M. Kaise, K. Sumiyama, K. Goda, H. Arakawa, and H. Tajiri, "A novel double-channel therapeutic endoscope ("R-scope") facilitates endoscopic submucosal dissection of superficial gastric neoplasms," *Endoscopy*, vol. 38, no. 10, pp. 1011–1015, 2006.

[18] L. L. Swanstrom, R. Kozarek, P. J. Pasricha et al., "Development of a new access device for transgastric surgery," *Journal of Gastrointestinal Surgery*, vol. 9, no. 8, pp. 1129–1137, 2005.

[19] R. A. Cahill, "Natural orifice transluminal endoscopic surgery—here and now," *The Surgeon*, vol. 8, no. 1, pp. 44–50, 2010.

[20] M. C. Meadows and R. S. Chamberlain, "A review on the status of natural orifice transluminal endoscopic surgery (NOTES) cholecystectomy: techniques and challenges," *Open Access Surgery*, vol. 3, pp. 73–86, 2010.

[21] G. Iddan, G. Meron, A. Glukhovsky, and P. Swain, "Wireless capsule endoscopy," *Nature*, vol. 405, no. 6785, pp. 417–418, 2000.

[22] X. Wang and M. Q. H. Meng, "A magnetic stereo-actuation mechanism for active capsule endoscope," in *Proceedings of the 29th Annual International Conference of IEEE-EMBS, Engineering in Medicine and Biology Society (EMBC '07)*, pp. 2811–2814, Lyon, France, August 2007.

[23] M. Quirini, A. Menciassi, S. Scapellato et al., "Feasibility proof of a legged locomotion capsule for the GI tract," *Gastrointestinal Endoscopy*, vol. 67, no. 7, pp. 1153–1158, 2008.

[24] M. E. Rentschler, J. Dumpert, S. R. Platt, S. M. Farritor, and D. Oleynikov, "Natural orifice surgery with an endoluminal mobile robot," *Surgical Endoscopy*, vol. 21, no. 7, pp. 1212–1215, 2007.

[25] M. E. Rentschler and D. Oleynikov, "Recent in vivo surgical robot and mechanism developments," *Surgical Endoscopy*, vol. 21, no. 9, pp. 1477–1481, 2007.

[26] A. C. Lehman, K. A. Berg, J. Dumpert et al., "Surgery with cooperative robots," *Computer Aided Surgery*, vol. 13, no. 2, pp. 95–105, 2008.

[27] J. Cadeddu, R. Fernandez, M. Desai et al., "Novel magnetically guided intra-abdominal camera to facilitate laparoendoscopic single-site surgery: initial human experience," *Surgical Endoscopy*, vol. 23, no. 8, pp. 1894–1899, 2009.

[28] B. C. Shah, S. L. Buettner, A. C. Lehman, S. M. Farritor, and D. Oleynikov, "Miniature in vivo robotics and novel robotic surgical platforms," *Urologic Clinics of North America*, vol. 36, no. 2, pp. 251–263, 2009.

[29] A. C. Lehman, J. Dumpert, N. A. Wood et al., "Natural orifice cholecystectomy using a miniature robot," *Surgical Endoscopy*, vol. 23, no. 2, pp. 260–266, 2009.

[30] S. R. Platt, J. A. Hawks, and M. E. Rentschler, "Vision and task assistance using modular wireless in vivo surgical robots," *IEEE Transactions on Biomedical Engineering*, vol. 56, no. 6, pp. 1700–1710, 2009.

[31] G. Dominguez, L. Durand, J. de Rosa, E. Danguise, C. Arozamena, and P. A. Ferraina, "Retraction and triangulation with neodymium magnetic forceps for single-port laparoscopic cholecystectomy," *Surgical Endoscopy*, vol. 23, no. 7, pp. 1660–1666, 2009.

[32] S. Park, R. A. Bergs, R. Eberhart, L. Baker, R. Fernandez, and J. A. Cadeddu, "Trocar-less instrumentation for laparoscopy: magnetic positioning of intra-abdominal camera and retractor," *Annals of Surgery*, vol. 245, no. 3, pp. 379–384, 2007.

[33] I. S. Zeltser, R. Bergs, R. Fernandez, L. Baker, R. Eberhart, and J. A. Cadeddu, "Single trocar laparoscopic nephrectomy using magnetic anchoring and guidance system in the porcine model," *The Journal of Urology*, vol. 178, no. 1, pp. 288–291, 2007.

[34] D. J. Scott, S. J. Tang, R. Fernandez et al., "Completely transvaginal NOTES cholecystectomy using magnetically anchored instruments," *Surgical Endoscopy*, vol. 21, no. 12, pp. 2308–2316, 2007.

[35] S. L. Best, W. Kabbani, D. J. Scott et al., "Magnetic anchoring and Guidance system instrumentation for laparo-endoscopic single-site surgery/natural orifice transluminal endoscopic surgery: lack of histologic damage after prolonged magnetic coupling across the abdominal wall," *Urology*, vol. 77, no. 1, pp. 243–247, 2011.

[36] T. Fukuda, S. Nakagawa, Y. Kawauchi, and M. Buss, "Structure decision method for self organising robots based on cell structures—CEBOT," in *Proceedings of the IEEE International Conference on Robotics and Automation*, pp. 695–700, Scottsdale, Ariz , USA, May 1989.

[37] E. Susilo, P. Valdastri, A. Menciassi, and P. Dario, "A miniaturized wireless control platform for robotic capsular endoscopy using advanced pseudokernel approach," *Sensors and Actuators A*, vol. 156, no. 1, pp. 49–58, 2009.

[38] K. Harada, E. Susilo, A. Menciassi, and P. Dario, "Wireless reconfigurable modules for robotic endoluminal surgery," in *Proceedings of the IEEE International Conference on Robotics and Automation*, pp. 2699–2704, Kobe, Japan, May 2009.

[39] K. Harada, S. Russo, T. Ranzani, A. Menciassi, and P. Dario, "Design of Scout Robot as a robotic module for symbiotic multi-robot organisms," in *Proceedings of the International Symposium on Micro-NanoMechatronics and Human Science*, November 2011.

[40] E. Diller, C. Pawashe, S. Floyd, and M. Sitti, "Assembly and disassembly of magnetic mobile micro-robots towards deterministic 2-D reconfigurable micro-systems," *The International Journal of Robotics Research*, vol. 30, no. 14, pp. 1667–1680, 2011.

11

Application of On-Board Evolutionary Algorithms to Underwater Robots to Optimally Replan Missions with Energy Constraints

M. L. Seto

Defence R&D Canada, Dartmouth, Nova Scotia, Canada B2Y 3Z7

Correspondence should be addressed to M. L. Seto, mae.seto@dal.ca

Academic Editor: Ivo Bukovsky

The objective is to show that on-board mission replanning for an AUV sensor coverage mission, based on available energy, enhances mission success. Autonomous underwater vehicles (AUVs) are tasked to increasingly long deployments, consequently energy management issues are timely and relevant. Energy shortages can occur if the AUV unexpectedly travels against stronger currents, is not trimmed for the local water salinity has to get back on course, and so forth. An on-board knowledge-based agent, based on a genetic algorithm, was designed and validated to replan a near-optimal AUV survey mission. It considers the measured AUV energy consumption, attitudes, speed over ground, and known response to proposed missions through on-line dynamics and control predictions. For the case studied, the replanned mission improves the survey area coverage by a factor of 2 for an energy budget, that is, a factor of 2 less than planned. The contribution is a novel on-board cognitive capability in the form of an agent that monitors the energy and intelligently replans missions based on energy considerations with evolutionary methods.

1. Introduction

Autonomous underwater vehicles (AUVs) are robots used for underwater tasks that range from surveys, inspection of submerged structures (e.g., pipelines), searching for downed aircraft, tracking oceanographic features, laying undersea cable, undersea mapping, and finding mines, to name a few. Such robots work in an unstructured dynamic environment with unique *perception*, *communication*, and *decision* issues compared to land, air, or space robots. Means for *perception* and detection of underwater targets include magnetic, optical, electric field, thermal (infrared), hydro-dynamic changes (pressure), and sound (acoustic). Sound is unsurpassed, compared to other means, for detection underwater. As an example, the sonar (sound navigation and ranging) is a popular underwater perception sensor that uses sound for detection, classification, and location of underwater targets. Having said that, there are acoustic propagation difficulties in the highly variable, noisy, and reverberant water medium. The ocean is a nonstationary and dynamic environment where the conductivity, temperature, and density of its water varies temporally and spatially and thus affects the propagation of acoustic signals within it. Add

to this multi-path, absorptive losses (high attenuation) [1], and low bandwidth for acoustic signal propagation that also occur in nonpredictable ways. Consequently, underwater *communication* issues stem from the variability and poorness of acoustic propagation in water.

These propagation limitations impact how AUVs are employed since reliable acoustic communications with their operators (or other AUVs) is not easily possible. Consequently, out of the land, ocean, space, and air robot environments, the ocean one is difficult for tasks that require persistent and reliable communications. Robotic autonomy, where the robot makes *decisions* and autonomously alters its mission plans *in situ* without operator intervention, in fulfillment of its mission, is necessary to exploit the potential of underwater robots. Autonomy is one way of coping with the poor underwater communications issue.

Autonomy or *decision* issues are addressed to make AUVs truly autonomous—able to operate long periods without operator intervention. It is desirable that the AUV have the autonomy for decision making or problem solving to deal with unexpected robot or mission events in a timely fashion. Mission autonomy is the ability to adapt a mission to unanticipated conditions in the environment or *in situ*

TABLE 1: Common energy sources for AUVs.

Battery type	Cost ($/Wh)	Energy density (Wh/liter)	Specific energy (Wh/kg)	Typical cell dimensions (mm)
lithium ion	4.27	300	130	$225 \times 212 \times 9.5$
silver zinc	1.75	240	130	$178 \times 161 \times 8$
lead acid	0.17	65	30	$330 \times 228 \times 152$

intelligence that can be exploited to better perform the mission. For example, in-situ environmental measurements can be applied to collect more optimal sonar images. Robot autonomy addresses issues that increase the robot's fault tolerance so it can adapt to unexpected robot events (e.g., more energy consumed earlier in the mission than planned precipitating a shortage for the rest of the mission). Wherever possible, if on-board autonomy can replan, or adapt, in light of unexpected events, the mission can be completed or better performed. Otherwise, the mission could be scrubbed (or in a drastic case, the robot is lost) due to an unexpected event. At the root of the required autonomy or decision-making are cognitive abilities for the robot. A timely scenario that would benefit from such autonomy or decision making ability is discussed next.

Large AUVs, of which the Explorer class [2] is an example, can have ranges on the order of hundreds of kilometers. Such missions can occur over a week or more which is a long time for an ocean environment to remain stationary. Consequently, energy shortages over long deployments are a real possibility. Energy shortages on missions can occur due to unplanned events, as the AUV travels through stronger currents, goes off-course and requires more energy to finish the survey *and* be at the recovery point on time, on-board hotel load increased because instrumentation use was higher (e.g., extra sensors employed), or the AUV was not trimmed for the local salinity conditions (final AUV trim only confirmed when AUV is underway) to name a few. The impact of such unplanned events cumulate over the course of long duration missions to the point where the planned mission may not be achievable anymore. Once the AUV is launched and underway, it is not (easily) possible to monitor these conditions in order to be forewarned of critical issues in time to have an operator recover the AUV to replan the survey. As well, the operator is not necessarily nearby on a support ship. With AUVs now used on longer missions, the issue of unexpected energy shortages is timely and relevant. The different types of AUV on-board energy is briefly discussed.

Like most mobile robots, AUVs carry all their energy on-board for a mission. This energy powers on-board equipment, sensors, computers, propulsion/locomotion, and so forth. On long-duration surveys, the bulk of the energy is for propulsion/locomotion. The on-board volume allocated to carry energy, in the form of batteries, is fixed, so the objective is to carry batteries of the highest specific energy and density possible. Table 1 (not an exhaustive list) shows properties of a few common AUV energy sources. Operating cost and endurance are correlated with the energy source selected. Consequently, endurance is also correlated with AUV size. A fair amount of effort has gone into investigating AUV energy sources [3–5]. They can range from the familiar lead acid battery to modern fuel cells [4, 5]. Fuel cells are an emerging technology that will impact how AUVs are tasked for long survey missions and for allowing on-board computationally intensive calculations. Presently, fuel cells are not commonly used in AUVs. The impact of energy on mission-planning is described next.

AUV path planning (or mission planning) is an active area of research [6–14]. As in air and land robots, underwater path planning objectives include obstacle (moving or stationary) avoidance [6, 7], optimal area coverage [8], and transiting accurately and safely between points with high variability in the water conditions [9–11]. Most related work considers the energy budget and the desire to minimize energy use [9, 10]. However, there is little work that studies replanning specifically due to an unexpected energy shortage on long survey missions. There is even less work on missions that employ side scan sonars.

A mission is planned with the AUV carrying 10% (or more) surplus energy. However, if unexpected events occur, and that surplus is insufficient, there is currently little contingent mission replanning. The usual response is to recover the AUV, replan the mission with an operator, download the mission to the AUV, and redeploy the AUV. Many such survey interruptions are undesirable on long missions as they require additional time, expense, and a nearby support ship. It also detracts from the value of using an AUV to begin with.

This paper's objective is to show that on-board autonomy to replan an AUV side scan sonar mission due to an unexpected AUV energy shortage enhances the mission's success.

This on-board autonomy is in the form of a novel knowledge-based autonomous agent developed to monitor and replan missions due to unexpected energy shortages. The agent replans the mission with a genetic algorithm (GA) that evaluates the fitness of proposed replanned missions to achieve the mission objectives (survey remaining area with less energy than planned and make the recovery schedule). It uses knowledge of the AUV's nonlinear hydrodynamics, dynamics, and control response to proposed replanned missions. Additionally, it takes into account the AUV's real-time attitudes, speed through water, hotel load, and so forth, to evaluate a replanned mission's fitness. The agent continually compares actual energy expended against anticipated energy to expend for the mission. This is achieved through on-line calculations of the remaining energy/endurance to predict and evaluate the AUV likelihood of completing the mission given the measured energy consumption rate. If it appears, at any point in the mission, that the AUV will have insufficient energy to complete its mission as planned, the agent initiates a behavior to replan its mission given the survey area left, the remaining power, and the original mission completion and AUV recovery times.

The rest of the paper is organized as follows. Section 2 defines the mission configuration followed by an overview of the autonomous agent. Then, details of the AUV model (Sections 2.1–2.3) used in the optimization to evaluate the AUV hydrodynamics, dynamics, control response, and subsequent energy consumption to missions proposed by the genetic algorithm are described. Section 2.4 describes the genetic algorithm implementation and how the afore-mentioned AUV models are used as the evaluation function for the GA. Section 3 describes the initial validation of the AUV energy model component in the agent. Section 4 analyzes an illustrative case by imposing a range of available energies, time constraints, and survey areas to evaluate optimal solutions proposed by the agent. It then compares the optimal solutions against what would have been achieved without autonomous replanning of the mission to highlight the agent's effectiveness. Section 5 concludes with a few remarks.

2. Autonomous Agent Description

The objective of this section is to define the mission config-uration, provide an overview of the knowledge-based agent, and then detail the AUV models (dynamics, hydrodynamics, controls, energy) to be used in the GA. This Section con-cludes with an overview description of the GA implemen-tation.

The AUV mission is to survey an area using long straight transits capped by 180-degree turns which put the AUV on a reciprocal heading—the "lawn mower tracks" whose lane spacing defines the mission geometry or configuration (Figure 1). Highlighted in yellow is the swath scanned by the AUV's side-scan sonar on a constant heading. Side-scan sonars scan from both sides to a specific sonar range (shown as 100 m). The objective is to survey/scan an area through side-by-side swaths (strips) with overlap between the swaths (similar to lawn mowing). The overlap is required to co-register features in the images, perform feature-based navigation, and so forth. These constraints govern operation with side scan sonars. For the case study of Figure 1, a projected AUV energy shortage is detected at $(X, Y) = (0, 0)$ m with the AUV recovery point at $(X, Y) = (2000, -2000)$ m to occur no later than t_R of 30,000 seconds later. The arrows show the AUV travel direction.

The swath width relative to the lane spacing determines the amount of area overlap across neighboring swaths, $A_{overlap}$. The area surveyed is the total area scanned by the sonar as it passes through. If all the area is scanned from the start to recovery point, then a total of 4 km² of area has been surveyed (area in turns do not contribute to area surveyed).

When the agent replans a mission, it calculates the energy to transit directly to the recovery point from every point along the mission. The optimal replanned mission surveys until there is just enough energy to branch off and transit to the recovery point. If the mission is optimally planned, the

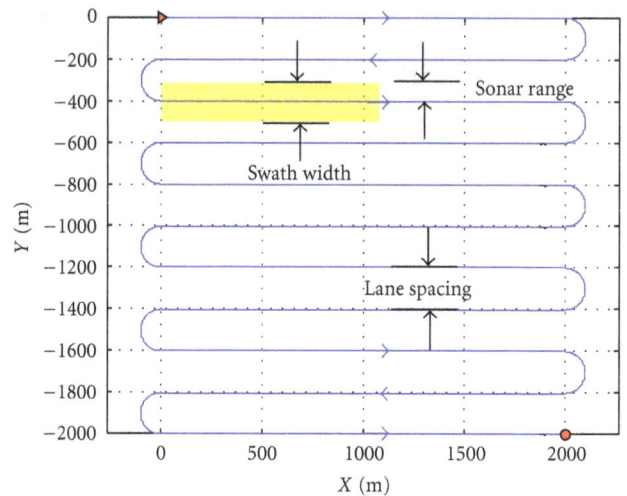

FIGURE 1: AUV side scan sonar survey mission (—) with 200 m lane spacing and 2 km survey leg lengths. Arrow heads indicate AUV travel direction (start point ▶; recovery point •).

branch off to the recovery point occurs near the end of the survey area.

The components of the agent are shown in Figure 2. Upon detection of a projected energy shortage, the agent replans a new mission by first determining dimensions of the remaining area to survey (agent knows survey area corners and AUV location within it) and checks the actual power on-board, and the time left to do the rest of the survey. These parameters are inputs to the genetic algorithm to generate the nominal optimal mission, $[\textbf{MIS}]_o$. $[\textbf{MIS}]_o$ consists of the desired AUV speed v_o, turn diameter D_o, and the total time t_o, to complete this mission. However, this is only a nominal optimal mission as it may or may not be realizable by the AUV. Function, N, takes v_o and D_o and uses them as set point values for the robot to achieve. Speed achieved, v_A, is determined based on the AUV propulsion, dynamics, hydrodynamics, and control as well as the water currents. v_A is then used to determine area surveyed in time t_A with the available energy. Note that there are three different time values, t_R, t_o, and t_A. The way the agent is implemented the recovery time, t_R is the largest of the three.

Finally, objective function, F, is applied to AUV response, $\textbf{R}$, to determine the fitness value, f, for proposed mission $[\textbf{MIS}]_o$. The details of the dynamic and control models used in the on-line evaluation of the AUV's response to a mission are briefly discussed next. This is followed by a description of the genetic algorithm implementation used.

2.1. Hydrodynamic and Dynamic Model. The on-board AUV hydrodynamics, dynamics and control models were imple-mented, and validated using the DRDC *Theseus* underwater vehicle [15] used in Arctic missions for laying cable under ice. The AUV equations of motion (1) for three rotational (yaw, pitch, roll) and three translational degrees-of-freedom are integrated. The notation used is shown in Figure 3.

X, Y, and Z, are the external forces due to the added masses, hydrodynamics, statics, and control fins. The control

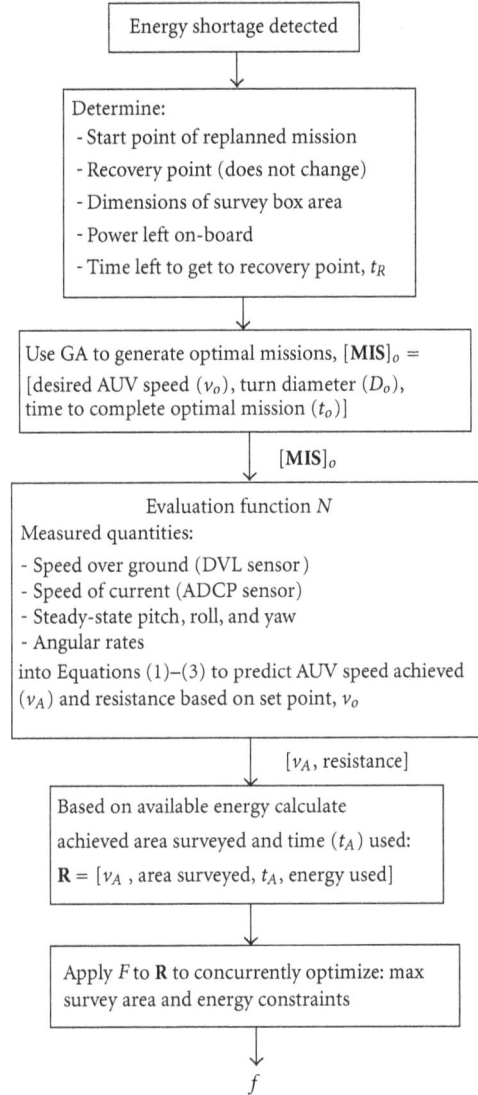

FIGURE 2: Overview of autonomous agent to replan AUV mission upon detection of an energy shortage.

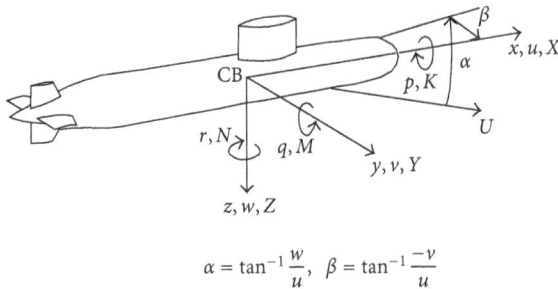

$$\alpha = \tan^{-1}\frac{w}{u}, \quad \beta = \tan^{-1}\frac{-v}{u}$$

FIGURE 3: Nomenclature for AUV motion in body frame axes.

fin damping, natural frequency, and maximum rates are described with a second order model as part of the fin response to commanded deflections (where δ_{bp} = bow port fin, δ_{ss} = stern starboard fin, and δ_r = rudder). K, M, and N are the moments of the roll, pitch, and yaw external forces. For brevity, only the X external forces are detailed in (2)

as it is most relevant to propulsive power (4). The other 5 external forces [16] are omitted here for brevity. Note ρ = water density, and the following pertain to the AUV: l = length, m = mass, W = vehicle weight, B = buoyancy, and X_{prop} = propulsive thrust. A more detailed description of the hydrodynamic terms is available [16]. The propulsion model, captured through X_{prop}, was validated to adequately capture the AUV behavior in level flight, diving, rising, and so forth [15].

$$X = m[\dot{u} - vr + wq - x_G(q^2 + r^2) + y_G(pq - \dot{r})$$
$$+ z_G(pr + \dot{q})],$$

$$Y = m[\dot{v} - wp + ur - y_G(r^2 + p^2) + z_G(qr - \dot{p})$$
$$+ x_G(qp + \dot{r})],$$

$$Z = m[\dot{w} - uq + vp - z_G(p^2 + q^2) + x_G(rp - \dot{q})$$
$$+ y_G(rq + \dot{p})],$$

$$K = I_x\dot{p} + \left(I_z - I_y\right)qr - (\dot{r} + pq)I_{xz} + (r^2 - q^2)I_{yz}$$
$$+ (pr - \dot{q})I_{xy}$$
$$+ m[y_G(\dot{w} - uq + vp) - z_G(\dot{v} - wp = ur)],$$
$$M = I_y\dot{q} + (I_x - I_z)rp - (\dot{p} + qr)I_{xy} + (p^2 - r^2)I_{zx}$$
$$+ (qp - \dot{r})I_{yz}$$
$$+ m[z_G(\dot{u} - vr + wq) - x_G(\dot{w} + uq + vp)],$$
$$N = I_z\dot{r} + \left(I_y - I_x\right)pq - (\dot{q} + rp)I_{yz} + (q^2 - p^2)I_{xy}$$
$$+ (rq - \dot{p})I_{zx}$$
$$+ m[x_G(\dot{v} - wp + ur) - y_G(\dot{u} - vr + wq)]; \tag{1}$$

$$X = -\frac{1}{2}\rho l^4\left[X_{pp}p^2 + X_{rr}r^2 + X_{qq}q^2 + X_{q|q|}q|q|\right]$$
$$+ \frac{1}{2}\rho l^3\left[X_{\dot{u}}\dot{u} + X_{\dot{v}}\dot{v} + X_{\dot{w}}\dot{w}\right]$$
$$+ \frac{1}{2}\rho l^4\left[X_{\dot{p}}\dot{p} + X_{\dot{q}}\dot{q} + X_{\dot{r}}\dot{r}\right]$$
$$+ \frac{1}{2}\rho l^3\left[X_{vr}vr + X_{wq}wq\right]$$
$$+ \frac{1}{2}\rho l^2\left[X_{uu}u^2 + X_{vv}v^2 + X_{ww}w^2 + X_{\delta r\delta r}u^2(\delta r)^2\right.$$
$$+ \frac{1}{2}X_{\delta b\delta b}u^2\left(\delta_{bp}^2 + \delta_{bs}^2\right)$$
$$+ \frac{1}{2}X_{\delta s\delta s}u^2\left(\delta_{sp}^2 + \delta_{ss}^2\right) + X_{\text{prop}}\Big]$$
$$+ (W - B)\sin\theta. \tag{2}$$

2.2. AUV Control Model.

The AUV in this study has two bow fins and a cruciform (+) stern fin configuration. The horizontal fins are used for pitch, roll, and depth control. Yaw (heading) control is achieved through the stern fins in the vertical plane. Both vertical stern fins are deflected together and constitute δ_r. This horizontal control is directly relevant to the mission/path-planning work in this study.

The fins are under closed-loop PID (proportional-integral-differential) control. Corrective fin deflections to minimize the attitude, heading, and depth errors are determined by applying the PID gains matrix to the AUV state vector: $[\psi, \theta, \varphi, \dot{\psi}, \dot{\theta}, \dot{\varphi}, z, \dot{z}]$ (yaw, pitch, roll, yaw rate, pitch rate, roll rate, depth, and depth rate) as shown in (3). $[\psi, \theta, \varphi]$ are the Euler angles that relate inertial frame axes to the body frame axes shown in Figure 3. The PID gains are arrived at through a combination of vehicle dynamic simulations and tuning during sea trials. In both cases, the AUV is commanded to do a variety of maneuvers that include holding and changing course, diving and rising, and turning at different speeds and rudder deflections, and so forth. In the case of the dynamic simulations, the AUV's hydrodynamic derivatives were both predicted and measured [17].

This controller has been studied in some detail and is near optimal in its control authority distribution to maintain the attitude and heading set points [18]:

$$\delta = \begin{bmatrix} \delta_{bs} \\ \delta_{bp} \\ \delta_{ss} \\ \delta_{sp} \\ \delta_r \end{bmatrix} = \begin{bmatrix} \text{PID} \\ \text{GAINS} \\ \text{MATRIX} \end{bmatrix} \times \begin{bmatrix} \psi \\ \theta \\ \phi \\ \cdot \\ \psi \\ \cdot \\ \theta \\ \cdot \\ \phi \\ Z \\ \cdot \\ Z \end{bmatrix}. \tag{3}$$

2.3. AUV Energy Consumption Model.

The energy consumption of the AUV is modeled as [19]:

$$e_t = \frac{\left(p_s + p_p + p_v\right) \times r}{3600 \times V}, \tag{4}$$

such that: e_t = total energy on-board (kWh) [fixed for mission], r = survey range (km) [changes with mission geometry, e.g., in Figure 1], V = AUV speed through water (m/s), p_v = on-board vehicle equipment power (W) [hotel load #1, changes with control plane usage, sensors on different power settings, use of ballast system, intensity of computations, etc. over mission], p_s = sonar power, (W) [hotel load #2, changes with sonar, sonar range, and mission geometry in step function manner], p_p = propulsion power (W) [~AUV resistance (2) × AUV water speed, predict to change as per (1)–(3)]. AUV propulsion power, p_p, is captured through time-varying models based on (1)–(3) which calculate the hydrodynamic resistance in order to predict energy requirements for the remaining mission. As shown in Figure 2, the inputs to these equations are the measured on-board AUV speeds (both speed through water and speed over ground), steady-state yaw, pitch, and roll angles, angular rates, and so forth. The on-board equipment and sensor energy usage is described in terms p_v and p_s, respectively. A running average of the hotel loads are used as measures of p_v and p_s for input into the autonomous agent.

As the AUV speed increases so does the AUV's hydrodynamic resistance ($\sim v^2$) and, consequently, the energy is consumed at a higher rate. This could result in a shorter mission time if it does not consume all the energy before the mission is complete. On long deployments, the propulsion power is often the largest term of the three. It varies as $\sim$ velocity^3 × time. If the AUV speed is fast, it consumes energy quickly and may not survey much area before having to transit to the recovery point—thus mission success is

TABLE 2: Bounds on optimizing parmeters. (sonar range = 50 m on one side → survey swath width = 100 m).

Optimizing parameter	Bounds imposed	
Speed over ground	[0.5–2.0] meters/second	AUV performance
Turn diameter	[50–80] m ~ [50–20] % overlap of neighbour swaths	sensor performance
Time	[80–100] % of max time	mission requirement

limited by energy. If the AUV speed is slow, it may also not cover much area before it has to transit to the recovery point (mission success limited by time). The optimal solution is somewhere in between.

Thus, the agent uses knowledge of the robot performance through the dynamic, hydrodynamic, and controller models with input from measured on-board AUV states, to monitor and predict energy usage to detect an energy shortfall. The energy consumption prediction is performed every τ seconds. Typically, τ is the time to complete 90% of a survey leg which is just prior to the AUV changing heading. It is a good point in a mission to replan and change a mission if required. These energy projections are also compared against a time-dependent battery energy-consumption curve since the consumption rate is nonlinear and specific to a battery.

2.4. Genetic Algorithm Implementation. The application to path-planning of a search procedure based on Darwin's theories of natural selection and survival has been recognized [20]. In these methods, referred to as genetic algorithms, a population of possible solutions is maintained and the paths are iteratively transformed by genetic operations like crossover and mutation [20]. GAs are applied to a variety of path-planning problems [6, 10, 13]. They are especially adept at solving problems with objective functions that are not continuous, differentiable, or possessing a closed tractable form.

This is exactly the case for the AUV energy problem here. The objective function acts on quantities that are obtained through integration of differential equations. As well, the propulsion energy varies as $\sim v^3 \times$ time so the three quantities are inextricably linked, yet they are optimized for a calculated quantity, the area surveyed. The application of genetic algorithms for the posed AUV energy problem, over other methodologies, is quite appropriate.

The optimization objective is to concurrently

(1) maximize area surveyed as shown in Figure 1 (or maximize range, r, in (4)) and

(2) stay within energy budget.

The optimizing parameters are

(i) AUV speed (AUV performance),

(ii) swath width overlap (or turn diameter for a given sonar range) (AUV sensor performance), and

(iii) replanned mission time (mission requirement).

Bounds over which the optimizing parameters can vary are imposed to confine the GA to search space regions with feasible solutions. This is so the agent does not waste time

looking in regions that are known to *not* yield physically achievable solutions. The bounds on the optimizing parameters are shown in Table 2. They are based on at-sea best practices. The speed bounds 0.5 m/s < AUV speed <2 m/s are achievable for the AUV used.

The lane spacing is a function of the sonar swath width and the area overlap between neighboring swaths. While high overlap is conducive to good target detection with side scan sonars, it does not make for an efficient survey. From just geometrical considerations (Figure 1) between overlapping rectangles, the lane spacing = sonar range × 2 × $(1 - A_{overlap})$.

The time to complete the survey and be at the recovery point is a mission requirement. It is desirable that the AUV be at the recovery point more-or-less on time where possible to meet the ship to ensure successful recovery of the AUV and data.

Formally, the multiple objectives are to optimize maximum coverage and to stay within the energy budget and time constraints. The optimization problem is thus posed as

given: objective function $F: A \rightarrow \mathfrak{R}$ from some set A of real numbers $\mathfrak{R}$

find: $[\mathbf{MIS}]_o \in A : F([\mathbf{MIS}]_o) \geq F([\mathbf{MIS}]) \forall [\mathbf{MIS}]$ in A,

where: A is the solution space spanned by solutions, $[\mathbf{MIS}]$ to F. $[\mathbf{MIS}]_o$ is the optimal solution and $\mathfrak{R}$ is the set of real numbers.

Applying function N takes mission parameter vector $[\mathbf{MIS}]_o$ as input to the AUV nonlinear equations of motion (1)–(3) to determine AUV response $\mathbf{R}$, that is,

$$\mathbf{R} = N([\mathbf{MIS}]_o)$$

$$\rightarrow \begin{bmatrix} \text{speed} & \text{area} & \text{time} & \text{energy} \\ \text{achieved} & \text{surveyed} & \text{needed} & \text{used} \end{bmatrix}. \quad (5)$$

Then, objective function F (6) assigns value, f, to response vector, $\mathbf{R}$, as a measure of a proposed $[\mathbf{MIS}]_o$ fitness to achieve the mission objectives. Weights w_i reflect the relative priorities of the different objectives and vary with mission, sensors, AUV, environmental conditions, and so forth:

$$F(\mathbf{R}) = f = w_1 \left(\frac{\text{area}}{\text{surveyed}} \right) + w_2 \left(\frac{\text{time}}{\text{required}} \right)$$

$$+ w_3 \left(\frac{\text{energy left-}}{\text{energy used}} \right) + w_4 \left(\frac{\text{hotel}}{\text{load}} \right) \quad (6)$$

$$+ w_5 \left(\frac{\text{speed achieved-}}{\text{speed desired}} \right).$$

Search space A is pruned to include solutions [**MIS**] whose:

(i) time for a complete area survey (based on proposed speed, lane spacing, etc.) is less than t_R, and

(ii) whose energy for a complete area survey (based on proposed speed, lane spacing, etc.) is less than e_t (4).

To account for operational requirements, the following heuristic constraints are imposed:

(i) energy consumed for a proposed solution (given proposed speed, lane spacing, and time) is less than $0.85 \times$ remaining power (though weight w_3)—as in practice, this saves energy for the AUV to get to the recovery point, and

(ii) the proposed mission time is less than $0.95 \times$ total time allocated (t_R), for the mission (through weight w_2), which ensures the AUV does not arrive at the recovery point too early.

With this scheme, the agent does not produce solutions exceeding energy budget e_t or maximum time t_R.

The genetic algorithm creates a population of solutions, [**MIS**] and uses cross-over and mutation [20] to generate better solutions. Then, it propagates the evolution of the better solutions by choosing the best solutions as parent solutions. Poor solutions, as in nature, can evolve and propagate. However, if sufficient generations are calculated, poor solutions do not survive. A reasonable solution that minimizes (or maximizes) the objective function is an optimal solution. Often there are several objectives to optimize, as in this case.

The genetic algorithm simulates the evolution of an [**MIS**] solution towards the optimal one where "survival of the fittest" is applied to a population of solutions. The steps in the genetic algorithm implementation are as follows:

(1) initialize space A spanned by a population of acceptable solutions, [**MIS**]—(5);

(2) evaluate each solution, perform $F(N[\mathbf{MIS}])$—(6);

(3) select a new population from the old population based on the fitness of the solutions;

(4) apply genetic operators cross-over and mutations to the new population to create new solutions;

(5) evaluate the newly-created solutions by applying $F(N[\mathbf{MIS}])$, and

(6) repeat steps 3–6 until termination criteria which is convergence of the fitness value f.

The initial solution space is seeded with similar historical in-water missions (i.e., good combinations of AUV velocity and swath spacing/turn diameter that surveyed a rectangular area within a prescribed time limit) that worked in the past. The GA achieves solutions with an initial population size of 40 solutions and a maximum 25 generations of evolution.

The chromosome representation to describe members of the solution population was real-valued over binary due to the greater (order of magnitude) computational efficiency of real valued representations [21]. This impacts the implementation of the GA and especially the way the nonuniform mutation and arithmetic cross-over operators are applied. Define parent solutions as vectors $\overline{X} = \{x_1, \ldots, x_i, \ldots, x_n\}$ and $\overline{Y} = \{y_1, \ldots, y_i \ldots, y_n\}$. The parameters of these vectors are AUV speed, swath spacing/turn diameter, and mission completion time. Nonuniform mutation changes one of the parameters of the parent solution based on a non-uniform probability distribution. This Gaussian distribution starts wide and narrows to a point distribution as the current generation approaches the maximum number of generations. Nonuniform mutation randomly selects one of the three (in this case) parameters, and sets it equal to a nonuniform random number:

$$x_i' = \begin{cases} x_i + (b_i - x_i)f(G) & \text{if } r_1 < 0.5 \\ x_i - (x_i + a_i)f(G) & \text{if } r_1 \geq 0.5 \\ x_i, & \text{otherwise,} \end{cases} \quad (7)$$

$$f(G) = \left(r_2 \left(1 - \frac{G}{G_{\max}} \right) \right)^b,$$

r_1, r_2 = uniform random numbers in range $[0, 1]$

G = the current generation number

$G_{\max}$ = maximum number of generations (set by user)

b = shape parameter (set by user)

a_i = minimum possible value for parameter x_i

b_i = maximum possible value for parameter x_i.

Note, there are two possibilities for the nonuniform random number depending on the value of random number r_1.

Arithmetic cross-over produces two complimentary linear combinations of the parent solutions $(\overline{X}, \overline{Y})$ where $r = a$ uniform random number between 0 and 1. The children of the cross-over operation, $(\overline{X}', \overline{Y}')$, are:

$$\overline{X}' = r\overline{X} + (1 - r)\overline{Y},$$
$$\overline{Y}' = (1 - r)\overline{X} + r\overline{Y}. \quad (8)$$

Normalizations on the new (children) solutions are not required—the results are immediately usable.

The posed optimization problem sets up the possibility of local extrema in solution space A. Given the objective is to survey as much area as possible with a fixed energy budget and to arrive at the recovery point by a given time, the optimization can be driven by energy, time, or both. The next section briefly discusses the validation of the AUV energy model.

3. Validation

A basic validation of the energy model in the agent was performed against sea trials data collected with DRDC's *Theseus* AUV [17]. The runs from that trial are maneuvers

TABLE 3: Optimized missions to survey 2 km × 2 km area with variable energy and time = 30,000 seconds until recovery.

Area surveyed (%)	Lane spacing (m)	AUV speed (m/s)	Energy used (kWh)	Energy avail (kWh)	Time-max 30 k sec (10 k sec)
76.34	78.29	1.46	133.0	150	29.837
83.58	79.83	1.59	179.3	200	28.112
91.61	80.00	1.72	231.0	250	28.600
94.98	**79.22**	**1.86**	**290.9**	**300**	**29.402**
98.59	75.91	1.92	335.9	350	29.741
98.76	76.05	1.95	270.3	400	29.652

TABLE 4: Optimized sonar lane spacing and resulting overlap with neighboring lanes (50 m Sonar Range).

Area surveyed (%)	Lane spacing (m)	Survey area overlap (%)
76.34	78.29	21.71
83.58	79.83	20.17
91.61	80.00	20.00
94.98	**79.22**	**20.78**
98.59	75.91	24.09
98.76	76.05	23.95

that are geometrically similar to those in Figure 1. While the resistance of the AUV in straight and level flight is understood and previously validated [17], the energy consumption from propulsion, in turn was not validated until this work. Energy consumption depends on whether the AUV is commanded to maintain closed-loop control over speed or power. In the validation runs, it maintained constant power which means the AUV decreased speed in a turn. The energy consumed was monitored and logged. With all the main components of the autonomous agent described the next section analyzes an illustrative example that highlights the agent's effectiveness.

4. AUV Path Planning Driven by Energy

To assess the agent behaviour, the dimensions of the survey area (X km $\times Y$ km) and the mission time, t_R, until recovery were fixed and the energy budget varied, from insufficient, to a surplus at the time the agent projects an energy shortage. The suspected appearance of local extrema in the solution space did materialize. Its manifestation depended on the energy budget. For the illustrative case, 300 kWh is known to be roughly enough to survey the area based on knowledge of the AUV from sea trials.

4.1. Agent Results and Discussions. The analysis on a representative case is used to highlight the agent's capabilities. The optimal mission parameters for a 2 km × 2 km survey area (sum of total distance travelled = r from (4)) is shown in Table 3. Around the AUV's known, energy budget of 300 kWh (e_t of (4)) is a solution, not necessarily optimal, that can survey the whole area which is both power and time limited. As the energy budget increases, the optimal solution becomes

time-limited, as desired. As the energy budget decreases, it can be time limited without all the energy consumed (150 kWh case) or it can be neither energy or time limited (200 kWh case).

The lane spacing is bound to have between a 20%–50% area overlap with neighboring swaths given the range of the sonar. Obviously, minimal overlap area is desired for a survey since it takes less time. However, a lane spacing that achieves that overlap may or may not give an optimal survey energy-wise given the AUV speed and time constraints. The agent does produce lane spacing that creates overlaps close to the minimum 20% as shown in Table 4.

As shown in Table 3, even with generous energy budgets, the agent surveys no more than 98% of the area. This is due to the way the optimizer works. 98% of the area surveyed is a completed survey of the area. From the results shown in Table 3, if an energy shortage is declared by the agent with less than 300 kWh of energy and only 30,000 seconds left for the remaining mission, the area will not be surveyed completely. However, the AUV will be at the recovery point having achieved the optimal maximal coverage possible shown in column 1 of Table 3.

The fitness values from a proposed [**MIS**] can congregate around two main values depending on whether the GA pursues a time, or energy-limited solution. The fitness f values from these two "sets" of solutions can be different by an order of magnitude (not shown). In that case, the agent is designed to choose the energetically favorable solution.

The trends observed in this case do not change with proportionately different sized survey areas or mission times (not shown). As a benchmark for the agent's performance, the AUV dynamic response component can be computed eight times faster than real time. When the genetic algorithm and energy model components are accounted for, the time to compute a solution is about three times faster than real time. Convergence to a solution typically takes less than 25 generations. Time wise, this is only a little (15%) slower than real-time. These are reasonable responses from an on-board-the-vehicle agent.

The effectiveness of these new cognitive capabilities was evaluated by comparing results of replanned missions against that achieved with a mission that did not adapt to energy shortfalls. The agent effectiveness was measured against the case of the optimal solution at 300 kWh in Table 3. Specifically, a mission speed of 1.86 m/s and a turn diameter of 79.22 m. Normally, the operator will use this mission for the speed, energy, and time constraints.

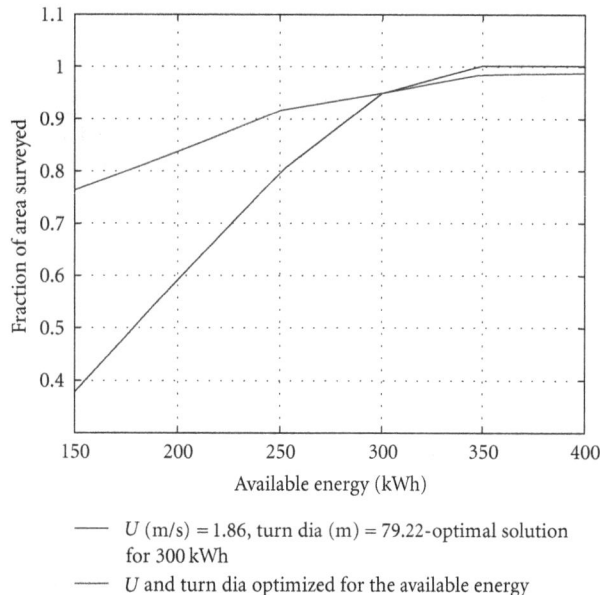

— U (m/s) = 1.86, turn dia (m) = 79.22-optimal solution for 300 kWh

— U and turn dia optimized for the available energy

FIGURE 4: Effectiveness of the autonomous agent to replan AUV missions to adapt to an energy shortage.

As shown in Figure 4, the autonomous agent effectiveness is apparent as the available energy budget drops. The area surveyed with the fixed inflexible mission, near optimal for 300 kWh, and decreases rapidly with decreased available energy (blue plot). The replanned missions (red plot) which optimizes the speed and turn diameter for the available energy perform better. At the lowest energy considered (factor of 2 below sufficient), the surveyed area *improves* over the near optimal 300 kWh mission, by almost a factor of 2.

4.2. Current Work. Encouraged by the results, the current work is implementing the autonomous agent within the DRDC Multi-Agent System framework designed for collaborative underwater vehicles [22]. New developments include collaborative vehicles replanning a mission given a vehicle(s) has an energy shortage. Parallel work uses the agent to replan a mission in the presence of currents measured with an on-board current profiler. At-sea trials are on-going for all developments.

5. Conclusions

With AUVs used on increasingly long deployments, the issue of unexpected energy shortages is timely and relevant. Scripted *a priori* missions based on subsumption architecture [23] cannot adapt to the unstructured dynamic ocean environment and evolving changes within the AUV. This is especially true for AUVs on long deployments. The paper objective was to show that on-board mission replanning for an AUV sensor coverage mission, based on available energy, increases mission success in the event of an energy shortage.

This objective was successfully achieved through on-board cognitive abilities in the form of a novel knowledge-based autonomous agent that replans its missions underway using on-line evolutionary methods to optimize the

replanned mission. The replanned missions take into account the AUV dynamic, hydrodynamic, and control performance, AUV-projected energy consumption, amount of survey area left, amount of available energy, time left for mission completion and AUV recovery, and amount of overlap desired in the side scan sonar images. The agent also makes use of on-line measurements of the AUV attitudes, speed through water, hotel load, and so forth and performs an on-going assessment of the AUV's ability to fulfil the mission energy wise.

This agent was tested on scenarios that varied the energy budget from below sufficient to well-above sufficient. With energy budgets that are insufficient to perform the survey mission, the agent can be either time- or energy-limited. With sufficient energy, the agent uses most of the energy and the time. With surplus energy, the agent is time limited. These are acceptable solutions for the replanned mission. For the illustrative case studied, the replanned mission can improve the survey area coverage by a factor of 2 for an energy budget that is a factor of 2 less than planned.

An effective on-board knowledge-based agent that can autonomously replan an optimal mission to intelligently adapt to an unexpected energy shortage has not been previously reported—especially one with considerations for side scan sonar requirements.

Acknowledgment

This work was supported in part by Defence R&D Canada.

References

[1] A. D. Waite, *SONAR for Practising Engineers*, John Wiley & Sons, New York, NY, USA, 3rd edition, 2002.

[2] T. Crees, C. Kaminski, J. Ferguson et al., "Preparing for UNCLOS—an historic AUV deployment in the Canadian high arctic," in *Proceedings of the Oceans / MTS Conference*, p. 8, 2010.

[3] J. G. Hawley and G. T. Reader, "A knowledge-based aid for the selection of autonomous underwater Vehicle Energy Systems," in *Proceedings of the Symposium on Autonomous Underwater Vehicle Technology (AUV '92)*, pp. 177–180, 1992.

[4] O. Hasvold, K. H. Johansen, and K. Vestgaard, "The alkaline aluminium hydrogen peroxide semi-fuel cell for the hugin 3000 autonomous underwater vehicle," in *Proceedings of the Workshop on Autonomous Underwater Vehicles (AUV '02)*, pp. 89–94, June 2002.

[5] I. Yamamoto, T. Aoki, S. Tsukioka et al., "Fuel cell system of AUV Urashima," in *Proceedings of the MTS/IEEE Oceans Conference*, pp. 1732–1737, November 2004.

[6] Z. Chang, Z. Tang, H. Cai, X. Shi, and X. Bian, "GA path planning for AUV to avoid moving obstalces based on forward looking sonar," in *Proceedings of the 4th International Conference on Machine Learning and Cybernetics (ICMLC '05)*, pp. 1498–1502, August 2005.

[7] H. Kawano and T. Ura, "Navigation algorithm for autonomous underwater vehicle considering cruising mission using a side scanning SONAR in disturbance," in *Proceedings of the MTS/IEEE Oceans*, vol. 1, pp. 403–440, November 2001.

[8] A. Kim and R. M. Eustice, "Toward AUV survey design for optimal coverage and localization using the Cramer Rao lower

bound," in *Proceedings of the MTS/IEEE Oceans Conference*, pp. 1–7, Biloxi, Miss, USA, October 2009.

[9] A. Alvarez, A. Caiti, and R. Onken, "Evolutionary path planning for autonomous underwater vehicles in a variable ocean," *IEEE Journal of Oceanic Engineering*, vol. 29, no. 2, pp. 418–429, 2004.

[10] D. Kruger, R. Stolkin, A. Blum, and J. Briganti, "Optimal AUV path planning for extended missions in complex, fast-flowing estuarine environments," in *Proceedings of the IEEE International Conference on Robotics and Automation (ICRA '07)*, pp. 4265–4270, 2007.

[11] G. Yang and R. Zhang, "Path planning of AUV in turbulent ocean environments used adapted inertiaweight PSO," in *Proceedings of the 5th International Conference on Natural Computation (ICNC '09)*, pp. 299–302, August 2009.

[12] K. Carroll, S. McClaran, E. L. Nelson et al., "AUV path planning: an A* approach to path planning with consideration of variable vehicle speeds and multiple overlapping, time-dependent exclusion zones," in *Proceedings of the Symposium on Autonomous Underwater Vehicle Technology (AUV '92)*, pp. 79–84, Washington, DC, USA, 1992.

[13] W. Hong-jian, A. Jie, B. Xin-qian, and S. Xiao-cheng, "An improved path planner based on adaptive genetic algorithm for autonomous underwater vehicle," in *Proceedings of the IEEE International Conference on Mechatronics and Automation (ICMA '05)*, pp. 857–861, Niagara Falls, Canada, 2005.

[14] Z. Chang, M. Fu, Z. Tang, and H. Cai, "Autonomous mission management for an unmanned underwater vehicle," in *Proceedings of the IEEE International Conference on Mechatronics and Automation (ICMA '05)*, pp. 1456–1459, August 2005.

[15] J. Thorliefson, T. Davies, M. Black et al., "The theseus autonomous underwater vehicle: a Canadian success story," in *Proceedings of the IEEE/MTS Oceans Conference and Exhibition*, pp. 1001–1008, 1997.

[16] J. Feldman, "DTNSRDC Revised standard submarine equations of motion," Tech. Rep. DTNSRDC/SPD-0393-09, p. 31, 1979.

[17] M. L. Seto and G. D. Watt, "Dynamics and control simulator for the THESEUS AUV," in *Proceedings of the 10th International Conference of Society of Offshore and Polar Engineers*, p. 6, Montreal, Canada, 2000.

[18] M. L. Seto, "An agent to optimally re-distribute control in an underactuated AUV," *International Journal of Intelligent Defence Support Systems*, vol. 4, no. 1, pp. 3–19, 2010.

[19] ISE Ltd., Design an Explorer, 2000.

[20] D. E. Goldberg, *Genetic Algorithms in Search, Optimization and Machine Learning*, Addison-Wesley, Boston, Mass, USA, 1989.

[21] A. Michaelewicz, *Genetic Algorithms + Data Structures = Evolution Programs*, AI, Springer, New York, NY, USA, 1994.

[22] H. Li, A. Popa, C. Thibault, and M. Seto, "A software framework for multi-agent control of multiple autonomous underwater vehicles for underwater mine counter-measures," in *Proceedings of the IEEE International Conference on Autonomous and Intelligent Systems (AIS '10)*, pp. 1–6, Povoa de Varzim, Portugal, June 2010.

[23] R. Brooks, "A robust layered control system for a mobile robot," *IEEE Transactions on Robotics and Automation*, vol. RA-2, no. 1, pp. 14–23, 1986.

12

Geometric Parameter Identification of a 6-DOF Space Robot Using a Laser-Ranger

Yu Liu,[1] Zainan Jiang,[1] Hong Liu,[1] and Wenfu Xu[2]

[1] *State Key Laboratory of Robotics and System, Harbin Institute of Technology, Harbin 150001, China*
[2] *Mechanical Engineering and Automation, Harbin Institute of Technology Shenzhen Graduate School, Shenzhen 518057, China*

Correspondence should be addressed to Yu Liu, lyu11@hit.edu.cn

Academic Editor: Zhuming Bi

The geometric parameters of a space robot change with the terrible temperature change in orbit, which will cause the end-effector pose (position and orientation) error of a space robot, and so weakens its operability. With this in consideration, a new geometric parameter identification method is presented based on a laser-ranger attached to the end-effector. Then, independence of the geometric parameters is analyzed, and their identification equations are derived. With the derived identification Jacobian matrix, the optimal identification configurations are chosen according to the observability index O_3. Subsequently, through simulation the geometric parameter identification of a 6-DOF space robot is implemented for these identification configurations, and the identified parameters are verified in a set of independent reference configurations. The result shows that in spite of distance measurement alone, pose accuracy of the space robot still has a greater improvement, so the identification method is practical and valid.

1. Introduction

Repeatability of the robot only represents the ability that the robot follows the same trajectory or gets to the same desired poses time after time, so it more indicates compactness of the robot. Comparatively, pose accuracy of the robot describes how close the end effector true pose is to desired pose. Good repeatability is the premise of high accuracy for a robot. Generally, for such simple tasks as conveying goods, spraying paint, or welding an automobile, high repeatability is already enough, because these jobs can be completed through teaching and playback. However, in some other occasions, for example, the medical robot bores a hole on the bone for a patient with the aid of X-ray image, or more typically the space robot guided by hand-eye vision maintains a faulty space vehicle, in this case it is necessary to map the end effector Cartesian coordinates into the joint coordinates, namely, the joint angles must be evaluated through inverse kinematics. However, subject to difference between the nominal geometrical parameters of the robot links and their true parameters, the calculated joint angles do not correspond with the desired ones, which cause the

end effector pose errors. At the same time, pose errors may result from nongeometrical errors, for example, joint and link deformation, transmission, and temperature.

Consequently, the robot kinematic parameter identification must be done to improve the end effector pose accuracy before it is used. Virtually, parameter identification is a software compensation algorithm, because it only seeks for the true kinematic parameters and does not physically change the links, joints, and controllers of the robot. It can be divided into two categories, that is, geometrical parameter identification and nongeometrical one. Most researchers concentrate on the former. Veitschegger and wu [1] developed a method of kinematic calibration and compensation, and with the least square algorithm calibrated the PUMA 560 experimentally. The experiment results showed that a greater than 70 times improvement in Cartesian pose errors resulted from the calibrated versus the nominal manipulator. Stone et al. [2, 3] modeled kinematics errors using a six-parameter "S-model" per link, then they introduced three features of the robot to estimate the 6n S-model parameters. Lukas Beyer and Wulfsberg [4] developed an ROSY calibration system with two CCD cameras and a reference sphere that enabled pose accuracy to be improved for conventional arms and parallel robots. Sun and Hollerbach [5] presented an active robot calibration algorithm using the determinant-based updating observability index and demonstrated it through simulation with a 6-DOF PUMA 560 robot. Kang et al. [6] introduced a new metrology method based on the product-of-exponential formula and the modified dyad kinematics to calibrate a modular robot, but there were no calibration results to be given.

Research on nongeometrical parameter identification has also made great progress. Chen and Chao [7] presented a six-parameter error model between two consecutive links in a general sense and developed a mathematical identification model composed of nongeometrical parameters, it considered the second, the third joint, and the link flexibility due to gravity. Judd and Knasinski [8] analyzed nongeometrical errors (gear train errors, joint, and link flexibility, etc.) and proposed an error model that can be identified with a common least squares procedure. Chunhe Gong et al. [9] built a comprehensive error model including geometric errors, position-dependent compliance errors, and time-variant thermal errors, and robot accuracy was improved by an order of magnitude after calibration. Lightcap et al. [10] applied a 30-parameter flexible geometric model to the Mitsubishi PA10-6CE robot, considering the flexibility in the harmonic drive transmission. Drouet et al. [11] decomposed the measured end-point error into generalized geometric and elastic errors and realized compensation for dynamic elastic effects. With a camera attached to the end effector, Radkhah et al. [12] used an extended forward kinematic model incorporating both geometric and nongeometric parameters to identify the KUKA KR 125/2 robot kinematic parameters.

Space robots lie in microgravity environments and move slowly, so nongeometrical errors due to joint and link flexibility will occupy a small proportion, and here they are omitted. However, subject to extreme temperature under space environment, the geometric parameters of space robots will have a great change. The extravehicular temperature scope in orbit is ±80°C or so, and the inner temperature scope of the space robot is −30°C~+50°C under the condition of temperature control. For a two-meter robotic arm, its maximum length variation is 2 mm or so. Besides, there is a temperature difference between the lighted surface of the space robot and the shady surface, which will cause deformation of the space robot. So, a space robot calibrated on the ground must be recalibrated on orbit to improve its pose accuracy. Sometimes, the space robot will carry a laser-ranger attached to its end-effector to detect the manipulated objects [13, 14], using it the paper will discuss geometric parameter identification of the on-orbit space robot, and give the simulation results.

2. Kinematic Model of the Space Robot

2.1. Outline of Identification Scheme. As shown in Figure 1, the space robot is fixed on the +Z surface (pointing to the center of the earth) of the satellite, and its end-effector carries a laser-ranger that is used to measure the distance from the

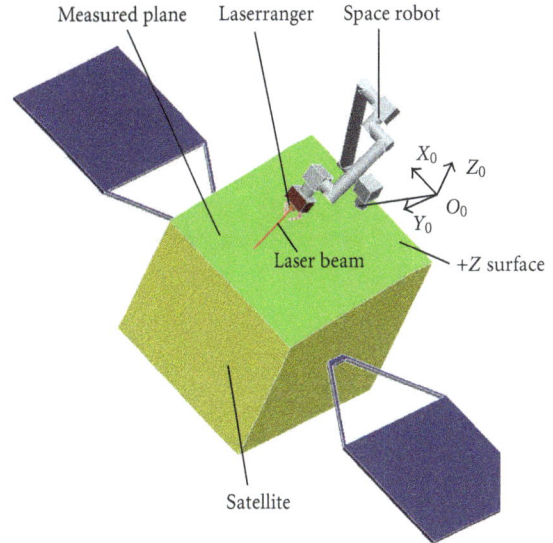

FIGURE 1: Sketch of parameter identification scheme of the space robot.

starting point of the laser beam to the measured declining plane. Because the equation of the plane with respect to the base coordinate frame is known and the starting point and the equation of the laser beam (line) with respect to the tool frame can be calibrated beforehand, so the distance can also be estimated according to the kinematic model. However, the model is inaccurate because of the geometric parameter errors of the space robot, so there exists the difference between the measured distance and that calculated with the nominal kinematic parameters, which is used to identify the geometric parameters of the space robot. Some other parameter identification methods [15–17] using a laser-ranger generally measured the distance from the robot end-point to a known object point, however, it was difficult to determine whether the laser beam just passed through the object point in practice. In the literature [15], the position-sensitive detector (PSD) was adopted, which increased complexity of parameter identification. Here, the known declining plane is chosen as the object measured by the laser-ranger, which simplifies the measurement scheme. In the literature [17], the laser spot was measured by a camera, which introduced measurement noise of the camera.

2.2. Kinematic Model. Commonly, with the D-H parameter method, the relative translation and rotation from the robot link frame $i - 1$ to the frame i can be described by a homogeneous transformation matrix ${}^{i-1}\mathbf{A}_i$ as

$$
{}^{i-1}\mathbf{A}_i = \begin{bmatrix} C\theta_i & -C\alpha_i S\theta_i & S\alpha_i S\theta_i & a_i C\theta_i \\ S\theta_i & C\alpha_i C\theta_i & -S\alpha_i \cos\theta_i & a_i S\theta_i \\ 0 & S\alpha_i & C\alpha_i & d_i \\ 0 & 0 & 0 & 1 \end{bmatrix}, \quad (1)
$$

where, $C\theta_i$ denotes $\cos(\theta_i)$, $S\theta_i$ represents $\sin(\theta_i)$, and the rest may be deduced by analogy. ${}^{i-1}\mathbf{A}_i$ includes four kinematic parameters, namely θ_i, d_i, a_i, and α_i. However, when a small angle variation creates between two consecutive parallel axes or near parallel axes, with the D-H method, it will lead to a large variation of the parameter d_i, in other words, in this case the axial offset d_i is very sensitive to the twist α_i. In view

of this, the matrix $^{i-1}\mathbf{A}_i$ is post-multiplied the matrix $^{i-1}\mathbf{A}_i$ by an additional rotational matrix $\text{Rot}(y,\beta_i)$ [18], namely the matrix $^{i-1}\mathbf{A}_i$ can be changed as

$$^{i-1}\mathbf{A}_i \leftarrow {}^{i-1}\mathbf{A}_i \cdot \text{Rot}(y,\beta_i) = \begin{pmatrix} C\theta_i C\beta_i - S\theta_i S\alpha_i S\beta_i & -S\theta_i C\alpha_i & C\theta_i S\beta_i + S\theta_i S\alpha_i S\beta_i & a_i C\theta_i \\ S\theta_i C\beta_i + C\theta_i S\alpha_i S\beta_i & C\theta_i C\alpha_i & S\theta_i S\beta_i - C\theta_i S\alpha_i C\beta_i & a_i S\theta_i \\ -C\alpha_i S\beta_i & S\alpha_i & C\alpha_i C\beta_i & d_i \\ 0 & 0 & 0 & 1 \end{pmatrix}, \tag{2}$$

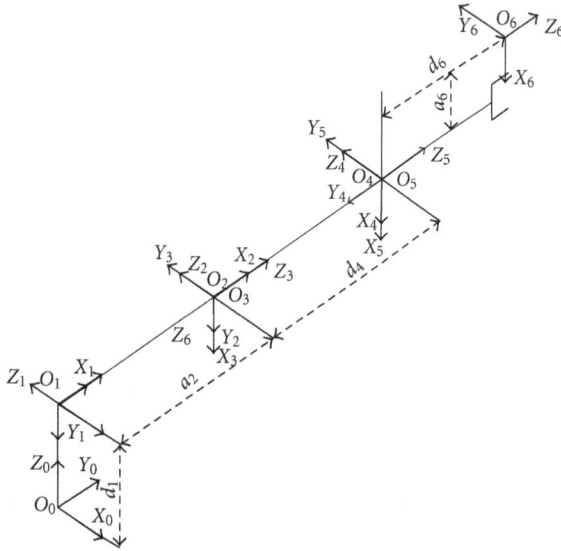

FIGURE 2: Representation of the coordinate frames of the space robot with D-H convention.

where

$$\text{Rot}(y,\beta_i) = \begin{bmatrix} C\beta_i & 0 & S\beta_i & 0 \\ 0 & 1 & 0 & 0 \\ -S\beta_i & 0 & C\beta_i & 0 \\ 0 & 0 & 0 & 1 \end{bmatrix}. \tag{3}$$

Supposed that $^0\mathbf{A}_1$ represents the transformation matrix from the base coordinate frame to the frame 1, in terms of Figure 2, the transformation matrix $\mathbf{T}_N$ from the base coordinate frame to the tool frame can be obtained from the well known loop closure equation:

$$\mathbf{T}_N = {}^0\mathbf{A}_n = {}^0\mathbf{A}_1\,{}^1\mathbf{A}_2 \cdots {}^5\mathbf{A}_6. \tag{4}$$

Further, the matrix $\mathbf{T}_N$ can be divided into the following submatrix:

$$\mathbf{T}_N = \begin{pmatrix} \mathbf{R}_N & \mathbf{p}_N \\ 0 & 1 \end{pmatrix}, \tag{5}$$

where $\mathbf{R}_N \in R^{3\times 3}$ is an orientation matrix of the tool frame with respect to the base frame, $\mathbf{p}_N \in R^3$ denotes the translational vector.

The configuration of the space robot is shown in Figure 2, the tool frame $O_6 - X_6 Y_6 Z_6$ of the space robot can be chosen

arbitrarily. Here, we might as well choose the laser-ranger coordinate frame fixed to the end-effector as the tool frame, namely, the starting point of the laser beam is located in the origin O_6 and the positive direction of the Z_6 axis acts as the emission direction of the laser beam, which helps to simplify identification process and decrease the complexity of robot identification.

3. Identification Model of the Geometric Parameters

3.1. Independent Parameters of the Identification Model. A complete kinematic model consists of a certain number of the independent parameters. If the model exceeds the scope, they will be relative. Therefore, extra increment of the number of the parameters is insignificant to improvement of pose accuracy. Everett et al. [19] gave the following calculative formula:

$$C = 4R + 2P + 6, \tag{6}$$

where, C denotes the number of independent parameters (also constrained equations), R is the number of the revolved joints, and P is the number of the translational joints. Besides, Figure 6 of (6) represents 6 constraints that determine the pose of the tool frame with respect to the link 5 frame $O_5 - X_5 Y_5 Z_5$. According to the above equation, the space robot shown in Figure 2 totally has 30 independent geometric parameters. However, different from a laser tracker to measure a 6-dimension pose of the robot, the laserranger only measures the distance from the origin of the laserranger coordinate frame to the objective point. Obviously, an arbitrary equal-distance rotation of the end-effector around the target point creates no significance to output of the laserranger, in other words, in the spherical surface whose spherical center is the target point, and its radius is the measured distance, however, the tool coordinate frame moves, the measured distance is same, and it means that the orientation of the end-effector cannot be constrained. In addition, a distance equation only constrains one of the three coordinates for a point, while the other two coordinates are free. Namely, compared with the laser tracker, the laserranger loses five constraints, and maximally there are 25 identifiable parameters for the space robot.

3.2. Identification Equation. According to (5) and the above laser-ranger coordinate frame, it is easy to know that the starting point $\mathbf{p}_s$ of the laser beam with respect to the base coordinate frame is equivalent to the translational vector $\mathbf{p}_n$.

Similarly, the laser beam unit vector $\mathbf{b}_l$ relative to the base coordinate frame is expressed as

$$\mathbf{b}_l = \mathbf{R}_N \begin{bmatrix} 0 \\ 0 \\ 1 \end{bmatrix}. \tag{7}$$

It is assumed that the measured plane equation in the base coordinate frame is

$$\mathbf{n}_l \cdot \mathbf{p} + f = 0, \tag{8}$$

where $\mathbf{n}_l(n_{lx}, n_{ly}, n_{lz})$ is the unit normal vector of the measured plane, its positive direction can be chosen arbitrarily, here, n_{lz} is given a positive value. $\mathbf{p}$ denotes the coordinate vector (p_x, p_y, p_z) of the arbitrary point in the plane, and f is a known scalar. Supposed that the laser beam vector $\mathbf{b}_l$ intersects the measured plane at the point $\mathbf{p}_j$, then according to the relation of the vectors, $\mathbf{p}_j$ can be written as

$$\mathbf{p}_j = \mathbf{p}_s + h\mathbf{b}_l, \tag{9}$$

where h denotes the distance from the starting point $\mathbf{p}_s$ of the laser beam to the intersectant point $\mathbf{p}_j$. As We know, $\mathbf{p}_j$. Also meets (8), then by substituting (9) into (8) h can be expressed as

$$h = -\frac{\mathbf{n}_l \cdot \mathbf{p}_s + f}{\mathbf{n}_l \cdot \mathbf{b}_l}. \tag{10}$$

The distance h in (10) is an estimated value based on the nominal geometric parameters of the space robot and the nominal plane equation. As stated previously, these geometric parameters on space orbit generally deviate from the nominal ones. The geometric errors in the link i are. respectively, written as $\Delta\theta_i$, Δd_i, Δa_i, $\Delta\alpha_i$, and $\Delta\beta_i$. Here, it is assumed that they are small amount, so a linear model can be developed for simplicity. If the true parameters of the link i are, respectively, given as θ_i^r, d_i^r, a_i^r, α_i^r, and β_i^r, there are the following relations:

$$\theta_i^r = \theta_i + \Delta\theta_i \qquad d_i^r = d_i + \Delta d_i,$$
$$a_i^r = a_i + \Delta a_i \cdots \alpha_i^r = \alpha_i + \Delta\alpha_i, \qquad \beta_i^r = \beta_i + \Delta\beta_i. \tag{11}$$

Differentiate (10), then

$$\Delta h = h^r - h$$
$$\approx \frac{\partial h}{\partial\theta_1}\Delta\theta_1 + \frac{\partial h}{\partial d_1}\Delta d_1 + \frac{\partial h}{\partial a_1}\Delta a_1 + \frac{\partial h}{\partial\alpha_1}\Delta\alpha_1$$
$$+ \frac{\partial h}{\partial\beta_1}\Delta\beta_1 + \cdots + \frac{\partial h}{\partial\theta_6}\Delta\theta_6 + \frac{\partial h}{\partial d_6}\Delta d_6 + \frac{\partial h}{\partial a_6}\Delta a_6$$
$$+ \frac{\partial h}{\partial\alpha_6}\Delta\alpha_6 + \frac{\partial h}{\partial\beta_6}\Delta\beta_6 + \frac{\partial h}{\partial n_{ly}}\Delta n_{ly} + \frac{\partial h}{\partial n_{lz}}\Delta n_{lz}, \tag{12}$$

where, h^r denotes the actual distance. Attentively, the above listed geometric parameters of the space robot amount to

32, but it does not mean that all these parameters can be identified, only for convenience. Equation(12) considers the influence of variation of the plane equation. Because $\mathbf{n}_l$ is a unit vector, the two of its three components are independent. Here we choose n_{ly} and n_{lz} as the parameters to be identified. Attentively, the parameter f is unidentifiable, because $-f/n_{lz}$ represents the intercept that the plane intersects the coordinate Z_0 axis, and obviously it is associate with the parameter d_1. Of course, the above explanation assumes that the measured plane is not parallel to the Z_0 axis. Besides, the roughness of the plane will also weaken accuracy of the measurement, a good choice is that it is classified as measurement noise. The number of the identification equation must be greater than that of the identified geometric parameters. Obviously, only (12) is not enough. Simply, the more identification configurations are chosen to obtain the more identification equations. Through combining these equations the following formula can be given:

$$\Delta\mathbf{h} = \mathbf{G}\Delta\mathbf{e}, \tag{13}$$

where $\Delta\mathbf{h}$ is the distance error vector, $\Delta\mathbf{h} = [\Delta h_1 \Delta h_2 \cdots \Delta h_m]$, m denotes the mth measurement configuration, $\Delta\mathbf{e}$ is the parameter error vector, $\Delta\mathbf{e} = [\Delta\theta_1, \Delta d_1, \Delta a_1, \Delta\alpha_1, \Delta\beta_1, \ldots, \Delta n_{ly}, \Delta n_{lz}]$, $\mathbf{G}$ is the identification Jacobian matrix. According to (13), through iteration, we can identify the geometric parameters of the space robot and the measured plane.

4. Simulation of Parameter Identification

4.1. Optimal Experimental Design. The different measurement configurations have a certain impact on identification results. So, the selection of the measurement configurations is also important. At present, there are several proposed observability indexes to evaluate a set of measurement configurations. Since E-optimality is the best criterion to minimize the uncertainty of the end-effector pose of a robot and the variance of the parameters [20], it is used as the observability index of the optimal experimental design. Its objective function is to maximize the minimum singular value of the identification Jacobian matrix, and it can be written as

$$O_3 = \max \sigma_{\min}(\mathbf{G}). \tag{14}$$

According to (14), when there are many sets of measurement configurations to be chosen, the set whose minimum singular value is maximal is the optimal experimental design.

4.2. Measurement Noise. There are usually some errors in the distance values measured by the laser-ranger, which will create disadvantageous effects on the geometric parameter identification of the space robot. In order to simulate the real case, measurement noise should be added to the error model so as to calibrate the space robot more exactly. Here, it is assumed that distance measurement noise follows a normal distribution with zero mean and standard deviation 0.2 mm.

For the same configuration, the more distance measurements will be taken to reduce disturbance of the stochastic

TABLE 1: Nominal D-H parameters of the space robot.

Link Number	θ_n/rad	α_n/rad	a_n/m	d_n/m	$\Delta\beta_n$/rad
1	$\pi/2$	$-\pi/2$	0	0.5	—
2	0	0	1	—	0
3	$-\pi/2$	$\pi/2$	0	0	—
4	0	$-\pi/2$	0	-0.8	—
5	π	$\pi/2$	0	0	—
6	0	0	-0.12	0.4	0

TABLE 2: Pre-assumed geometrical Parameter Errors.

Link Number	$\Delta\theta_n$/mrad	$\Delta\alpha_n$/mrad	Δa_n/mm	Δd_n/mm	$\Delta\beta_n$/mrad
1	-7.23	-3.22	0.23	0.73	—
2	0.52	0.13	1.94	—	1.45
3	0.56	-2.23	0.11	0.34	—
4	0.36	1.92	0.18	1.35	—
5	-5.52	-4.83	0.27	0.29	—
6	-0.34	0.62	0.47	0.85	-3.36

measurement noise, then the average of these measurements is provided as the measurand. On the other hand, the more redundant measurement configurations are used to identify the geometric parameters of the space robot, which has also an effect on filtering measurement noise.

4.3. Simulation Approach. According to the description above, the simulation approach of parameter identification of the space robot can be summarized as shown in Figure 3. Because the identification method is verified through simulation, a distance value calculated with the preassumed true parameters and the above-mentioned measurement noise will be used as a measurement value, and it is equal to the sum of the real value plus measurement noise. Besides, here, the estimated distance denotes the distance calculated with the nominal geometrical parameters.

4.4. Initial Condition. The nominal D-H parameters of the space robot are shown in Table 1 and its preassumed geometrical parameter errors are shown in Table 2.

In view of the space robot working on orbit lighted by the sun, the above length errors Δa_n and Δd_n are given a positive number in relation to their lengths, while the angle errors $\Delta\alpha_n$, $\Delta\beta_n$, and $\Delta\theta_n$ are given based on a normal distribution with zero mean and standard deviation 3.49 mrad. Attentively, the geometric parameters marked "—" in Table 2 are unidentifiable, so the identifiable parameters of the space robot amount to 25.

Besides, the measured plane equation is chosen as

$$y + 4.6z - 0.69 = 0. \tag{15}$$

Attentively, as shown in Figure 1, the equation cannot be given such the form as $z + f = 0$, or it will make three

geometric parameters of the space robot unidentifiable, that is, θ_1, a_1, d_3. Obviously, if the measured plane is perpendicular to the Z_0 axis, the three parameters will make no difference to the measured distance, which will weaken completeness of the identified geometric model. Because the measured plane expressed by (15) is parallel to the X_0 axis, for simplicity, here we only give the coefficient 4.6 an error, it is 0.005.

4.5. Simulation Result. Subsequently, the above geometric parameter identification algorithm will be verified through simulation. Here, we have chosen 101 measurement configurations in all where the space robot is nonsingular. Then, the two cases will be simulated, namely, 50 configurations 10 repetitions (the first case) and 100 configurations 10 repetitions (the second case), x repetitions denote the number of repeated measurements for a same measurement configuration. As stated in the Section 4.1, according to the optimal experimental design criterion, we will calculate C_{101}^{100} minimum singular values of **G** for the first case, similarly for the second case, it is C_{101}^{50} ones which are a huge number, and the task is difficult to come true. In fact, with the observability index O_3, we calculate a part of the minimum singular values for the first case and all of them for the second case in simulation. According to the calculation results, the observability indexes of the above two cases are equal to 0.048 and 0.180, respectively.

Besides, a set of independent validation configurations (20 configurations) distributing in the whole workspace of the space robot are selected to evaluate the identification effect. In nature, parameter identification is a fit for the measured data in the measurement configurations, so the extra validation configurations are necessary.

Figure 4 represents the distance errors in the measurement configurations, respectively, with the nominal parameters, the identified parameters for the first and second cases. It is easy to find that, after parameter identification, the maximum distance error in the measurement configurations decreases to less than 0.4 mm for the first case and to less than 0.2 mm for the second case, compared with more than 40 mm prior to parameter identification, so the parameter identification is a very good fit for the distance measurement values. At the same time, the maximum distance error with the identified parameters for the second case is less than that for the first case, which reflects the importance of more identification configurations. Of course, after identification, there still exist some fractional residual distance errors, which mainly come from measurement noise.

The position errors in the measurement configurations with the nominal, and the identified parameters for the first and second cases, are depicted in Figure 5, and the orientation errors are in Figure 6. Correspondingly, the position errors in the validation configuration are depicted in Figure 7, and the orientation errors are in Figure 8. In general, after parameter identification, pose accuracy of the space robot has a great improvement, for example, the position errors in the identification configuration are reduced from more than 15 mm to less than 1.5 mm for the second case and the orientation errors from 15 mrad or so

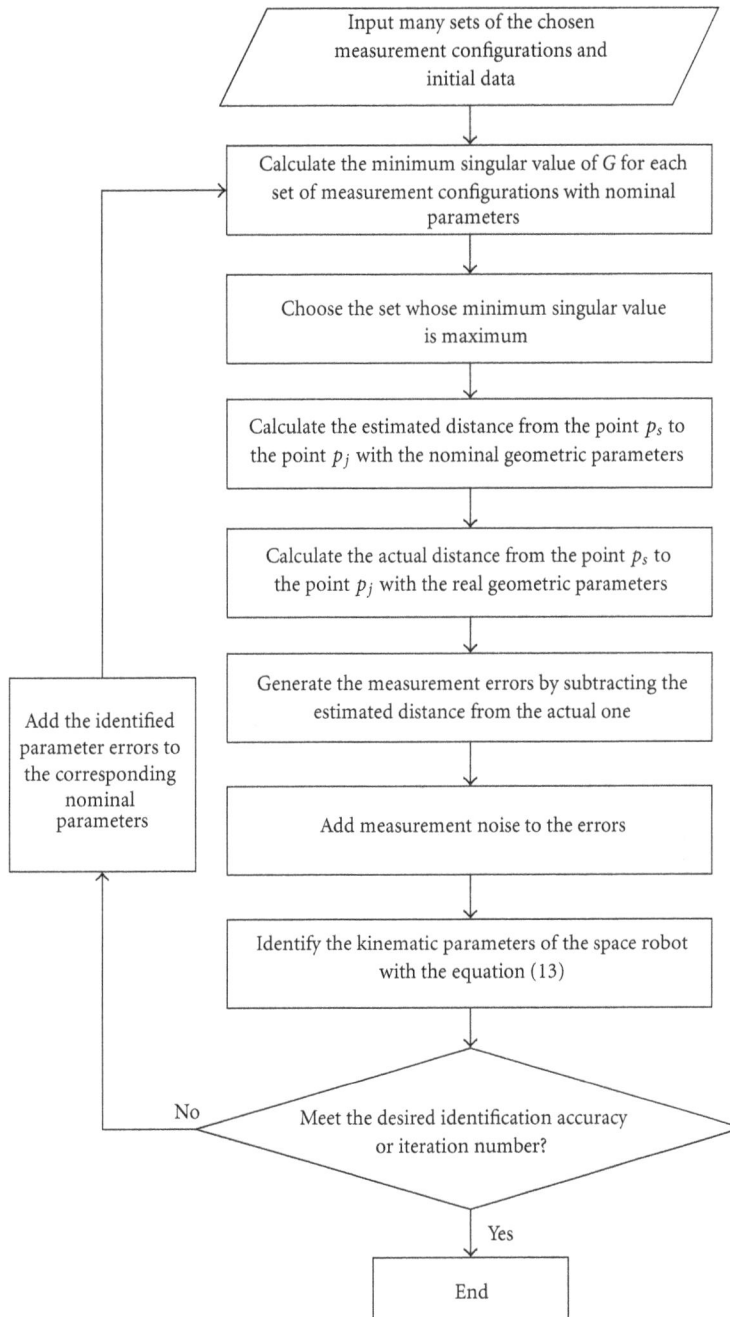

FIGURE 3: Simulation flowchart of parameter identification.

to 1.6 mrad or so, especially in the validation configurations, it can be found that the position errors are reduced from 20 mm or so to less than 2 mm and the orientation errors from 20 mrad or to less than 2.5 mrad. Besides, we noticed a law, namely, the pose errors in the identification configuration are fewer than those in the validation configuration, and the more the number of the identification configurations is, the higher the pose accuracy after identification is. In nature, parameter identification is a fit for measurement data in the identification configurations. However, it is an extrapolation in the validation configuration. So, the results

in the identification configuration are better than those in the validation configurations. The observability index O_3 in the more identification configurations is greater than that in the fewer identification configurations, so the identification results in the more identification configurations are better.

Tables 3 and 4, respectively, give the identified geometric parameter errors for the first and second case. In the two tables, the identified coefficient errors of the plane equation are not listed, and they are, respectively, 0.00243 and 0.00386 for the first and second case. Under the disturbance of measurement noise, these identified geometric parameter errors

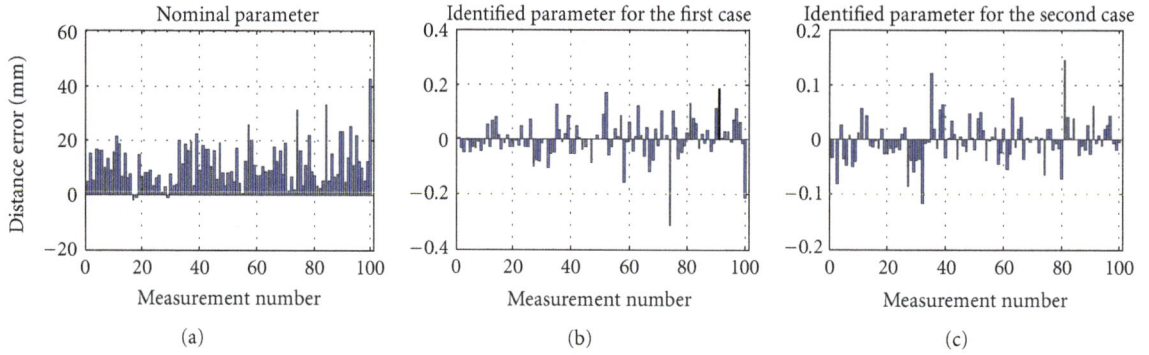

FIGURE 4: Distance errors prior to and after parameter identification.

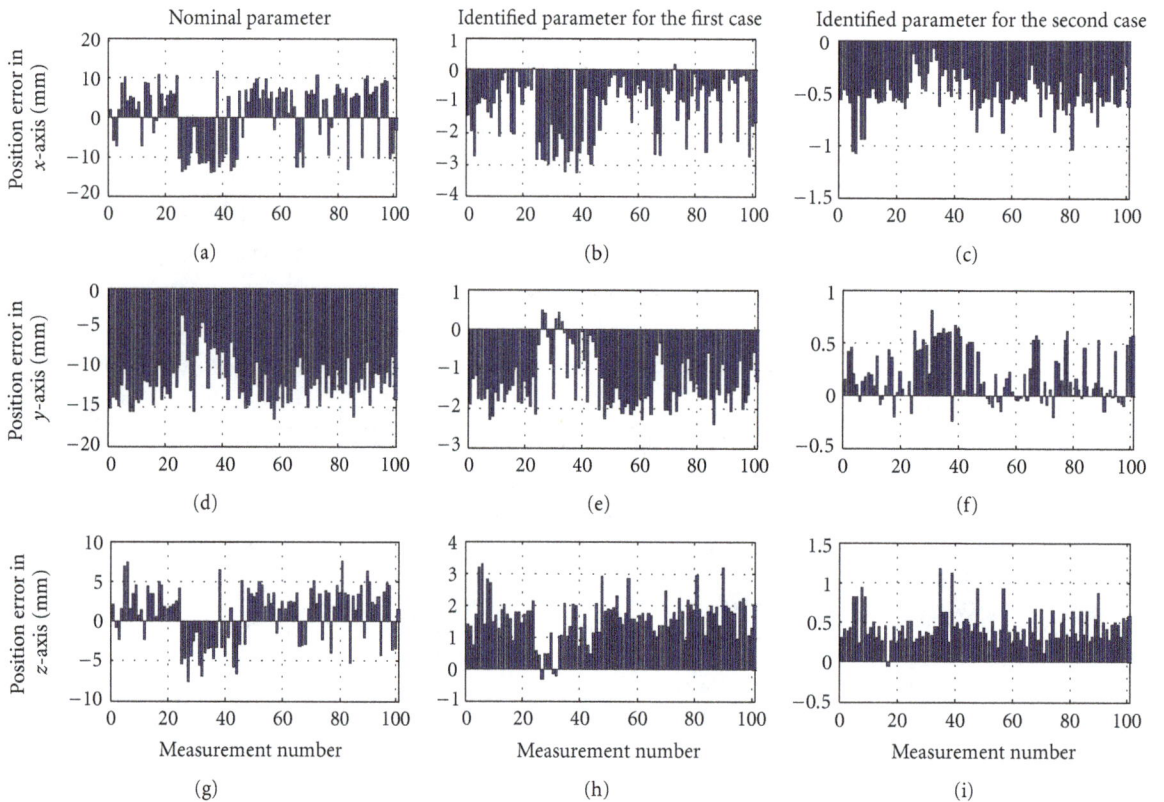

FIGURE 5: Position errors in the 101 identification configurations with the nominal, and the identified parameters for the first and second cases.

are inconsistent with the preassumed ones, but the identified parameter errors for the second case more approach them than those for the first case, which reflects that more measurement configurations can filter measurement noise better. If measurement noise is not added to the simulation, the identified parameters can match the preassumed parameters perfectly, which has been verified in the simulation.

Table 5 gives a statistical comparison of position and orientation errors calculated, respectively, with the nominal parameters, and the identified parameters for the first and second cases in the validation configurations. Here, RMS

represents root mean square of pose errors, with respect to position or orientation error in the x axis, it is written as

$$\text{RMS\_pose} = \sqrt{\frac{1}{m}\sum_{i=1}^{m}(\mathbf{p}_{rx} - \mathbf{p}_x)^2}, \quad (16)$$

where $\mathbf{p}_{rx}$ denotes the real position or orientation vector in the x axis, and $\mathbf{p}_x$ is an estimated position or orientation vector with the nominal or identified parameters in the x axis. The maximum position error denotes the maximum absolute position error value in the x, y, and z axes,

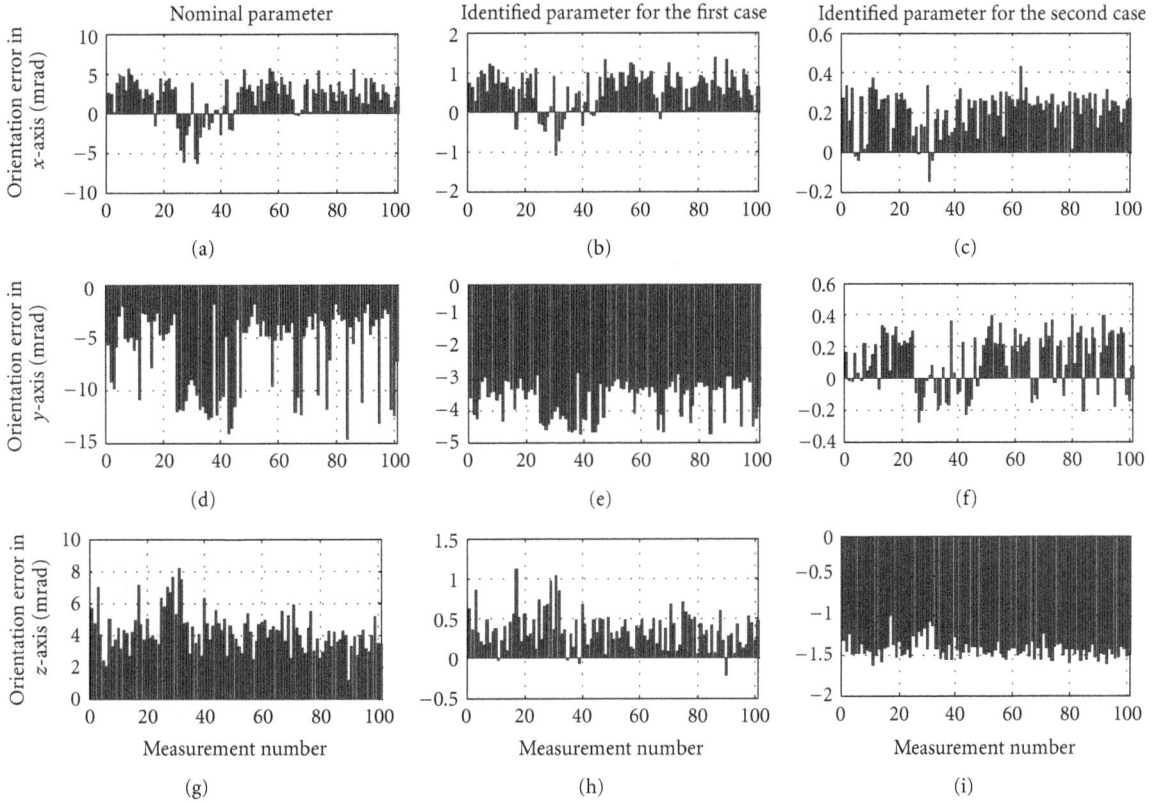

FIGURE 6: Orientation errors in the 101 identification configurations with the nominal, and the identified parameters for the first and second cases.

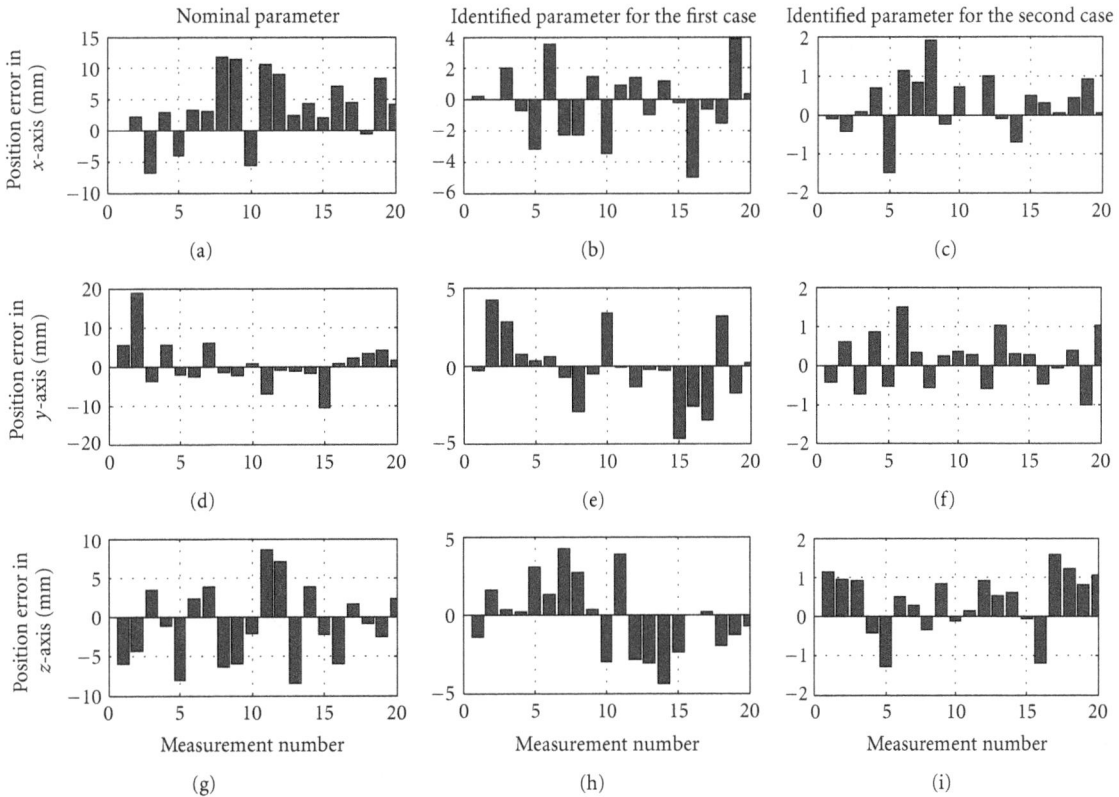

FIGURE 7: Position errors in the 20 validation configurations with the nominal, and the identified parameters for the first and second cases.

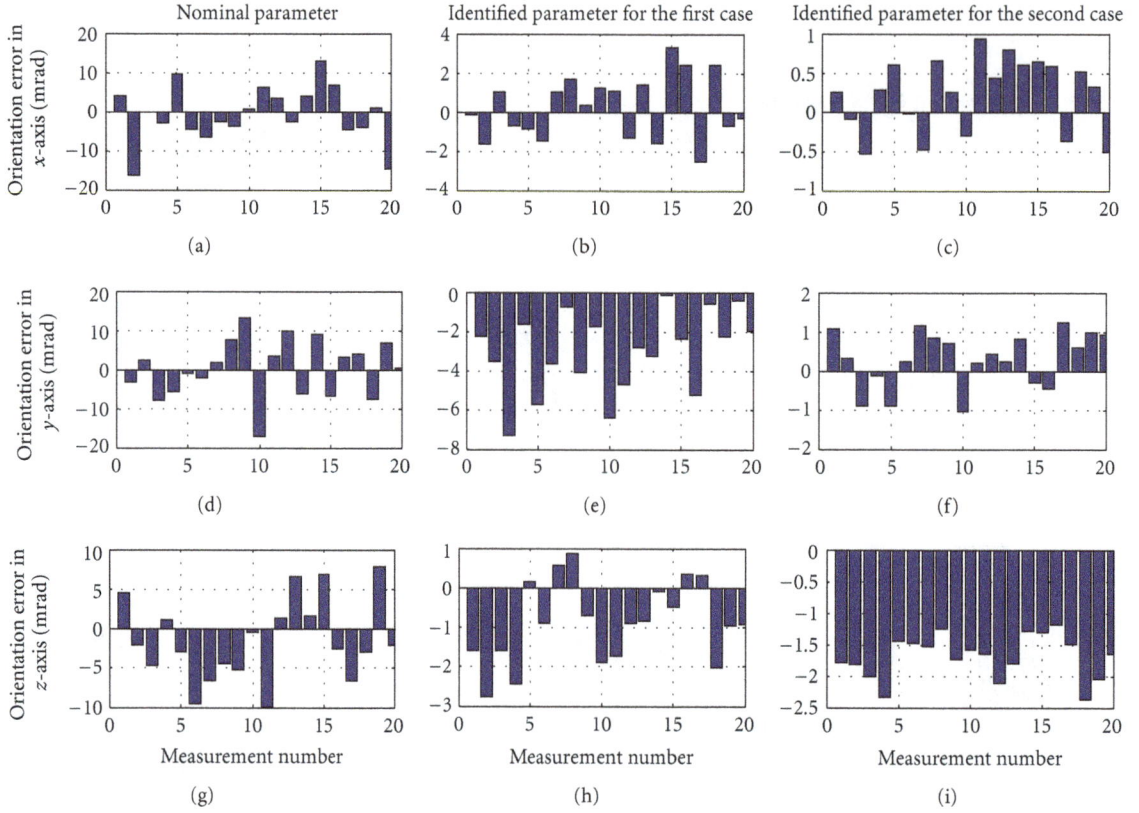

FIGURE 8: Orientation errors in the 20 validation configurations with the nominal, and the identified parameters for the first and second cases.

TABLE 3: Identified geometrical parameter errors for the first case.

Link Number	$\Delta\theta_n$/mrad	$\Delta\alpha_n$/mrad	Δa_n/mm	Δd_n/mm	$\Delta\beta_n$/mrad
1	−7.8525	−2.4425	−2.1491	0.5724	—
2	1.2119	0.0553	1.6291	—	1.1436
3	0.0949	−1.5013	1.3863	0.2838	—
4	0.2936	2.6439	0.3531	1.2179	—
5	−5.2130	−3.8485	−0.0797	0.2895	—
6	0.5628	0.5207	0.8714	0.7422	−3.7951

TABLE 4: Identified geometrical parameter errors for the second case.

Link Number	$\Delta\theta_n$/mrad	$\Delta\alpha_n$/mrad	Δa_n/mm	Δd_n/mm	$\Delta\beta_n$/mrad
1	−7.0607	−2.8193	−0.1191	0.1555	—
2	0.8234	0.2574	1.6439	—	1.4527
3	0.3428	−2.0967	1.0172	−0.5597	—
4	0.5364	2.2958	0.0014	1.4096	—
5	−5.6786	−4.2081	0.3607	0.6214	—
6	1.4004	0.4974	0.4545	0.4274	−3.3556

and also for orientation errors. According to Table 5, it is found that improvement of pose accuracy after parameter identification is significant, and the maximum position error in the y axis is reduced from 18.6857 mm to 1.4779 mm and the maximum orientation error from 17.0006 mrad to 1.2271 mrad. Comparatively, the identification results for the second case are better than those for the first case as a whole, which shows that increment of the redundant measurement configurations can weaken disadvantageous influence of measurement noise and enhance identification effect. If more measurement configurations are added, better identification results can be expected.

5. Conclusions

(1) With the laser-ranger carried by the end effector the paper presents a geometric parameter identification method, and the 25 independent parameters of the space robot are identified through simulation. In the process of identification, independence of the parameters is discussed to avoid parameter dependence.

(2) Because space temperature environment also causes change of the measured plane, its coefficient needs also to be identified. In view of selection of the optimal measurement configurations, the observability index is used to evaluate the combinations of the measurement configurations, which

TABLE 5: Comparison of position and orientation errors in the validation configuration.

Error item		RMS position error/mm	RMS orientation error/mrad	Maximum position error/mm	Maximum orientation error/mrad
Nominal parameter	x	2.7612	3.1491	11.7347	16.1688
	y	2.5917	3.2196	18.6857	17.0006
	z	2.2119	2.3480	8.5899	9.9512
Identified parameter for the first case	x	0.9921	0.7003	4.9654	3.2981
	y	1.0148	1.6199	4.6951	7.3035
	z	1.0524	0.5910	4.3151	2.7741
Identified parameter for the second case	x	0.3427	0.2273	1.9067	0.9378
	y	0.2974	0.3374	1.4779	1.2271
	z	0.3785	0.7671	1.5664	2.3669

reduces the possibility of inferior configurations to be introduced. At the same time, measurement noise of the laser-ranger is simulated to meet the actual state as much as possible.

(3) The simulation results show that in spite of distance measurement alone, the identification technique significantly improves pose accuracy of the space robot, which verifies the feasibility of the method.

Acknowledgment

This work is supported by National Nature Science Foundation of China (Nos. 60775049 and 60805033).

References

[1] W. K. Veitschegger and C. H. Wu, "Robot calibration and compensation," *IEEE Journal of Robotics and Automation*, vol. 4, no. 6, pp. 643–656, 1988.

[2] H. W. Stone, A. C. Sanderson, and C. P. Neuman, "Arm signature identification system," in *Proceedings of the IEEE International Conference on Robotics and Automation*, pp. 41–48, San Francisco, Calif, USA, 1986.

[3] H. W. Stone, A. C. Sanderson, C. P. Neuman et al., "A prototype arm signature identification system," in *Proceedings of the IEEE International Conference on Robotics and Automation*, pp. 175–182, Raleigh, NC, USA, 1987.

[4] L. Beyer and J. Wulfsberg, "Practical robot calibration with ROSY," *Robotica*, vol. 22, no. 5, pp. 505–512, 2004.

[5] Y. Sun and J. M. Hollerbach, "Active robot calibration algorithm," in *Proceedings of the IEEE International Conference on Robotics and Automation (ICRA '08)*, pp. 1276–1281, May 2008.

[6] S. H. Kang, M. W. Pryor, and D. Tesar, "Kinematic model and metrology system for modular robot calibration," in *Proceedings of the IEEE International Conference on Robotics and Automation*, pp. 2894–2899, New Orleans, Fla, USA, May 2004.

[7] J. Chen and L. M. Chao, "Positioning error analysis for robot manipulators with all rotary joints," *IEEE Transactions on Robotics and Automation*, vol. 3, no. 6, pp. 539–545, 1987.

[8] R. P. Judd and A. B. Knasinski, "A technique to calibrate industrial robots with experimental verification," in *Proceedings of the IEEE International Conference on Robotics and Automation*, pp. 351–357, Raleigh, NC, USA, 1987.

[9] C. H. Gong, J. X. Yuan, and J. Ni, "Nongeometric error identification and compensation for robotic system by inverse calibration," *International Journal of Machine Tools and Manufacture*, vol. 40, no. 14, pp. 2119–2137, 2000.

[10] C. Lightcap, S. Hamner, T. Schmitz, and S. Banks, "Improved positioning accuracy of the PA10-6CE robot with geometric and flexibility calibration," *IEEE Transactions on Robotics*, vol. 24, no. 2, pp. 452–456, 2008.

[11] P. Drouet, S. Dubowsky, S. Zeghloul, and C. Mavroidis, "Compensation of geometric and elastic errors in large manipulators with an application to a high accuracy medical system," *Robotica*, vol. 20, no. 3, pp. 341–352, 2002.

[12] K. Radkhah, T. Hemker, and O. V. Stryk, "A novel self-calibration method for industrial robots incorporating geometric and nongeometric effects," in *Proceedings of the IEEE International Conference on Mechatronics and Automation (ICMA '08)*, pp. 864–869, Takamatsu, Japan, August 2008.

[13] G. Hirzinger, K. Landzettel, B. Brunner et al., "DLR's robotics technologies for on-orbit servicing," *Advanced Robotics*, vol. 18, no. 2, pp. 142–144, 2004.

[14] G. Hirzinger, B. Brunner, J. Dietrich, and J. Heindl, "Sensor-based space robotics-ROTEX and its telerobotic features," *IEEE Transactions on Robotics and Automation*, vol. 9, no. 5, pp. 649–661, 1993.

[15] Y. Liu, Y. Shen, N. Xi et al., "Rapid robot/workcell calibration using line-based approach," in *Proceedings of the 4th IEEE Conference on Automation Science and Engineering (CASE '08)*, pp. 510–515, Arlington, Va, USA, August 2008.

[16] C. S. Gatla, R. Lumia, J. Wood, and G. Starr, "Calibration of industrial robots by magnifying errors on a distant plane," in *Proceedings of the IEEE/RSJ International Conference on Intelligent Robots and Systems (IROS '07)*, pp. 3834–3841, San Diego, Calif, USA, November 2007.

[17] I. W. Park, B. J. Lee, S. H. Cho, Y. D. Hong, and J. H. Kim, "Laser-based kinematic calibration of robot manipulator using differential kinematics," *IEEE/ASME Transactions on Mechatronics*, pp. 1–9, 2011.

[18] S. A. Hayati, "Robot arm geometric link parameter estimation," in *Proceedings of the IEEE International Conference on Decision and Control*, pp. 1477–1483, San Antonio, Tex, USA, 1983.

[19] L. Everett, M Driels, and B. Mooring, "Kinematic modeling for robot calibration," in *Proceedings of the IEEE International Conference on Robotics and Automation*, pp. 183–190, Raleigh, NC, USA, 1987.

[20] Y. Sun and J. M. Hollerbach, "Observability index selection for robot calibration," in *Proceedings of the IEEE International Conference on Robotics and Automation (ICRA '08)*, pp. 831–836, Pasadena, Calif, USA, May 2008.

13

Partition Learning for Multiagent Planning

Jared Wood[1] and J. Karl Hedrick[2]

[1] Vehicle Dynamics Lab and Center for Collaborative Control of Unmanned Vehicles, Department of Mechanical Engineering,
University of California, Berkeley, 6141 Etcheverry Hall, Berkeley, CA 94720-1740, USA
[2] Vehicle Dynamics Lab, Department of Mechanical Engineering, University of California, Berkeley, 6141 Etcheverry Hall,
Berkeley, CA 94720-1740, USA

Correspondence should be addressed to Jared Wood, jared.jwood@gmail.com

Academic Editor: Duško Katić

Automated surveillance of large geographic areas and target tracking by a team of autonomous agents is a topic that has received significant research and development effort. The standard approach is to decompose this problem into two steps. The first step is target track estimation and the second step is path planning by optimizing directly over target track estimation. This standard approach works well in many scenarios. However, an improved approach is needed for the scenario when general, nonparametric estimation is required, and the number of targets is unknown. The focus of this paper is to present a new approach that inherently handles the task to search for and track an *unknown* number of targets within a *large* geographic area. This approach is designed for the case when the search is performed by a team of autonomous agents and target estimation requires general, nonparametric methods. There are consequently very few assumptions made. The only assumption made is that a time-changing target track estimation is available and shared between the agents. This estimation is allowed to be general and nonparametric. Results are provided that compare the performance of this new approach with the standard approach. From these results it is concluded that this new approach improves search and tracking when the number of targets is unknown and target track estimation is general and nonparametric.

1. Introduction

The advancement of computing technology has enabled the practical development of intelligent autonomous systems. Intelligent autonomous systems can be used to perform difficult sensing tasks. One such sensing task is to search for and track targets over large geographic areas. Much research has gone into this task resulting in a standard approach. This standard approach decomposes the problem into two steps.

(1) Target track estimation.
(2) Agent path optimization based on target track estimation.

Significant research has been accomplished for each of these steps. Target track estimation has largely been solved [1–5] and this paper proposes no new methods for target track estimation. Agent path optimization based on target track estimation has been solved for many scenarios. However, the general scenario of when the number of targets is unknown still requires more development.

The standard approach in general works particularly well when it can be assumed that there is a single target [6–14]. And in many scenarios the standard approach works well even when there are multiple targets [15, 16]. However, when there are multiple targets, methods following the standard approach start requiring limiting assumptions on the problem. For example, many methods require that the geographic area be easily scanned so that there are frequent target detections throughout the geographic area. When this can be assumed it allows for simpler estimation, such as Gaussian distributions, and consequently agent paths can be more readily optimized. However, these methods do not extend well when the geographic area is too large to scan quickly. When the geographic area is large, agents frequently do *not* detect targets. These no-detection events must be utilized to estimate the target tracks [6, 10, 12]. General recursive Bayesian estimation methods are required to accomplish estimation for this case. This results in a general, nonparametric estimate of the target tracks. Yet, because of this generality, as the number of targets increases it becomes very difficult to optimize sensor paths over the target track estimates.

The focus of this paper is then on the problem of search and tracking an *unknown* number of possibly multiple targets in a *large* geographic area utilizing a team of autonomous, sensing agents. As such, very few assumptions are placed on the problem. All that is assumed is that time-changing target track estimates are provided and shared by the team of autonomous agents [10, 11, 13, 15, 16]. The target track estimates are completely general and nonparametric and the geographic area is too large to scan quickly.

A solution to this general problem is provided by proposing a new approach to perform autonomous search and tracking that inherently handles the case of an unknown number of targets in a general, nonparametric estimation setting. The performance of this new approach is then compared with the standard approach. The specific method that will be used for comparison is direct optimization over target estimation distribution [6, 10, 14].

2. Problem Structure: New Approach

In this paper a new approach is presented to perform autonomous search and tracking over large geographic areas. The union of such large geographic areas will be referred to as the surveillance area S in this paper. This new approach is designed to inherently handle the case of the number of targets being unknown and target track estimation being general and non-parametric. Figure 1 provides an overview of this approach. Notice, instead of the standard two-step approach there are three steps. These three steps are

(1) target density estimation [17],

(2) partition learning based on target density estimation,

(3) agent path planning based on partitions.

With this decomposition of the problem, separate subproblems are defined for each step. Each step will be described in the following sections. Two of these steps leverage existing work by extending existing methods to fit the structure of the problem presented in this paper. These steps are target density estimation and agent path planning based on partitions. The required extensions will be presented in this paper. Partition learning requires further development. Consequently most discussion provided in this paper will focus on the partition learning step.

Also note, these steps are repeated at each time instance. As the target estimation changes with time, so do the partitions and the agent paths.

3. Step 1: Target Density Estimation

The first step in the problem decomposition is target density estimation [18]. Recall that the first step of the standard approach is target track estimation. Target track estimation is

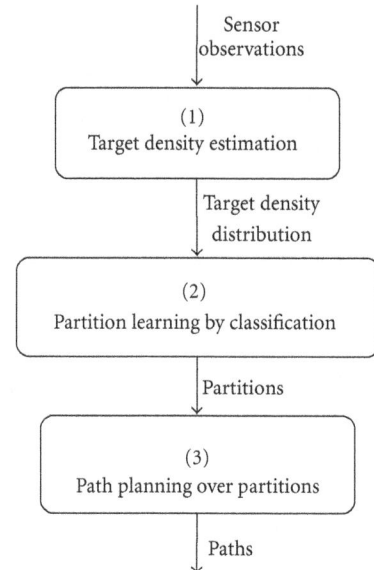

FIGURE 1: Problem structure of new approach for autonomous search and tracking. The problem is decomposed into three steps. Each of these steps is performed at each instance in time. As the target estimation changes with time, so do the partitions and the agent paths.

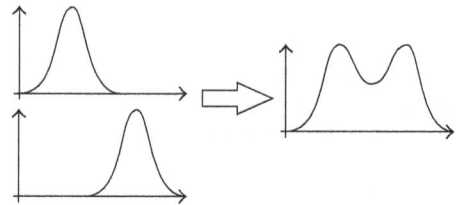

FIGURE 2: Depiction of capturing the complexity of the high-dimensional target track estimation space and mapping it to the surveillance area. This forms the target density distribution.

not identical to target density estimation. In order to understand the difference, consider a search and tracking application that estimates the position of N targets utilizing the standard approach. The first step of the standard approach is target track estimation. Assume that target positions are within some region S of the plane (R^2). Call this region the surveillance area. The estimation space is then $N \times S$. The dimension of this space is potentially very high. In order to estimate the position of the targets a probability distribution is determined which is defined over this high-dimensional space. The second step of the standard approach is then agent path planning by optimizing over this high-dimensional space. However, from the perspective of agent path planning it would be beneficial to optimize paths over the significantly lower dimensional space S, instead of $N \times S$. This is the purpose of the target density distribution.

The target density distribution captures the complexity of the high-dimensional target track estimation space and maps it to the single planar space of the surveillance area. This process is depicted in Figure 2. This figure shows multiple single-target distributions combining to form a single target density distribution. A sample target density distribution, defined over a planar surveillance area, is provided in Figure 3. In this figure the target density distribution is represented by contour lines. Red lines have high density and blue lines have low density. The details of this distribution's shape are not important. What is important is to note that a target density distribution captures the complexity of combining the high-dimensional target track estimation.

The space of the target density distribution is the surveillance area S. Now consider some subspace $A \subset S$ of the surveillance area. The target density can be defined as a distribution $f(x)$ that is defined over the surveillance area as

$$\text{EN}_A = \int_{x \in A \subset S} f(x) d\mu(x), \qquad (1)$$

where N_A is the number of targets in the subspace A and $\mu(x)$ denotes that measure on which the integral is performed. For example, if the space is discretized then the integral represents a summation.

Notice that the target density distribution provides the estimated number of targets within regions of the surveillance area. The expected total number of targets in the surveillance area is then

$$\text{EN}_S = \int_{x \in S} f(x) d\mu(x). \qquad (2)$$

The target density distribution can be computed from target track estimation. For example, consider the case when the positions of N targets are estimated independently. Target track estimation then provides a set of target distributions $\{P(X^1), \dots, P(X^N)\}$. The target density distribution can be computed as $f(x) = \sum_i P(x^i)$. Consequently there is no need to develop new methods for estimating the target density distribution. Instead, the vast body of target track estimation work can be leveraged.

However, it is not necessary to obtain the target density distribution from target track estimation. Instead, it can be estimated directly from sensor observations. One approach [5] that accomplishes this utilizes random set theory [19] to obtain an approximate target density distribution.

4. Step 2: Partition Learning

Recall that the information contained in a target density distribution can be very complex. This is because it combines the information of all target track estimates and maps them onto the surveillance area. To aid in distributing agents across the surveillance area, the complexity of the target density distribution can be used to partition the surveillance area into disjoint regions. For example, recall the sample target density distribution depicted in Figure 3. One possible set of partitioned regions is depicted in Figure 4. Note that in this figure the target density distribution is represented by contour lines and the boundaries of the partitions are represented by thick, straight lines. The particular choice of partitions displayed is of little significance. What is significant is to note that the partitions are determined based on the information content provided by the target density distribution.

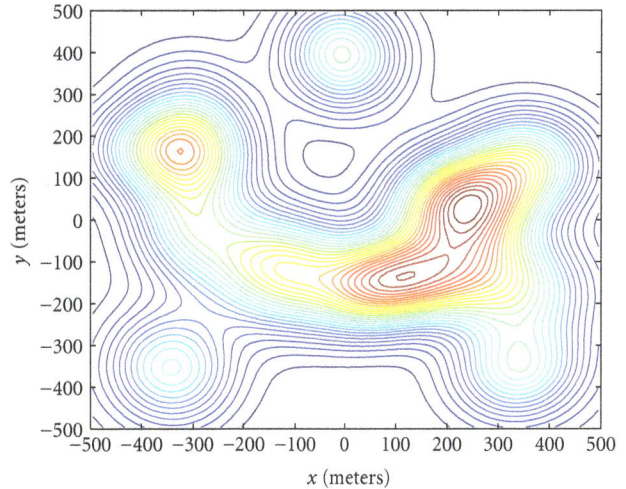

FIGURE 3: Sample target density distribution, represented by contour lines, defined over a surveillance area. The details of this distribution's shape are not important. What is important is to note that a target density distribution captures the complexity of the high-dimensional target track estimation space.

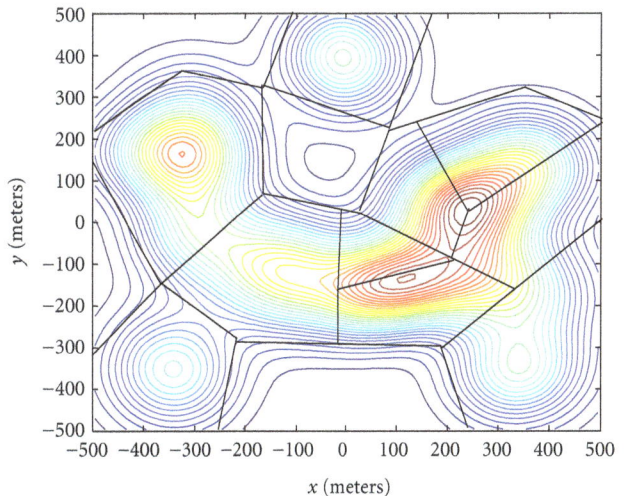

FIGURE 4: Example partitioning of the surveillance area based on a target density distribution. The target density distribution is represented by contour lines. The boundaries of the partitions are represented by thick, straight lines. The space of this plot represents the surveillance area over which the target density distribution is defined. The particular choice of partitions displayed is of little significance. What is significant is to note that the partitions are determined based on the information content provided by the target density distribution.

Instead of partitioning the surveillance area into arbitrary regions the approach taken in this paper is to partition the surveillance area into regions that correspond to

(1) a null target partition,

(2) an exploration partition,

(3) a set of search and tracking partitions.

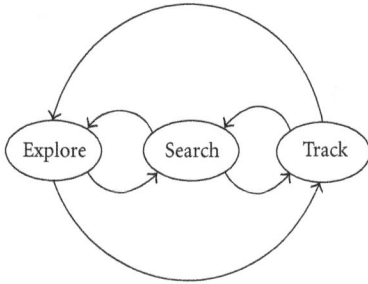

FIGURE 5: The path planning modes that may exist when surveillance is performed by region-based planning.

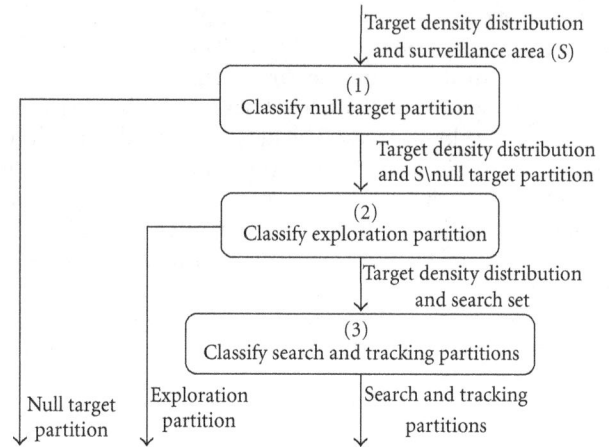

FIGURE 6: Cascade of classifiers that partition the surveillance area into (1) a null target partition, (2) an exploration partition, and (3) a set of search and tracking partitions. Note that at each instance of time that the target density distribution changes, this classifier is processed on the new target density distribution. In this sense partitions change with time.

By partitioning the surveillance area into these types of regions it gives agent path planning algorithms the flexibility to switch between modes of surveillance exploration, target search, and target tracking. These modes are depicted in Figure 5. In order to compute each of these partitions a different classifier is designed. These partition classifiers along with corresponding modes will be described further below.

The combination of these partition classifiers constructs the overall partition learning classifier. The structure of this classifier is presented in Figure 6. The general steps are to classify (1) the null target partition, (2) the exploration partition, and (3) a set of search and tracking partitions.

Before describing the details of these steps, the computational flow of the overall classifier can be understood by considering a set Γ and how it changes as it moves through the classifier. Let Γ be the set of all points in the surveillance area. As such, the target density distribution is defined over Γ. The flow of the classifier can then be understood as

(1) the null target partition S_{null} is classified and removed from Γ ($\Gamma \leftarrow \Gamma \setminus S_{\text{null}}$) in block (1) of Figure 6,

(2) Γ is now a subset of the surveillance area. Within Γ, the exploration partition S_{explore} is classified and then removed from Γ ($\Gamma \leftarrow \Gamma \setminus S_{\text{explore}}$) in block (2) of Figure 6,

(3) Γ now consists of only the subset of the surveillance area that will be partitioned into search and tracking partitions. In block (3) of Figure 6 an ordered set of search and tracking partitions are classified within Γ.

Each of these partition learning classifier steps will now be discussed in more detail.

4.1. Partitioning Step 1: Null Target Partition. Over time some regions will be repeatedly observed. Much of the observed regions will never have targets detected. Dependent on anticipated possible target mobility, it may be concluded that no targets exist in these areas. It is necessary to maintain a partition that classifies regions in which no targets exist. These regions form the null target partition.

The first step of partition learning is to classify this null target partition. The importance of this partition is seen by considering two scenarios. The first is when there is no

target in the surveillance area. At some point in time the conclusion should be reached that there is no target. The second scenario is when there is a vast exploration partition. As regions become fully observed, but no targets have been detected, these observed regions should cease to be explored.

The method of classification for this step of partition learning will now be described. To do this the features used for classification will be described first. Then the classifier will be described.

4.1.1. Features. Low values for target density are what define the null target partition. The only feature required to classify this partition is then simply values of the target density distribution. The target density distribution was defined previously in (1).

4.1.2. Classifier. Recall that Γ represents the set over which the classifier operates. As such, Γ is initially the entire surveillance area ($\Gamma \leftarrow S$). The first step (block (1) of Figure 6) of partition learning is to determine the null target partition S_{null} and remove it from Γ ($\Gamma \leftarrow \Gamma \setminus S_{\text{null}}$). This step of the classifier is visualized in Figure 7 for a simple one-dimensional target density distribution.

To determine which points in Γ belong to the null target partition, a target nullity threshold ϵ_{null} is required. This threshold specifies the value of target density below which it is assumed no targets exist. With this target nullity threshold given, all points in the surveillance area that correspond to regions of essentially no targets can then be defined by

$$S_{\text{null}} := \{x \in S : f(x) < \epsilon_{\text{null}}\}. \tag{3}$$

The set of points in S_{null} then form the null target partition. This set of points is removed from Γ and the classifier continues by classifying the exploration partition.

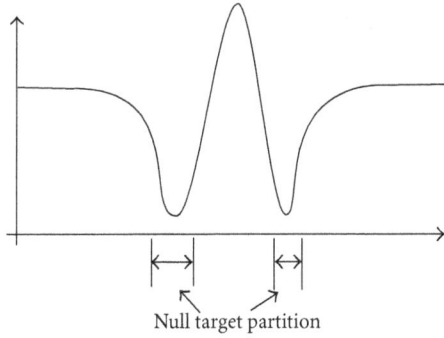

FIGURE 7: Visualization of a simple target density distribution with corresponding null target partition noted.

4.2. Partitioning Step 2: Exploration Partition.

4.2. Partitioning Step 2: Exploration Partition. The second step of partition learning is to classify the exploration partition. The exploration partition consists of areas within the surveillance area that have low-to-no information bias. For example, a region over which there is defined a uniform probability distribution would be included in the exploration partition. Because there is no bias in information, an exploration-oriented mode of path planning may be preferred for these regions [20, 21]. To allow this exploration-oriented mode of path planning, these types of regions are classified into a separate partition.

The method of classification for this step of partition learning will now be described. To do this the features used for classification will be described first. Then the classifier will be described.

4.2.1. Features. In order to classify points in Γ into the exploration partition, two features are required. These features are

(1) local uncertainty,

(2) target density.

Local uncertainty is used to determine regions of locally uniform value. Initially it may appear that only local uncertainty is required to define the exploration partition completely. However, there is a subtle aspect that requires the addition of target density in order to completely capture the entire exploration partition.

This subtle aspect can be understood by considering the case when the entire surveillance area is initially uniformly distributed. An agent makes imperfect no-detection observations. As such, some regions will have low-target density (regions that have been observed well), completely unobserved regions will have an unchanged uniform value, and others will have value somewhere in between the low value and the unchanged uniform value (due to poor observation in these areas). These in-between-valued areas will not have locally uniform value, yet will still belong to the exploration partition. This case suggests that target density, in addition to local uncertainty, is required to catch the complete exploration partition.

Before defining local uncertainty a definition for local area is required. The local area $S_r(x_0) \subseteq S$ of some point

$x_0 \in S$ in the surveillance area (where S is the space of the surveillance area) is defined as

$$S_r(x_0) := \{x \in S : d(x, x_0) < r\}, \tag{4}$$

where $d(x, x_0)$ is some measure of distance.

Local uncertainty is defined by first selecting a measure for uncertainty. In this paper local uncertainty is based on entropy. As such, local uncertainty is computed by evaluating the local entropy defined as

$$H_r(x_0) := \int_{x \in S_r(x_0)} f_r(x, x_0) \log\left(\frac{1}{f_r(x, x_0)}\right) d\mu(x), \tag{5}$$

where $f_r(x, x_0)$ is the locally normalized target density function defined by

$$f_r(x, x_0) := \frac{f(x)}{\int_{y \in S_r(x_0)} f(y) d\mu(y)}. \tag{6}$$

Note, if local uncertainty were not defined on a locally normalized target density distribution there would not be a well-established maximum value for local uncertainty. A locally normalized density is then required so that the maximum value for local uncertainty can be referenced during classification. This maximum value allows the classifier to determine if a particular region contains significantly biased information of target density. To aid in understanding local uncertainty Figure 8 presents the computation of local entropy over a surveillance area when the underlying target density distribution is a simple Gaussian probability distribution centered in the middle of the surveillance area. In this figure, local uncertainty is represented by shade value where white is high value and black is low value.

4.2.2. Classifier. At this point of the classifier Γ consists of a subset of the surveillance area defined by $\Gamma := S \setminus S_{\text{null}}$. In this step (block (2) of Figure 6) of partition learning the exploration partition S_{explore} is classified and removed from Γ ($\Gamma \leftarrow \Gamma \setminus S_{\text{explore}}$). Define this resulting state of Γ to be the search partition (or search set) S_{search}. The process of this stage of the classifier is visualized in Figure 9 for a simple one-dimensional target density distribution.

To determine which regions in Γ are approximately locally uniform, a new set S_{HE}, called the high-entropy set, is computed. The high-entropy set is defined as

$$S_{\text{HE}} := \{x \in \Gamma : |H_r(x) - H_{\max}| < \epsilon\}, \tag{7}$$

where $H_r(x)$ is the local entropy feature as defined in (5) and $H_{\max}$ is the maximum local entropy possible defined as

$$H_{\max} := \log|S_r|, \tag{8}$$

where $|S_r| := \int_{x \in S_r} d\mu(x)$. Take, for example, the case when the target density distribution is discretized on a fixed grid defined over the surveillance area. Recall the definition of S_r in (4). In (4), note that in order to define S_r it is required to define a measure of distance $d(x, x_0)$ between two points x

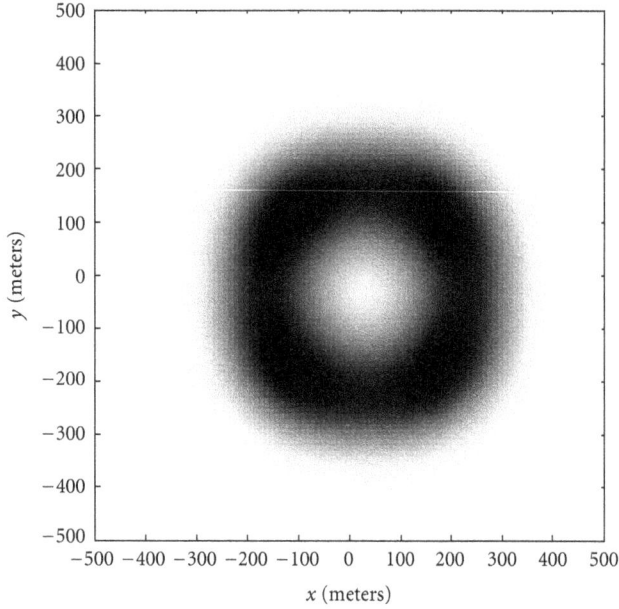

FIGURE 8: A sample computation of local uncertainty over a surveillance area when the underlying target density distribution is a simple Gaussian probability distribution. Local uncertainty is represented by shade value where white is high-local uncertainty and black is low-local uncertainty. In this sample, notice that the peak and the tails of the Gaussian have high-local uncertainty whereas the regions in between the peak and tails have low uncertainty.

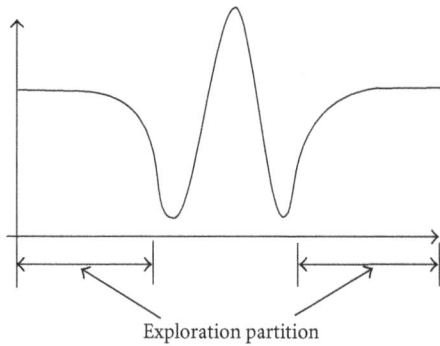

FIGURE 9: Visualization of a simple target density distribution with corresponding exploration partition noted.

and x_0 in the surveillance area. For a fixed grid, this distance is defined as

$$d(x, x_0) := \max(\{|x(1) - x_0(1)|, |x(2) - x_0(2)|\}), \quad (9)$$

where the numbers 1 and 2 specify the indices of the points. Then, according to the definition of S_r, the maximum local entropy is $H_{\max} = \log N^2$, where N is the number of rows/columns in the square local area S_r.

To catch regions for which the subtlety mentioned above applies another set is computed. This set is called the low-density set S_{NI}. The definition of this set is simple, however, it requires explanation. At the beginning of a surveillance task an initial target density (or prior distribution) is constructed

that expresses prior belief in possible target locations. At a minimum this prior consists of two pieces. These pieces are

(1) a prior distribution $f_{\text{prior}}(x)$ of previously known target positions,

(2) an estimated number of additional targets $EN_{\text{additional}}$ that may exist in the surveillance area.

The prior target density distribution $f_{\text{prior}}(x)$ provides a target density bias based on where targets have most recently been observed and where they might be now. For example, f_{prior} may consist of a summation of Gaussian distributions with each Gaussian representing the possible location of a particular target whose position was once known or whose position is simply guessed. The estimated number of additional targets $EN_{\text{additional}}$ affects the initial target density distribution by defining a uniform target density distribution $U(x)$ such that

$$\int_{x \in S} U(x) d\mu(x) = \frac{EN_{\text{additional}}}{\int_{x \in S} d\mu(x)}. \quad (10)$$

The resultant prior target density distribution is then

$$f(x) = f_{\text{prior}}(x) + U(x). \quad (11)$$

Note that some regions of the prior target density distribution will have a characteristic uniform value U. This characteristic low-information value can then be used to define the low-density set. The regions that must be captured in the low-density set are those regions that have target density in between the locally uniform density and the null target threshold. Yet, because $\Gamma = S \setminus S_{\text{null}}$ at this point of the classifier, the low-density set can be defined simply as

$$S_{NI} := \{x \in \Gamma : f(x) < U + \epsilon\}. \quad (12)$$

Then, combining the low-density set with the high-entropy set, the exploration partition is defined as

$$S_{\text{explore}} := S_{HE} \cap S_{NI}. \quad (13)$$

S_{explore} can then be removed from Γ as $\Gamma \leftarrow \Gamma \setminus S_{\text{explore}}$. Let this state of Γ be called the search partition (or search set) S_{search}. S_{search} is then defined as

$$S_{\text{search}} := S \setminus \{S_{\text{null}} \cup S_{\text{explore}}\}. \quad (14)$$

After removing the exploration partition from Γ the classifier then continues on to classifying the search and tracking partitions.

4.3. Partitioning Step 3: Search and Tracking Partitions. The third step of partition learning is to classify a set of search and tracking partitions. Both search and tracking partitions are classified by the same classifier.

In terms of information content, the opposite type of region to the exploration partition is a tracking partition. Tracking partitions are small spatially and are partitions in which there is strong bias of target density and high certainty.

These are partitions in which targets have been detected consistently. Consequently, tracking partitions define locations of known targets. These partitions must continue to be tracked according to the mobility of the tracked targets. The search strategy then becomes that of keeping observance of the known position of the targets. For example, the points of maximum density within tracking partitions are kept in observance. The search strategy for tracking partitions is then the most constraining on agent motion. For example, if an agent is a fixed-wing aircraft it will have to fly orbit-like paths encircling the known position of the target [14, 22–25].

Similar to tracking partitions are search partitions. The similarity that the search partitions have with the tracking partitions is that both consist of an information set that provides a bias to aid in optimizing search plans. However, search partitions are different from tracking partitions in that the information content is not very certain. Consequently, not much can be said about exactly where a target may be located. However, there is bias over which parts of the regions have high possibility of target existence. It then becomes the duty of a search plan to optimize paths based on the information content in order to yield a new distribution with higher certainty. The search strategy then becomes to maximize some type of information gain, and a searcher's paths are guided to improve the information content in order to ultimately observe a target [6–8, 12, 26, 27].

The method of classification for this step of partition learning will now be described. To do this the features used for classification will be described first. Then the classifier will be described.

4.3.1. Features. In order to classify points in Γ into search and tracking partitions, two features are required. These features are

(1) normalized position within the surveillance area,

(2) local expected number of targets.

Normalized position is computed for some point $x \in S$ by dividing by the size of the surveillance area S. Local expected number of targets comes readily from the target density distribution. To understand this, recall the definition of the target density distribution. From this definition it is apparent that the expected number of targets within some region A is just the integration of the target density distribution over that region. Recalling the definition of local area, local expected number of targets is then computed by integrating the target density distribution over the local area as

$$\mathrm{I}_r(x_0) := \mathrm{EN}_{S_r(x_0)}$$
$$= \int_{x \in S_r(x_0)} f(x) d\mu(x). \tag{15}$$

To aid understanding of local expected number of targets, Figure 10 presents a sample computation over a surveillance area when the target density distribution is a simple Gaussian probability distribution. In this figure the value of local expected number of targets is represented by shade value, white being high and black being low. From this figure it can

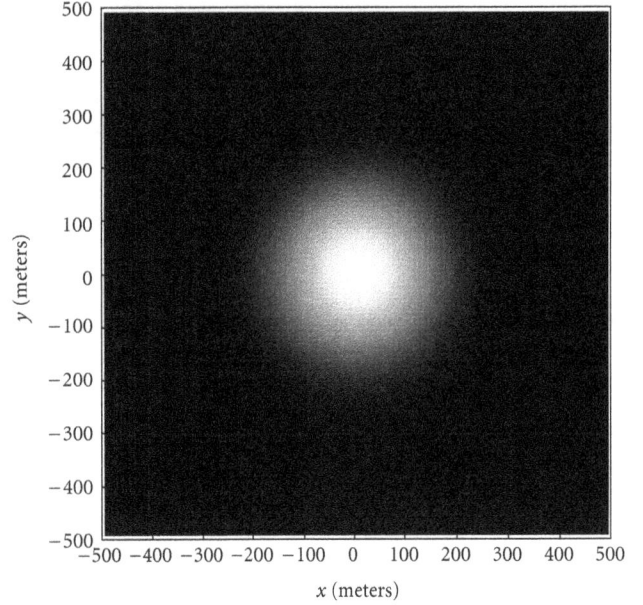

FIGURE 10: A sample computation of local expected number of targets I_r over a surveillance area when the underlying target density distribution is a simple Gaussian probability distribution. Notice that local expected number of targets acts as a smoothing filter over the target density distribution.

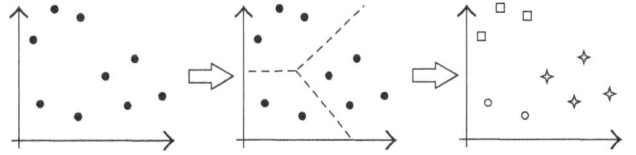

FIGURE 11: Visualization of how points in the target density distribution feature space are convexly partitioned. In this example three partitions are classified. The resulting three partitions are represented by circles, stars, and rectangles.

be observed that local expected number of targets acts as a smoothing filter over the target density distribution.

4.3.2. Classifier. At this point in partition learning (block 3 in Figure 6) the set on which classification operates is $\Gamma = S_{\mathrm{search}} = S \setminus \{S_{\mathrm{null}} \cup S_{\mathrm{explore}}\}$. Γ then consists of regions in the surveillance area with some level of biased information of possible target locations. In this step of partition learning Γ is partitioned into a set of search and tracking partitions. This step cannot be performed by a simple set computation as was done for the previous two steps of partition learning. Instead, points in Γ are clustered according to the target density feature space. In order to help visualize the action that occurs at this stage of the classifier, observe Figure 11. This figure shows how a set of points in the target density distribution feature space are convexly partitioned.

There are many convex clustering methods [28] that would work for this level of the classifier. The approach taken in this paper is to perform classification by utilizing both K-means and Gaussian Mixture Model EM [28]. K-means is used to initialize search and tracking partitioning at the beginning of the surveillance task. Gaussian Mixture Model EM is then used at subsequent steps in time, where the Gaussian Mixture Model EM is seeded with previous partition means [17]. Both K-means and Gaussian Mixture Model EM operate in the target density distribution feature space consisting of normalized position within the surveillance area and local expected number of targets.

In order to perform this classification, the number of partitions to classify must be initialized. The initial number of partitions is determined from the total expected number of targets in the surveillance area

$$N_{\text{initial}} := \text{ceil}(EN_S)$$
$$= \text{ceil}\left(\int_{x \in S} f(x) d\mu(x) \right). \quad (16)$$

After performing K-means or Gaussian Mixture Model EM to convexly partition $\Gamma = S_{\text{search}}$ into a set of search and tracking partitions, these partitions are then ordered. The partitions are ordered so that tracking partitions appear first and uncertain searching partitions appear last. This enables path planning algorithms to prioritize the various search and tracking partitions. To accomplish this ordering the partition density ρ_{P_i} of each partition P_i is computed. Partition density is defined as

$$\rho_{P_i} := \frac{\int_{x \in P_i} f(x) d\mu(x)}{\int_{x \in P_i} d\mu(x)}. \quad (17)$$

At this point in this step of partition learning an ordered set of search and tracking partitions have been classified. Some of these partitions may correspond to regions with many densely located targets. It may be beneficial to allocate more searching resource to these types of partitions. Accordingly, these partitions are further subpartitioned.

In order to determine if some partition P_i should be subpartitioned its expected number of targets EN_{P_i} is computed. EN_{P_i} is easily computed from the target density distribution $f(x)$ as

$$EN_{P_i} := \int_{x \in P_i} f(x) d\mu(x). \quad (18)$$

If $EN_{P_i} > 1$ then it is expected that there is more than one target within P_i. In order to track all of these possible targets multiple agents may be required. To account for the possible need of multiple agents, any P_i with $EN_{P_i} > 1$ is subpartitioned into $\text{ceil}(EN_{P_i})$ new partitions. And this is where the classifier ends. The end result is one null target partition, one exploration partition, and a set of ordered search and tracking partitions.

5. Step 3: Path Planning over Partitions

The final step of the approach presented in this paper for autonomous search and tracking is path planning. This path planning is performed over the set of partitions. In order to

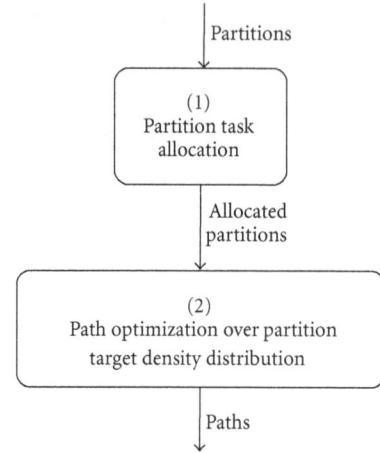

FIGURE 12: Structure of path planning decomposed into task allocation over partitions and path planning by optimizing directly over the target density distribution.

plan paths over partitions, path planning is decomposed into two steps as depicted in Figure 12. These steps are

(1) partition task allocation,

(2) target density distribution based path optimization.

In the first step partitions are allocated to the team of agents [29, 30]. In the second step agent paths are determined within allocated partitions by optimizing directly over partition level target density distributions [6–8, 10, 14].

By decomposing path planning in this manner the vast amount of work that has been developed for vehicle routing (for partition task allocation) and receding horizon path optimization (for target density-distribution-based path optimization) can be leveraged. Now, all that is required is extensions of existing methods where necessary. As such, this section refers the reader to the body of work that is leveraged and then presents any required extensions.

5.1. Path Planning Step 1: Partition Task Allocation. The partitions generated by the classifier define areas over which subsets of the target density distribution can be extracted. This suggests the application of some kind of task allocation algorithm that takes each of the partitioned search areas as tasks with varying level of certainty or priority. The exact method of task allocation is beyond the scope of this paper since it has been well developed by researchers already. Refer specifically to [29, 30] for methods that directly apply. For task allocation algorithms that have been developed for projects at the Center for Collaborative Control of Unmanned Vehicles refer to [29].

5.2. Path Planning Step 2: Distribution-Based Optimization. Optimizing a path over a target density distribution is almost identical to optimizing a path over a probability distribution. Fortunately, much work has been done to develop path optimization over probability distributions [6–8, 10, 14]. These existing methods are leveraged in this paper. The general approach of these methods is to define a function for measuring the utility of a path based on an underlying probability distribution. Then, this utility function is used to optimize paths through the surveillance area. These methods are extended by defining a utility function based on target density distributions. Defining this utility function is the focus of this section.

In order to define a path's utility the utility of a point in the surveillance area must be defined. However, before defining the utility of a point, an agent's sensor observation coverage $f_C(x, x_0)$, about a point x_0 in the surveillance area, must be determined. $f_C(x, x_0)$ essentially specifies how applicable some point x in the surveillance area is to a particular agent when the agent is located at x_0. For example, consider the case of a fixed sensor. This sensor is free to rotate in order to observe its surroundings. However, it cannot see beyond r meters. Consequently, any point farther away than r meters is of little significance to this sensor.

An agent's observation coverage is determined by the properties of the agent's sensor. For example, if an agent can make observations perfectly within a radius r, then the observation coverage is an indicator function defined by

$$f_C(x, x_0) = \begin{cases} 1, & \text{if } \|x - x_0\| < r, \\ 0, & \text{otherwise.} \end{cases} \quad (19)$$

However, in general, the observation coverage is determined by the sensor's capable field of view as well as the resolution of observable points within the field of view and the probability of missed detection [31]. In order to further visualize possible sensor observation coverages, consider two cases.

(1) An agent can view it's surroundings perfectly within 25 meters. Beyond that, the agent's view linearly degrades until it cannot make any observation at 70 meters.

(2) An agent cannot view anything near it until a distance of 45 meters away. After that, observations quickly become perfect but then start to fade around 65 meters. By 90 meters, observations are no longer possible.

The first of these cases is similar to what is true for many sensing agents [32]. They are designed such that their observations improve with proximity. Figure 13 depicts this sensor coverage. In this figure the quality of a point's coverage by the agent's sensor is represented by a shaded value where white is high utility and black is low utility.

The second of these cases may seem odd, but is actually similar to what was used in an experiment performed with autonomous aircraft equipped with visual spectrum cameras [33]. In this experiment, a camera on-board an aircraft was zoomed in to detect features of a pedestrian. The zoom was designed so that good resolution would be provided when the aircraft orbited the pedestrian. Consequently, it was

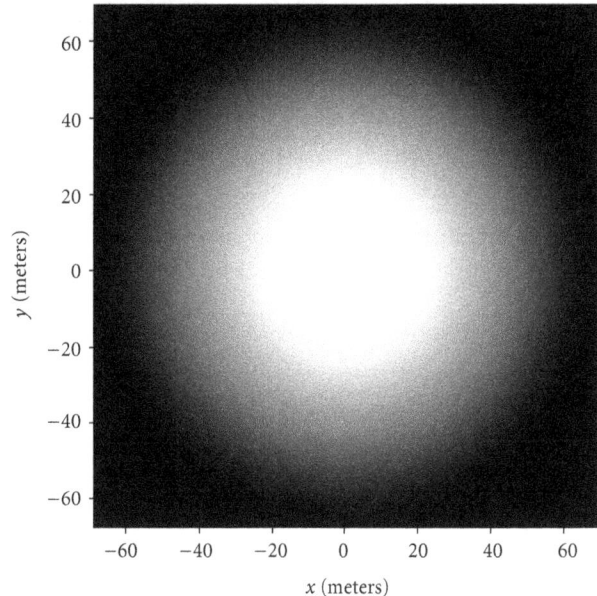

FIGURE 13: A sample sensor observation coverage where the quality of coverage is represented by shaded value, white being high quality and black being low quality. This type of coverage is applicable when a sensor's observations are improved with close proximity.

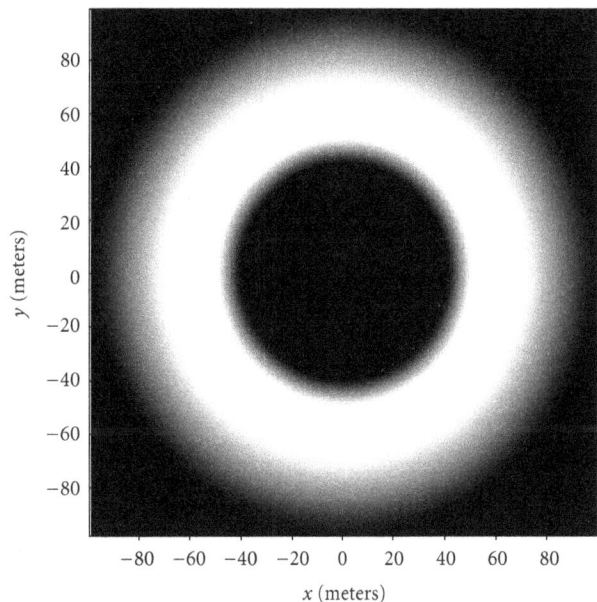

FIGURE 14: A sample sensor observation coverage where the quality of coverage is represented by shaded value, white being high quality and black being low quality. This type of coverage is applicable when a sensor makes good observations at some specified distance away.

designed to make good observations at an orbit's radius away from the aircraft. Figure 14 depicts this sensor coverage. In this figure the quality of a point's coverage by the agent's sensor is represented by a shaded value where white is high utility and black is low utility.

The utility of an agent's point x_0 in the surveillance area can then be defined utilizing sensor coverage. First, consider a zero horizon path. The utility of a point x_0 is

$$V_0(x_0) := \int_{x \in S} f(x) f_C(x, x_0) d\mu(x), \qquad (20)$$

where $f(x)$ is the target density distribution. Extending this to finite horizon planning, define the H-step horizon observation coverage over the path $x_{0:H} = (x_0, \ldots, x_H)$ as

$$f_C(x, x_{0:H}) := 1 - \prod_{t=0:H} (1 - f_C(x, x_t)). \qquad (21)$$

The utility of a point $x_0 \in S$, and consequently the path $x_{0:H}$, is then defined as

$$V_H(x_0) := \max_{\substack{x_i \in \mathrm{R}(x_{i-1}) \\ i=1,\ldots,H}} \int_{x \in S} f(x) f_C(x, x_{0:H}) d\mu(x), \qquad (22)$$

where $\mathrm{R}(x)$ is the set of all points within the reach set of x [14]. Intuitively, (22) represents the expected number of targets within the sensor coverage over a H-step path originating from the point x_0. Maximizing (22) then corresponds to choosing the point $x_0^\star$ that yields the maximum expected number of targets within the observation coverage of a path originating from $x_0^\star$.

This definition of path utility was based on the entire target density distribution. In order to optimize paths through specific partitions the utility must be defined for partition level target density distributions. Fortunately, this extension is easily accomplished. First, let the set of partitions defined over the surveillance area be $P = \{P_1, \ldots, P_n\}$. The partition level target density distribution $f_{P_i}(x)$ for partition $P_i \in P$ is then defined as

$$f_{P_i}(x) := \begin{cases} f(x) & \text{if } x \in P_i, \\ 0 & \text{otherwise.} \end{cases} \qquad (23)$$

And then replacing $f(x)$ with $f_{P_i}(x)$, the partition level utility of a path starting at x_0 is then defined as

$$V_H^{P_i}(x_0) := \max_{\substack{x_i \in \mathrm{R}(x_{i-1}) \\ i=1,\ldots,H}} \int_{x \in S} f_{P_i}(x) f_C(x, x_{0:H}) d\mu(x). \qquad (24)$$

This equation fully specifies partition level target density-distribution-based path optimization. Further development and application-specific details can be found in [14, 17].

6. Results

In this paper a new approach for autonomous search and tracking was presented. This new approach was designed for the case when the surveillance area is large, the number of targets is unknown, and target estimation is general and nonparametric. In this section, results of the performance of this approach are presented.

The performance of partition learning to aid in agent path planning was tested by constructing a simulation environment. In this environment the team of agents consisted of autonomous aircraft equipped with visual spectrum

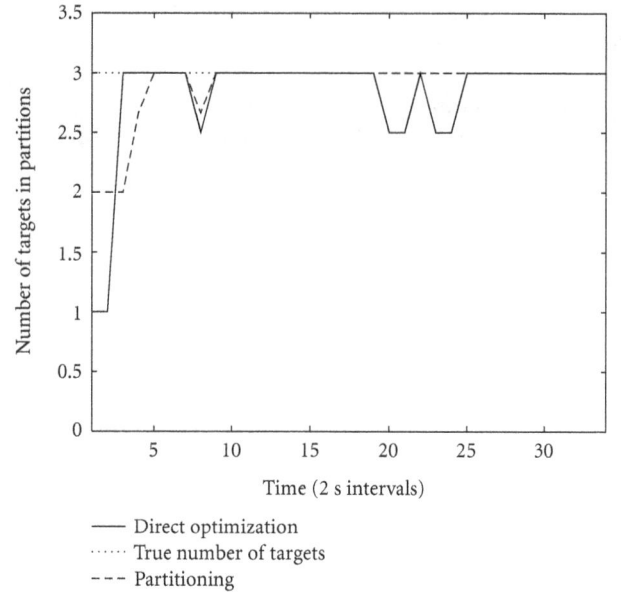

— Direct optimization
······ True number of targets
--- Partitioning

FIGURE 15: Comparison of sample mean number of targets within search or tracking partitions over simulation time between state-of-the-art direct distribution optimization and partition learning classification path planning approaches.

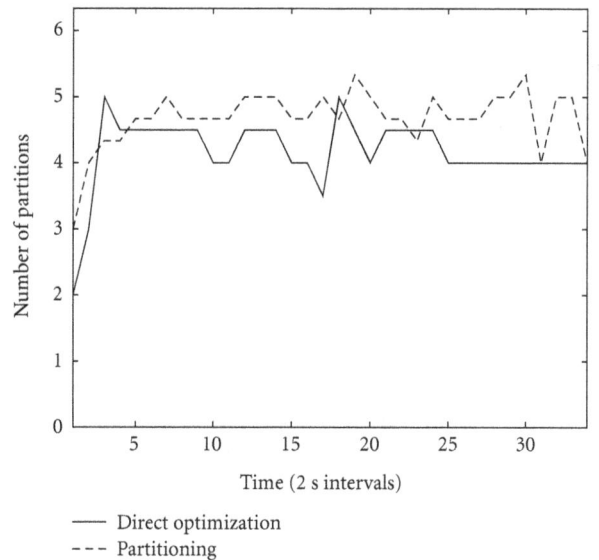

— Direct optimization
--- Partitioning

FIGURE 16: Comparison of sample mean number of search and tracking partitions over simulation time between state-of-the-art direct distribution optimization and partition learning classification path planning approaches.

gimballed camera sensors. The capabilities of these agents were designed to closely represent behaviors observed in flight experiments [32, 33]. The camera characteristics were designed to represent a field of view resulting from a 0.9273 rad view angle. Effects of resolution were included by limiting the distance of observations to 250 m. The agents were designed to fly at 25 m/s and 100 m altitude with a

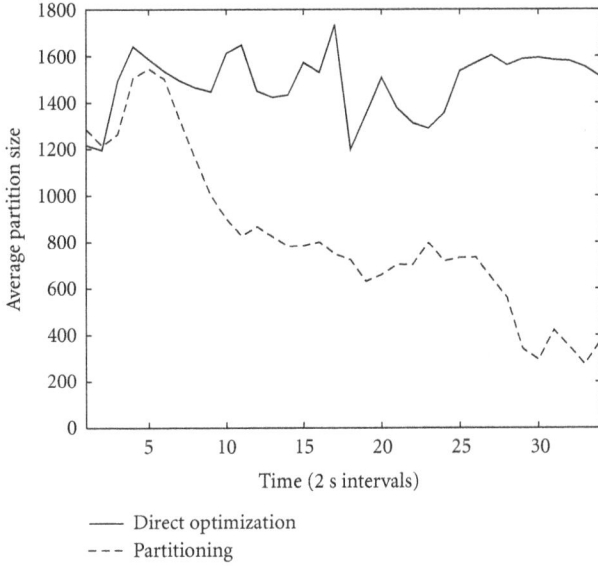

FIGURE 17: Comparison of sample mean average search and tracking partition size over simulation time between state-of-the-art direct distribution optimization and partition learning classification path planning approaches.

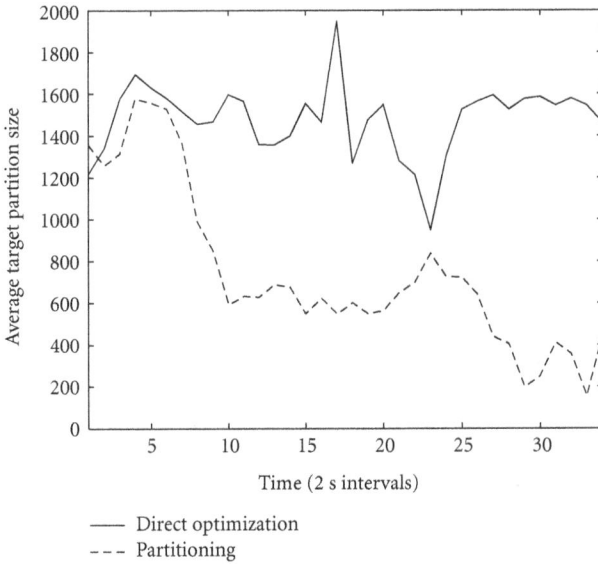

FIGURE 18: Comparison of sample mean average partition size of partitions containing targets over simulation time between state-of-the-art direct distribution optimization and partition learning classification path planning approaches.

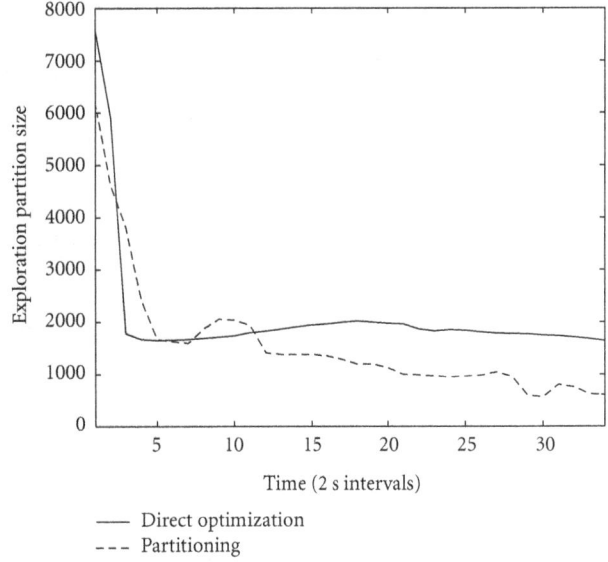

FIGURE 19: Comparison of sample mean exploration partition size over simulation time between state-of-the-art direct distribution optimization and partition learning classification path planning approaches.

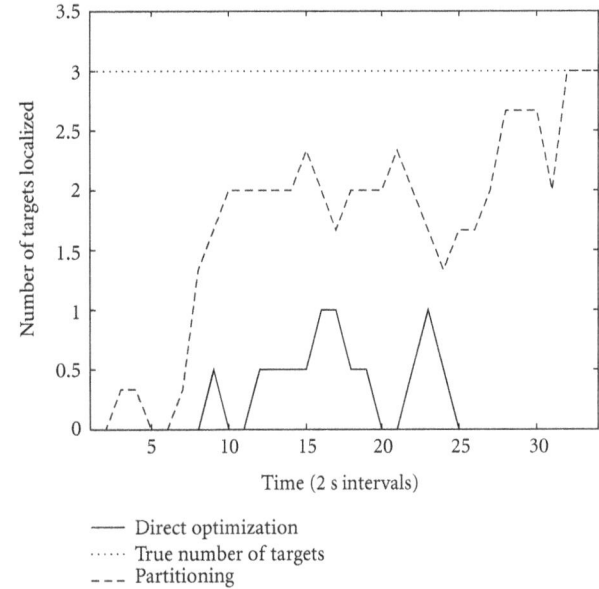

FIGURE 20: Comparison of sample mean number of targets localized over simulation time between state-of-the-art direct distribution optimization and partition learning classification path planning approaches.

maximum turn rate of 0.2 rad/s. The targets were allowed to move according to a transition model defined by

$$x_{T,t} = x_{T,t-1} + r \begin{bmatrix} \cos(\theta) \\ \sin(\theta) \end{bmatrix}, \tag{25}$$

where r and θ were distributed as

$$r \sim \text{Gaussian } (\mu, \sigma^2),$$
$$\theta \sim \text{Uniform } ([0, 2\pi)), \tag{26}$$

with $\mu = 2\,\text{m}$ in one second and $\sigma^2 = 10\,\text{m}^2$. The time interval of each simulation iteration was 4 seconds. Several

simulation samples were performed with various initial target and agent positions as well as various prior density distributions.

The performance of partition classification is affected by the quality and diversity of observations made over the surveillance area. Partition classification should perform well according to the observations it receives. To measure this performance several metrics were used. These metrics are

(1) number of targets in search and tracking partitions,

(2) number of search and tracking partitions,

FIGURE 21: Sample sequence of partition learning classification path planning for a scenario involving six agents and six targets (the number of targets was unknown to the agents). Agents are represented by blue/green circles with protruding lines that represent the direction of the agents. The approximate field of view of each agent is represented by green connected lines. Targets are represented by black/red circles. The entire surveillance area is partitioned. These partitions are represented by different colors. Follow the sequence from left to right and from top to bottom. In (a) initial partitions are formed. In (b) tracking partitions appear. By (f) all targets are within tracking partitions.

(3) average search and tracking partition size,

(4) average size of search and tracking partitions containing targets,

(5) exploration partition size,

(6) number of targets localized.

It is additionally necessary to provide a comparison in order to see how well the presented methods perform. Recall the standard approach for autonomous search and tracking. The standard approach optimizes agent paths directly over the distribution. The comparison provided in this section is then between using partition learning to aid path planning versus optimizing paths directly over the target distribution. This standard approach will be referred to as the state-of-the-art. Note, however, in order to compute some of the metrics above, it is necessary to run the partition classification algorithm for both cases. The partition learning classifier was run for both path planning approaches (that presented in this paper and the state-of-the-art). Yet, only the path planning approach presented in this paper utilized the partitions for path planning purposes.

In this paper, results for the scenario in which there are six agents and three targets is provided. Additionally, a sample sequence of partition learning is provided as a visual aid to understand how these partitions may look for the case when there are six agents and six targets. This sample sequence is found in Figure 21. This figure spans an entire page so it is provided after all other figures. Additional scenarios are provided in [17].

The results provided in this section demonstrate the scenario when there are sufficient resources to perform surveillance. From these results it is concluded that the the approach presented in this paper performed well. This conclusion is determined by observing Figures 15 and 16. Figure 15 presents the number of targets in search or tracking partitions over time. Figure 16 presents the number of search or tracking partitions over time.

From Figures 15 and 16 it is apparent that all targets are quickly captured within search or tracking partitions and the number of partitions is bounded. However, the performance of the two path planning approaches is very different. From Figure 17, it can be seen that the average partition size does not decrease for state-of-the-art path planning. However, the average partition size decreases substantially for partition learning classification path planning.

A similar result is also true for the average size of partitions containing targets, plotted in Figure 18. Additionally, according to Figure 19, the exploration size continually decreases for partition learning classification, but tends to level off for state-of-the-art path planning. Furthermore, the state-of-the-art path planning did not perform well to localize targets in this scenario. In contrast to this, partition learning classification path planning performed well to eventually localize all targets. This can be seen in Figure 20. From the results presented here, it is then apparent that partition learning classification path planning performs well to find and localize targets for this scenario, as compared to state-of-the-art path planning.

7. Conclusions

In this paper a new approach for autonomous search and tracking was presented. This new approach was designed for the case when the geographic area is large, the number of targets is unknown, and target track estimation is general and nonparametric. This is a challenging problem because very little is assumed. All that was assumed is that some form of target track estimation is available and shared among the team of autonomous agents performing the search. This new approach decomposes the search and tracking problem into three steps. The first step is target density distribution estimation. The second step is partition learning classification based on the target density distribution. The third step is path planning based on the partitions.

The vast body of work available for target track estimation and path planning over probability distributions was leveraged to provide solutions for the first and third steps. As such, the main focus of this paper was on partition learning. In order to determine the performance of this new approach, it was compared with the standard approach of directly optimizing paths over target estimation distributions. From this comparison, it is concluded that the approach presented in this paper performs well and provides an improved solution for this very general form of the autonomous search and tracking problem.

References

[1] Y. Bar-Shalom, *Tracking and Data Association*, Academic Press, San Diego, Calif, USA, 1987.

[2] Y. Bar-Shalom and X. R. Li, *Multitarget-Multisensor Tracking: Principles and Techniques*, YBS, Urbana, Ill, USA, 1995.

[3] S. S. Blackman and R. Popoli, *Design and Analysis of Modern Tracking Systems*, Artech House, Norwood, Mass, USA, 1999.

[4] L. Stone, *Theory of Optimal Search*, Academic Press, New York, NY, USA, 1975.

[5] R. P. S. Mahler, *Statistical Multisource-Multitarget Information Fusion*, Artech House, Norwood, Mass, USA, 2007.

[6] F. Bourgault, T. Furukawa, and H. Durrant-Whyte, "Optimal search for a lost target in a bayesian world," in *Field and Service Robotics*, S. Yuta, H. Asama, E. Prassler, T. Tsubouchi, and S. Thrun, Eds., vol. 24 of *Springer Tracts in Advanced Robotics*, pp. 209–222, Springer, Berlin, Germany, 2006.

[7] C. M. Kreucher, A. O. Hero, K. D. Kastella, and M. R. Morelande, "An information-based approach to sensor management in large dynamic networks," *Proceedings of the IEEE*, vol. 95, no. 5, pp. 978–999, 2007.

[8] B. Grocholsky, A. Makarenko, and H. Durrant-Whyte, "Information-theoretic coordinated control of multiple sensor platforms," in *Proceedings of the IEEE International Conference on Robotics and Automation (ICRA '03)*, vol. 1, pp. 1521–1526, September 2003.

[9] G. M. Hoffmann and C. J. Tomlin, "Mobile sensor network control using mutual information methods and particle filters," *IEEE Transactions on Automatic Control*, vol. 55, no. 1, pp. 32–47, 2010.

[10] J. Tisdale, Z. W. Kim, and J. K. Hedrick, "Autonomous UAV path planning and estimation: an online path planning framework for cooperative search and localization," *IEEE Robotics and Automation Magazine*, vol. 16, no. 2, pp. 35–42, 2009.

[11] F. Bourgault, T. Furukawa, and H. F. Durrant-Whyte, "Coordinated decentralized search for a lost target in a bayesian world," in *Proceedings of the IEEE/RSJ International Conference on Intelligent Robots and Systems*, vol. 1, pp. 48–53, October 2003.

[12] A. D. Ryan, H. Durrant-Whyte, and J. K. Hedrick, "Information-theoretic sensor motion control for distributed estimation," in *Proceedings of the International Mechanical Engineering Congress and Exposition (IMECE '07)*, November 2007.

[13] G. M. Mathews, H. Durrant-Whyte, and M. Prokopenko, "Decentralised decision making in heterogeneous teams using anonymous optimisation," *Robotics and Autonomous Systems*, vol. 57, no. 3, pp. 310–3320, 2009, selected papers from IEEE International Conference on Multisensor Fusion and Integration (MFI '06).

[14] J. G. Wood, B. Kehoe, and J. K. Hedrick, "Target estimate pdf-based optimal path planning algorithm with application to UAV systems," in *Proceedings of the Dynamic Systems and Control Conference (DSCC '10)*, pp. 749–756, ASME, September 2010.

[15] E. M. Wong, F. Bourgault, and T. Furukawa, "Multi-vehicle Bayesian search for multiple lost targets," in *Proceedings of the IEEE International Conference on Robotics and Automation (ICRA '05)*, pp. 3169–3174, April 2005.

[16] T. Furukawa, F. Bourgault, B. Lavis, and H. F. Durrant-Whyte, "Recursive Bayesian search-and-tracking using coordinated UAVs for lost targets," in *Proceedings of the IEEE International Conference on Robotics and Automation (ICRA '06)*, pp. 2521–2526, May 2006.

[17] J. Wood, *Search and tracking of an unknown number of targets by a team of autonomous agents utilizing time-evolving partition classification [Ph.D. thesis]*, University of California, Berkeley, Calif, USA, 2011.

[18] J. Wood and J. K. Hedrick, "Space partitioning and classification for multi-target search and tracking by heterogeneous unmanned aerial system teams," in *Proceedings of the Infotech@Aerospace*, American Institute of Aeronautics and Astronautics, 2011.

[19] I. R. Goodman, R. P. S. Mahler, and H. T. Nguyen, *Mathematics of Data Fusion*, Kluwer Academic, Boston, Mass, USA, 1997.

[20] G. Mathew, A. Surana, and I. Mezić, "Uniform coverage control of mobile sensor networks for dynamic target detection," in *Proceedings of the 49th IEEE Conference on Decision and Control (CDC '10)*, pp. 7292–7299, December 2010.

[21] G. Mathew and I. Mezi, "Metrics for ergodicity and design of ergodic dynamics for multi-agent systems," *Physica D*, vol. 240, no. 4-5, pp. 432–442, 2011.

[22] H. Chen, K. Chang, and C. S. Agate, "Tracking with UAV using tangent-plus-lyapunov vector field guidance," in *Proceedings of the 12th International Conference on Information Fusion*, pp. 363–372, July 2009.

[23] M. Shanmugavel, A. Tsourdos, B. White, and R. Zbikowski, "Co-operative path planning of multiple UAVs using Dubins paths with clothoid arcs," *Control Engineering Practice*, vol. 18, no. 9, pp. 1084–1092, 2010.

[24] J. Lee, R. Huang, A. Vaughn et al., "Strategies of path-planning for a UAV to track a ground vehicle," in *Proceedings of the 2nd Annual Symposium on Autonomous Intelligent Networks and Systems*, 2003.

[25] S. C. Spry, A. R. Girard, and J. K. Hedrick, "Convoy protection using multiple unmanned aerial vehicles: organization and coordination," in *Proceedings of the American Control Conference (ACC '05)*, pp. 3524–3529, June 2005.

[26] A. Sinha, T. Kirubarajan, and Y. Bar-Shalom, "Autonomous ground target tracking by multiple cooperative UAVs," in *Proceedings of the IEEE Aerospace Conference*, pp. 1–9, March 2005.

[27] P. Skoglar, *Planning methods for aerial exploration and ground target tracking [Licentiate thesis]*, Linköping University, Linköping, Sweden, 2009.

[28] R. O. Duda, P. E. Hart, and D. G. Stork, *Pattern Classification*, John Wiley & Sons, New York, NY, USA, 2001.

[29] M. Godwin and J. K. Hedrick, "Stochastic approximation of an online vehicle routing problem for autonomous aircraft," Infotech@Aerospace Conference, 2011.

[30] M. Alighanbari and J. P. How, "A robust approach to the UAV task assignment problem," *International Journal of Robust and Nonlinear Control*, vol. 18, no. 2, pp. 118–134, 2008.

[31] Z. Kim and R. Sengupta, "Target detection and position likelihood using an aerial image sensor," in *Proceedings of the IEEE International Conference on Robotics and Automation (ICRA '08)*, pp. 59–64, May 2008.

[32] R. Sengupta, J. Connors, B. Kehoe, Z. Kim, T. Kuhn, and J. Wood, "Autonomous search and rescue with ScanEagle," Tech. Rep., 2010, prepared for Evergreen Unmanned Systems and Shell International Exploration and Production Inc.

[33] J. Garvey, B. Kehoe, B. Basso et al., "An autonomous unmanned aerial vehicle system for sensing and tracking," in *Proceedings of the Infotech@Aerospace Conference*, 2011.

14

Smart Localization Using a New Sensor Association Framework for Outdoor Augmented Reality Systems

F. Ababsa, I. Zendjebil, J.-Y. Didier, and M. Mallem

Laboratoire IBISC, EA 4526, Université d'Evry-Val-d'Essonne, 40 rue du Pelvoux, 91020 Evry, France

Correspondence should be addressed to F. Ababsa, ababsa@iup.univ-evry.fr

Academic Editor: Huosheng Hu

Augmented Reality (AR) aims at enhancing our the real world, by adding fictitious elements that are not perceptible naturally such as: computer-generated images, virtual objects, texts, symbols, graphics, sounds, and smells. The quality of the real/virtual registration depends mainly on the accuracy of the 3D camera pose estimation. In this paper, we present an original real-time localization system for outdoor AR which combines three heterogeneous sensors: a camera, a GPS, and an inertial sensor. The proposed system is subdivided into two modules: the main module is vision based; it estimates the user's location using a markerless tracking method. When the visual tracking fails, the system switches automatically to the secondary localization module composed of the GPS and the inertial sensor.

1. Introduction

The idea of combining several kinds of sensors is not recent. The first multi-sensors system appeared with robotic applications where, for example, in [1] Vieville et al. proposed to combine a camera with an inertial sensor to automatically correct the path of an autonomous mobile robot. This idea has been exploited these last years by the community of Mixed Reality. Several works proposed to fuse vision and inertial data sensors, using a Kalman filter [2–6] or a particular filter [7, 8]. The strategy consists in merging all data from all sensors to localize the camera following a prediction/correction model. The data provided by inertial sensors (gyroscopes, magnetometers, etc.) are generally used to predict the 3D motion of the camera which is then adjusted and refined using the vision-based techniques. The Kalman filter is generally implemented to perform the data fusion. Kalman filter is a recursive filter that estimates the state of a linear dynamic system from a series of noisy measurements. Recursive estimation means that only the estimated state from the previous time step and the current measurement are needed to compute the estimate for the current state. So, no history of observations and/or estimates is required.

In [2] You et al. developed a hybrid sensor combining a vision system with three gyroscopes to estimate the orientation of the camera in an outdoor environment. Their visual tracking allows refining the obtained estimation. The system described by Ababsa [5] combines an edge-based tracking with inertial measurements (angular velocity, linear acceleration, magnetic fields). The visual tracking is used for accurate 3D localization while the inertial sensor compensates errors due to sudden motion and occlusion. The measurements of gravity and magnetic field are used to limit the drift problem. The gyroscope is employed to automatically reset the tracking process. Data provided by the two sensors are combined with an extended Kalman filter using a constant velocity model. More recently [9], the same authors proposed to use the GPS positions to re-initialize visual tracking when it fails. Thus, initialization of the visual tracking is obtained by defining a search area represented by an ellipse centered on the GPS position.

Recently, Bleser and Stricker [6] proposed to combine a texture-based tracking with an inertial sensor. The camera pose is predicted from data provided by the accelerometers using an Extended Kalman filter (EKF). In order to estimate the pose, the EKF fuse the 2D/3D correspondences obtained from the image analysis and the inertial measurements acquired from the inertial sensor. A rendering of CAD model (textured patches) is made using the predicted poses. This allows aligning iteratively the textured patches in the current image to estimate the 2D motion and to update the estimate given by the filter. Natural feature points are tracked by a KLT (Kanade Lucas Tomasi) tracker. The motion model assumes constant acceleration and constant angular velocity. This approach needed offline preparation for generating a textured CAD model of the environment. Hu et al. [10] proposed to combine a camera, a GPS and an inertial gyroscope sensor. The fusion approach is based on PPM (Parameterized model matching algorithm). The road shape model is derived from the digital map with respect to GPS position, and matches with road features extracted from the real images. The fusion is based on a predictor-corrector control theory. After checking data integrity, GPS data will start a new loop and reset gyro's integrated. Gyro's prediction will be feedback into the gyro integration module as a dynamical correction factor. When the image feature tracking fails, gyro's prediction data is used for the camera pose estimation. Ababsa and Mallem [8] proposed a particle filter instead of the Kalman filter. Particle filters (PF), also known as methods of Monte-Carlo sequential, are sophisticated techniques for estimating models based on simulation. PFs are generally used to estimate Bayesian models. They represent an alternative to extended Kalman filter, their advantage is that they approach the optimal Bayesian estimation using enough samples. Ababsa et al. merged data from fiducial-based method with inertial data (gyros and accelerometers). Their fusion algorithm is based on a particle filter with sampling importance resampling (SIR). As the two sensors have different sampling frequency, the authors implemented two complementary filters. Thus, if there is no data of vision (e.g., occlusion), the system uses only data from the inertial sensor and vice versa. Aron et al. [11] used the inertial sensor to estimate the orientation of the camera only when the visual tracking fails. The orientation allows tracking the visual primitives by defining a search area in the image to perform the features matching. A homography is estimated from this set of matched features to estimate the camera pose. The errors of the inertial sensor are taken into account to optimize the search area. Unlike the approach proposed by Aron et al. [11] which only estimates the camera orientation, Maidi et al. [12] used an inertial sensor to estimate both the position and the orientation. Their multimodal system allows tracking fiducials and handling occlusions by combining several sensors and techniques depending on the existing conditions in the environment. When the target is partially occluded, the system uses a point-based tracking. In presence of a total occlusion of the fiducials, inertial sensor helps to overcome the vision failure. However, the estimation of position from acceleration produces drift over time resulting in a tracking failure. Combining sensors following the assistance scheme seems more interesting than the data fusion. Indeed, assistance approach makes the system more intelligent so that it can adapt itself to different situations and uses at each time only the data provided by the available sensors.

In this research work, we are interested in developing an original localization system combining three heterogeneous sensors (Camera, GPS and IMU) in order to increase the accuracy and the robustness. Our objective is to carry out a generic solution for the 3D localization, 3D visualization and interaction adaptable to several outdoor environments. The remainder of the paper is organized as follows: Section 2 is devoted to the overview of our Hybrid localization system. In Section 3, we give the formulation of the camera pose estimation problem when using point features. Sections 3 and 4, will describe in details our proposed hybrid localization system. Section 5 presents some performed experiments and results obtained in outdoor environments under real conditions. We conclude in Section 5 and suggest future works.

2. Hybrid Localization System Overview

Our localization system is wearable and composed of a Tablet PC, an Inertial Measurement Unit (IMU), a camera and a GPS receiver (see Figure 1). The IMU is rigidly coupled with the camera and used to estimate the camera rotation. We used an Xsens MTi sensor which contains gyroscopes, accelerometers and magnetometers. The advantage of the MTi is that it incorporates an internal digital signal processor which runs a real-time sensor fusion algorithm providing a reliable 3D orientation estimate. Data from MTi are synchronously measured at 100 Hz. For the vision, we opted for the uEye UI-2220-RE-C CCD camera with 6 mm lens which is extremely compact, low-cost and well adapted for outdoor environments. Color images with resolution of 768×576 pixels at a frame rate of 25 Hz are streamed to a PC using a USB 2.0 connection. A Trimble GPS Pathfinder ProXT receiver mounted on a user provides GPS measurements. The ProXT receiver integrates a multipath rejection technology providing a submeter accuracy. Its rate update is about 1 Hz. The ProXT receiver uses a Bluetooth wireless connection, to communicate with the computer. The three sensors providing measurements to the system are synchronized in hardware and runs at different rates.

This localization system uses an assistance scheme between the several sensors, and thus it is subdivided in two subsystems: a main subsystem and an auxiliary one. The main subsystem corresponds to the vision based localization. The auxiliary subsystem is used only when the visual tracking fails; it is composed of the GPS and the inertial sensors. This hybrid system estimates continuously the position and orientation of the point of view, even when the vision fails. Figure 2 provides a flow chart to describe the 3D localization process using the proposed assistance scheme.

3. Vision-Based Localization System

3.1. Camera Pose Problem Formulation. Throughout this paper, we assume a calibrated camera and a perspective projection model. If a point has coordinates $(x, y, z)^t$ in the coordinate frame of the camera, its projection onto the image plane is $(x/z, y/z, 1)^t$. In this section, we present the

FIGURE 1: Sensor components of our Hybrid tracker.

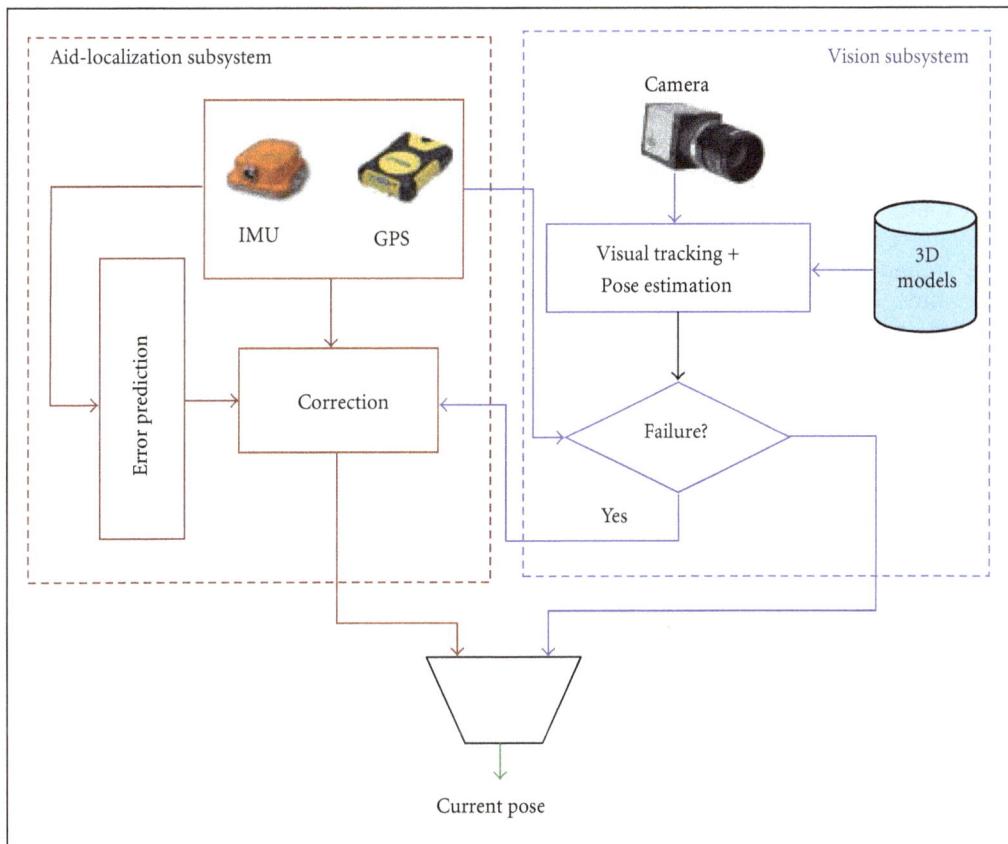

FIGURE 2: The system data flow scheme.

constraints for camera pose determination when using point and line features.

3.2. *Problem Definition.* Let $\mathbf{p}_i = (x_i, y_i, z_i)^t, i = 1, \ldots, n, \ n \geq 3$ a set of 3D non-collinear reference points defined in the world reference frame, the corresponding camera-space coordinates $\mathbf{q}_i = (x_i', y_i', z_i')$ are given by:

$$\mathbf{q}_i = R\mathbf{p}_i + T, \qquad (1)$$

where $R = (\mathbf{r}_1^t, \mathbf{r}_2^t, \mathbf{r}_3^t)^t$ and $T = (t_x, t_y, t_z)^t$ are a rotation matrix and a translation vector, respectively. R and T describe

the rigid body transformation from the world coordinate system to the camera coordinate system and are precisely the parameters associated with the camera pose problem.

Let the image point $\mathbf{g}_i = (u_i, v_i, 1)^t$ be the projection of $\mathbf{p}_i$ on the normalized image plane. Using the camera pinhole model, the relationship between $\mathbf{g}_i$ and $\mathbf{p}_i$ is given by:

$$\mathbf{g}_i = \frac{1}{\mathbf{r}_3^t \mathbf{p}_i + t_z}(R\mathbf{p}_i + T) \qquad (2)$$

which is known as the colinearity equation.

The point constraint corresponds to the image space error, it gives a relationship between 3D reference points, their corresponding 2D extracted image points and the camera pose parameters as follows:

$$E_i^p = \sqrt{\left(\hat{u}_i - \frac{\mathbf{r}_1^t \mathbf{p}_i + t_x}{\mathbf{r}_3^t \mathbf{p}_i + t_z}\right)^2 + \left(\hat{v}_i - \frac{\mathbf{r}_2^t \mathbf{p}_i + t_y}{\mathbf{r}_3^t \mathbf{p}_i + t_z}\right)^2}, \quad (3)$$

where $\hat{\mathbf{m}}_i = (\hat{u}_i, \hat{v}_i, 1)^t$ are the observed image points.

The pose estimation problem is to find the rigid transform (R, t) that best fits the known 3D reference points with the observed 2D image points. Usually this is achieved by minimizing some form of accumulation of errors (least squares methods) based on (3). Typically Gauss-Newton or Levenberg-Marquardt methods are used for this purpose [13, 14].

3.3. Discussion. 3D-2D feature matching is critical for the camera pose estimation and still a difficult unsolved problem in computer vision. The tracking system needs an initialization that provides a set of good 3D-2D matched points. In practice, the accuracy of the matching process depends on the relevance of the information associated to the 3D points for their recognition. One interesting approach is to define a reference patches around the image points corresponding to the 3D model points. Matching is then performed by aligning these references patches within those extracted from the current frame. The correlation can be used to measure the similarity between the patches as in [15]. However, this method is not robust against illumination variation. Other approaches use the SIFT descriptors [16] for their robustness to changes in viewing conditions. The main disadvantage of the SIFT is its complexity and its high time consumption, this makes it not suitable for real-time applications. In this work, we propose two vision-based initialization approaches. The first one is semi-automated and requires the user intervention to guide the matching process; it is used to start the tracking system. The second approach is fully automated; it is executed when the tracking is lost. The next sections give an overview of these approaches.

3.4. Semi-Automated Initialization Approach. The proposed semi-automated initialization approach is performed in two steps: the wireframe model of the environment (here the building frontage) is first render, in real time, on the video flow coming from the camera, using a set of predefined poses (see Figure 3(a)). At the same time the user moves the camera in order to align the projected model within its image. Once this alignment is achieved (see Figure 3(b)) the user validates the corresponding pose and the system switch to the matching step in order to perform, with high accuracy, the 3D-2D points matching. Aligning the rendered model allows to limit the search area of the 2D points in the current image. This makes the approach faster and robust against outliers.

The 3D-2D matching is performed as follows. A search box is defined around each projected 3D model point on the aligned image. The interest points are then extracted from these image regions. As 3D points represent corners in the model, we use the Harris detector [17] in order to extract the 2D interest points. Then, a SIFT descriptor is computed and associated to each extracted Harris point. We choose to use the SIFT descriptor because it is scale-rotation invariant and allows real-time tracking. The distances between the reference descriptor associated to the 3D point and the descriptors of the extracted 2D points are measured and compared. The 2D point that minimizes this distance is selected as the corresponding 2D point. We have also used a RANSAC algorithm [18] in order to detect and remove outliers in the matching set, and thus increasing the accuracy of the initialization.

In order to validate the whole matching 3D-2D points, we introduce a coherence test which is used as a quality measurement for the estimated camera pose. We assume that this pose is close to that selected when the wireframe model is aligned within the image. Let, $P_a = [R_a \mid T_a]$ be the predefined camera pose that is used for the model/image alignment, and $P = [R \mid T]$ the camera pose estimated using the set of candidate matched points. As the two matrices are identical, we can write then:

$$P_a \cdot P^{-1} = I, \quad (4)$$

where I is a 4×4 identity matrix.

So, the trace of the matrix $P_a \cdot P^{-1}$ tends to 4. Out coherence test can be formulated as:

$$\delta < \text{Trace}(P_a \cdot P) < 4, \quad (5)$$

where δ is a threshold below which the two matrices are considered different.

3.5. Automated Initialization Approach. Unlike the semi-automated approach described above, in automated initialization procedure, the user intervention is not required. The system switches automatically to this mode every time when the tracking fails because of noise, occlusion or image blurring. This approach is performed as follows (see Figure 4).

Let F_i be the reference frame that corresponds to the last captured image before the tracking failed. Several information are associated to this frame namely: the camera poses P_i and the set of matched 3D-2D points. The idea is to generate new 3D-2D matched points between the reference and the current frames.

For that, we first project the 3D points on the current frame using the reference camera pose P_i to generate predicted research areas. These image areas, named patches, are centred on the projected 3D points and have rectangular shape. The interest points corresponding to the SIFT features are then extracted inside these patches and matched with those extracted from the reference frame. We also use a RANSAC algorithm in order to discard the outliers. To find the 3D-2D matched points for the current frame, we only need to identify the transformation that maps interest points defined in the reference image to those extracted in the current image. We assume that this transformation is a

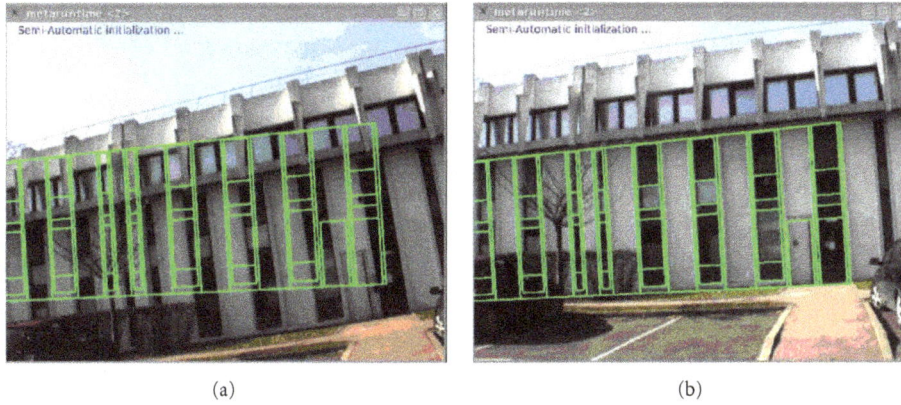

FIGURE 3: (a) the model environment rendering. (b) manual alignment between the projected model and its image.

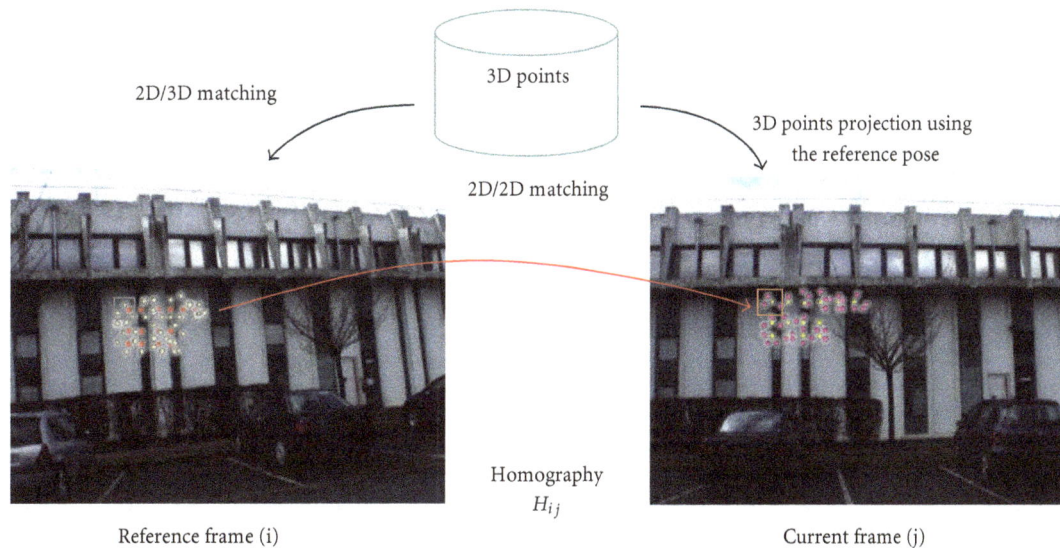

FIGURE 4: Automated initialization approach.

homography because the movement between the two frames is not meaningful.

Let H_{ij} be this homography, m_i and m_j the interest points extracted from reference and current frames F_i, and F_j, respectively. The relationship between m_i, m_j and H_{ij} is defined as:

$$m_j = H_{ij} \cdot m_i. \tag{6}$$

Once the homography is estimated, we apply it to the 2D image points associated to the 3D model points for the reference frame, in order to transform them in the current image. This allows updating correspondence between 3D and 2D points for the current image, and hence restarts automatically the tracking process.

3.6. Visual Tracking. Once the vision system is initialized the visual tracking can start. To estimate the camera pose, we must keep the 2D/3D matching for each current view. This can be achieved by using a frame-to-frame 2D points tracking. Tracking consists in following features from one frame t-1 to another frame t. Several approaches can be

used such as correlation matching methods; however they are very expensive in computing time. To track 2D features in real time, the chosen method must be fast and accurate. For that the KLT Tracker [19] can be adopted. This algorithm used an optical flow computation to track features points or a set of predefined points from the previous image I_{t-1} to the current image I_t. Therefore, this algorithm tracks a set of 2D points associated to visible 3D points. Briefly, 2D points are searched in the neighborhood of its position in view t-1 based on the minimization of brightness difference. To minimize the computation time, the KLT tracker uses a pyramid of images for the current view. Therefore, tracking is done at the coarsest level and then propagate to the finest. This allows following the features over a long distance with great precision. The approach is fast and accurate, but it requires that the tracked points are always visible. So the approach does not handle occlusions.

4. IMU-GPS Localization System

This subsystem, replaces the vision-based localization system when this one fails. The position and orientation given by the vision subsystem are substituted by the absolute position provided by the GPS receiver and the orientation given by the inertial sensor. The use of the Auxiliary subsystem is not limited only to replace the vision subsystem. The Auxiliary subsystem is also used to initialize the vision subsystem. Moreover, from the position and orientation given by this subsystem, we can measure the accuracy of the 3D localization estimated by the vision subsystem by defining some confidence intervals. The Auxiliary subsystem is composed of two modules: prediction and correction. The prediction module is used to predict accuracy errors of the localization system. It is based on online training of the error between the two subsystems. Once the localization system switches to the Auxiliary subsystem, the error is predicted following a Gaussian model and used to improve the position and the orientation provided by the GPS and the inertial sensor. The two parts composing the system interact continuously with each other. Also, the use of GPS for position estimation solves the problem of inertial sensor's drift, which is used only for orientation estimation.

4.1. Sensors Calibration. In our hybrid system, each sensor provides data in its own reference frame. The inertial sensor computes the orientation between a body reference frame attached to it and a local level reference frame. Also, the GPS position is expressed in an earth reference frame defined by WGS84 (World Geodetic System) standard. For registration, we need to estimate continually the camera pose which relates the world reference frame to the camera reference frame. Thus, the 3D localization provided by the IMU-GPS system must be aligned with the camera reference frame. The several sensors must be aligned in a unified reference frame in order to have the same position and orientation of the point of view. So, the hybrid sensor must be calibrated to determine the relationships between the several sensors and thus to unify the measurements. The accuracy of the IMU-GPS system depends on the accuracy of the calibration processes. We have developed an original calibration method in order to compute the several transformations with high accuracy. Our method is divided in two calibration processes. The first one consists in estimating the relationship between inertial sensor and camera (Inertial/Camera calibration). The second one estimates the transformation which maps the GPS position to the camera position (GPS/Camera transformation).

4.1.1. Inertial/Camera Calibration. In order to deduce the camera orientation from the orientation given by the inertial sensor, we need to estimate the transformation between the references frames attached to the camera and the inertial sensor. The used reference frames are illustrated in Figure 5.

Let R_{CW} be the rotation of the world frame R_W with respect to the camera frame R_C. R_{CI} represents the rotation of R_I with respect to the camera frame R_C. The rotation between the body frame R_I and R_G is noted R_{IG}. Finally, R_{GW} represents rotation of the world frame R_W with respect to R_G. The rotation R_{IG} is given by the inertial sensor and R_{CW} is obtained from the camera pose estimation. Thus, we need

to compute rotation R_{CI} and R_{GW}. The relationship between the several frames is expressed by:

$$R_{CW} = R_{CI} \cdot R_{IG} \cdot R_{GW}. \tag{7}$$

In this case, the Inertial/Camera calibration process consists in estimating the rotation matrix R_{CI} and deducing the matrix R_{GW}. We assume that Z-axis of the R_G frame is pointing up along the vertical and is collinear with the Z-axis of the world coordinate frame R_W (Figure 6).

Therefore, this configuration implies that the matrix R_{GW} will correspond to a single rotation around the Z-axis with an angle θ. So, the matrix R_{GW} can be expressed as follows:

$$R_{GW} = \begin{pmatrix} \cos(\theta) & -\sin(\theta) & 0 \\ \sin(\theta) & \cos(\theta) & 0 \\ 0 & 0 & 1 \end{pmatrix}. \tag{8}$$

Assuming that $R_{IG} = (r_i^{IG})$, $i = 1, 2, 3$, where r_i^{IG} is the ith column of the R_{IG} matrix. Equation (8) can be written as:

$$R_{CW}$$
$$= \begin{pmatrix} R_{CI} \cdot r_1^{IG} & R_{CI} \cdot r_2^{IG} & R_{CI} \cdot r_3^{IG} \end{pmatrix} \cdot \begin{pmatrix} \cos(\theta) & -\sin(\theta) & 0 \\ \sin(\theta) & \cos(\theta) & 0 \\ 0 & 0 & 1 \end{pmatrix} \tag{9}$$

and then,

$$R_{CW}$$
$$= \begin{pmatrix} R_{CI} \cdot r_1^{IG} \cdot \begin{pmatrix} \cos(\theta) \\ \sin(\theta) \end{pmatrix} R_{CI} \cdot r_2^{IG} \cdot \begin{pmatrix} -\sin(\theta) \\ \cos(\theta) \end{pmatrix} R_{CI} \cdot r_3^{IG} \end{pmatrix} \tag{10}$$

From (10) we can deduce that:

$$r_3^{CW} = R_{CI} \cdot r_3^{IG} \tag{11}$$

We rewrite this equation in the linear form $A \cdot X = b$, so:

$$\begin{pmatrix} (r_3^{IG})^T & 0 & 0 \\ 0 & (r_3^{IG})^T & 0 \\ 0 & 0 & (r_3^{IG})^T \end{pmatrix} \cdot X = r_3^{CW}, \tag{12}$$

where $X = \begin{pmatrix} r_{11}^{CI} & r_{12}^{CI} & r_{13}^{CI} & r_{21}^{CI} & r_{22}^{CI} & r_{23}^{CI} & r_{31}^{CI} & r_{32}^{CI} & r_{33}^{CI} \end{pmatrix}^T$. We need at least 3 matrixes R_{IG} and R_{CW} in order to estimate R_{CI} using the least mean squares algorithm. Once the matrix R_{CI} is computed we can deduce the matrix R_{GW} using (1), so:

$$R_{GW} = R_{IG}^T \cdot R_{CI}^T \cdot R_{CW}. \tag{13}$$

The estimation of the two matrixes R_{CI} and R_{GW}, which is done off line, allows the system to estimate the rotation of the camera in real-time using only the orientation matrix given by the inertial sensor. This is very important to recover the camera pose when the visual tracking fails.

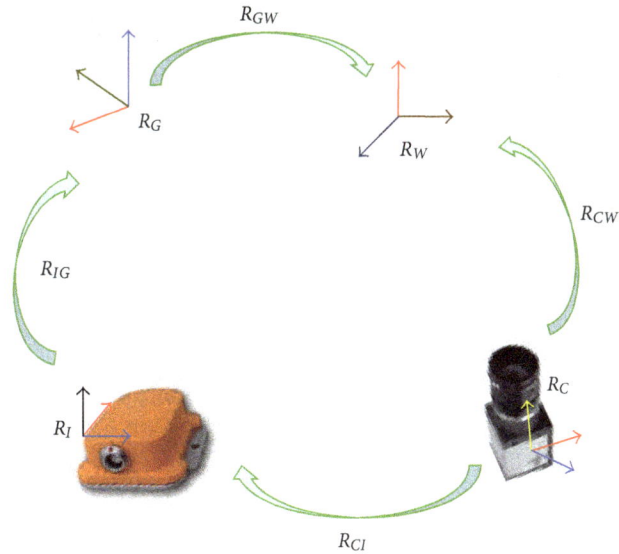

FIGURE 5: Inertial/Camera: Reference frames configuration.

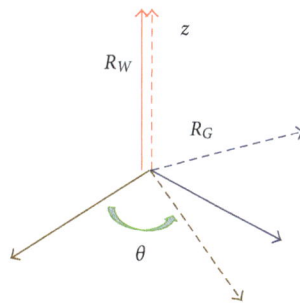

FIGURE 6: The relationship between the world reference frame R_W and the global inertial reference frame R_I.

4.1.2. GPS/Camera Calibration. This second calibration process estimates the rigid transformation (rotation + translation) which maps the GPS position to the local world reference frame R_W (see Figure 6). In our case, the position provided by the GPS receiver is expressed in degrees. A Lambert conic projection is necessary to express the GPS positions in meters. This projection minimizes the distortions and is adapted to the region where the experiments are done (i.e., France). The Lambert conic projection superimposes a cone over the sphere of the Earth. Let p_{gps} the camera position given by the GPS receiver and expressed in the earth reference frame and p_{cam} the camera position in the local world reference frame. The relationship between p_{gps} and p_{cam} is then given by:

$$p_{cam} = R \cdot p_{gps} + t, \qquad (14)$$

where p_{cam} is computed using the camera calibration parameters as follows: $p_{cam} = -R_{CW}^T \cdot t_{CW}$.

Finally, the rigid transformation is obtained by minimizing the following criterion:

$$\sum_{i=1}^{n} \left\| p_{cam}^i - \left(R \cdot p_{gps}^i + t \right) \right\|^2. \qquad (15)$$

In order to simplify the problem we have used the axis and the angle $(\vec{n}, \theta)$ to represent the rotation matrix R [20]. This allows subdividing the optimization problem into two sub-problems, one for the rotation estimation and one for the translation vector.

4.2. Localization Error Prediction. The estimation of the 3D localization provided by the combination of the GPS and the inertial sensor is less accurate then the vision-based estimation. The computation of the produced error is important in the localization process. Indeed, it allows quantifying the quality of measurements in order to improve the 3D localization estimation provided by the IMU-GPS localization system. The error represents the offset between the camera pose and the position and orientation deduced from GPS and inertial sensor. When the vision fails, this error must be predicted. For that, the error is modeled as a regression with a Gaussian process [21]. The Gaussian process is a stochastic process which generates samples and can be used as a prior probability distribution over functions in Bayesian inference.

Let be $x_1, x_2, \ldots, x_n$ a set of training data associated to $y_1, y_2, \ldots, y_n$ where $y_i = f(x_i)$. The goal is to predict the value y_{n+1} associated to the data x_{n+1}. We consider $(Y_1, Y_2, \ldots, Y_{n+1})$ a set of $(n + 1)$ random variables which have a Gaussian distribution with zero mean and covariance matrix Σ_{n+1} such as:

$$\Sigma_{n+1} = \begin{pmatrix} \Sigma_n & \kappa \\ \kappa^T & \kappa_{n+1} \end{pmatrix}, \tag{16}$$

Where κ_n is $n \times n$ matrix, κ is a n-column vector and κ_{n+1} is a scalar. If $y_1, y_2, \ldots, y_n$ are the observed variables associated to $x_1, x_2, \ldots, x_n$ then the conditional distribution $P(Y_{n+1} | Y_1, \ldots, Y_n)$ yielding a Gaussian distribution with:

$$\begin{aligned} \mu_{Y_{n+1}} &= \kappa^T \Sigma_n^{-1} y^n \\ \sigma^2_{Y_{n+1}} &= \kappa_{n+1} - \kappa^T \Sigma_n^{-1} \kappa, \end{aligned} \tag{17}$$

where $y^n = (y_1, y_2, \ldots, y_n)^T$, $\kappa_{n+1} = \mathrm{cov}(y_{n+1}, y_{n+1})$, $\kappa_i = \mathrm{cov}(y_{n+1}, y_i)$ and $\Sigma_{ij} = \mathrm{cov}(y_i, y_j)$. The covariance between y_i and y_j is the same as the covariance between x_i and x_j whish is given by:

$$\mathrm{cov}(x_i, x_j) = \frac{1}{N - |i - j|} \sum_{n=1}^{N - |i - j|} x_n \cdot x_{n+|i-j|}. \tag{18}$$

In our case x_i represents either the GPS position or the orientation matrix give by the inertial sensor, and y is the error localization that we want to predict. So, during the visual tracking, the offset between the IMU-GPS localization system and the vision-based localization system is recorded for the online training step. Training data are used to learn sampled covariance function. When the visual tracking fails, the Gaussian process predicts the offset made by GPS and the inertial sensor. This offset, which is represented by the mean error, is used to correct the estimated 3D localization. Indeed, the Gaussian Process allows computing $p(y \mid x)$ the likelihood of the error localization. This likelihood function is Gaussian, with mean and variance at the point x given by the Gaussian Process based map. In our case, mean and variance are used to compute the error ellipse of the estimated localization.

5. Real-Time Operating

Our localization system operates using a finite state machine scheme (see Figure 7). A finite state machine is an abstract model composed of a finite number of states, transitions between those states, and actions. This formalism is mainly used in the theory of computability and formal languages and allows running in real-time with high accuracy.

The proposed state machine is composed of three states: the Auxiliary predominance state, the initialization state and the visual predominance state. The transitions between different states are as follows: At the initialization state, the Auxiliary subsystem provides an estimation of the pose (1). This estimation is refined with vision subsystem (2). When the visual tracking fails, the Auxiliary subsystem takes over to estimate the 3D localization (3). Since the Auxiliary subsystem is less accurate than the vision subsystem, the

estimation is corrected taking into account the predicted error. Thereafter, the estimation is used to re-initialize the visual tracking (4).

We have defined three criteria in order to quantify the quality of the estimated pose using the visual tracking. If one of these criteria is not verified, the pose is rejected and the system switches to the Auxiliary subsystem.

5.1. Number of Tracked Points. The number of 2D/3D matching points affects the accuracy of the minimization process used to estimate the camera pose. Indeed, the more we have a large set of 2D/3D matched points, the more the estimated pose is accurate and vice versa. For this, we define a minimum number of matching. Below this threshold, it is considered impossible to estimate the pose with the vision subsystem.

5.2. Projection Error. The number of matched points is not sufficient to ensure the accuracy of the pose estimation; the projection error criterion can also be used. This error represents the average square of the difference between the projection of 3D points using estimated pose and the 2D points. If the error is large, greater than an empirical threshold, the pose is considered wrong.

5.3. Confidence Intervals. The data provided by the Auxiliary subsystem can also be used as an indicator of the pose validation. In fact, from the position and orientation given by the Auxiliary subsystem, confidence intervals are defined. They are represented by an ellipsoid centered by the orientation provided by the inertial sensor and an ellipse which center is determined by the 2D position given by GPS. The axes of the ellipse or the ellipsoid can be defined $3 * \sigma$ (standard deviation of the offset between the camera pose and Auxiliary estimation) or empirically. If the pose computed by the vision subsystem is included in these confidence intervals (position in the ellipse and the orientation in ellipsoid), the pose is considered correct.

6. Experiments

Several experiments have been achieved to study the behavior of the proposed localization system when used in outdoor environments. The first experiment points out the performances of the semi-automated initialization approach in several conditions. Figure 8 shows that this approach performs good matching in spite of the illumination change.

Furthermore, in order to analyze the error in the matching process, we estimate the mean distance between the 2D points obtained after the semi-automated initialization step and the 2D points extracted from the images after refining the camera pose. We found a mean error equal to 3.8216 pixels with a standard deviation about 1.0873 pixels. This means that semi-automated approach is very efficient to generate a rough estimate of 2D-3D correspondences and really helps the tracking system to rapidly converge to the optimal solution.

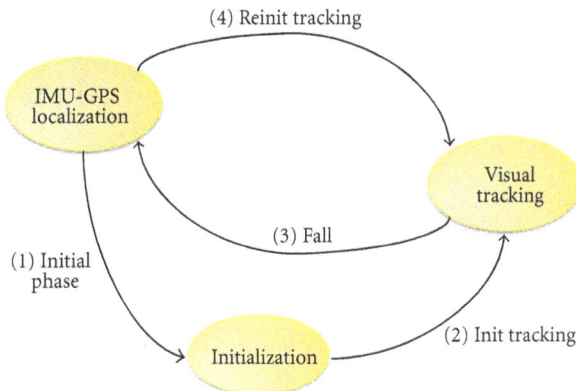

FIGURE 7: The state machine scheme of 3D localization system's operation.

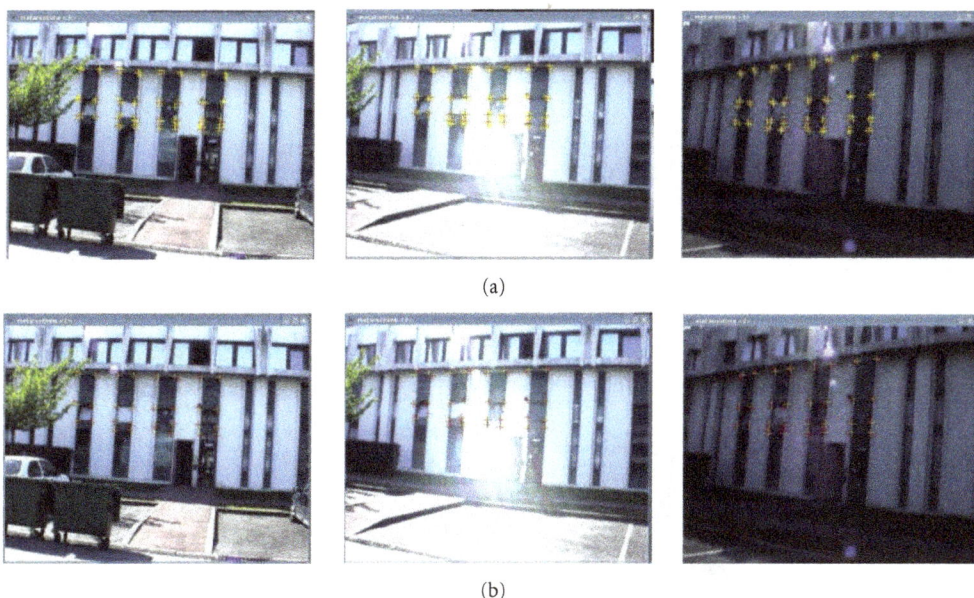

(a)

(b)

FIGURE 8: Performance of semiautomated approach.

TABLE 1: Computation times for the semiautomated initialization approach.

Steps	Times
Manual alignment	unknown
Extraction and matching	50 ms
RANSAC	100 ms
Total (without manual alignment)	150 ms

We also analyzed the execution time by carefully evaluating the processing time needed to achieve each step in the semi-automated initialization procedure. An example of these computation times is given in Table 1.

This table shows that the computation time needed to achieve the semi-automated initialization matching is quite fast and makes this approach particularly efficient for the initialization stage of the tracking system.

In addition, we also tested the performances of our automated initialization approach. For that we have considered several images taken under different viewpoints. Figure 9 shows some obtained results. We can see that, for all the considered cases, the reference points (taken on the frontage of the building) are well matched in the current frames. In this example we have considered coplanar points.

We have also tested our approach for non coplanar points chosen on the tower of the castle (Figure 10). Obtained results in this case are satisfying. Indeed, combining the SIFT points with the RANSAC algorithm provides an accurate and robust homography estimation, and thereby allows good points matching. The matching process in this case gives a mean error about 1.7823 pixels with a standard deviation of 0.6634 pixels.

The computation time of this approach depends mainly on the SIFT features extraction and matching. Introducing a prediction stage in order to limit the research area of the interest points in the current frame has significantly reduced

(a) Reference Image

(b) 1st result

(c) 2nd result

(d) 3rd result

FIGURE 9: Matching results for autometed initialization approach, case of the coplanr points.

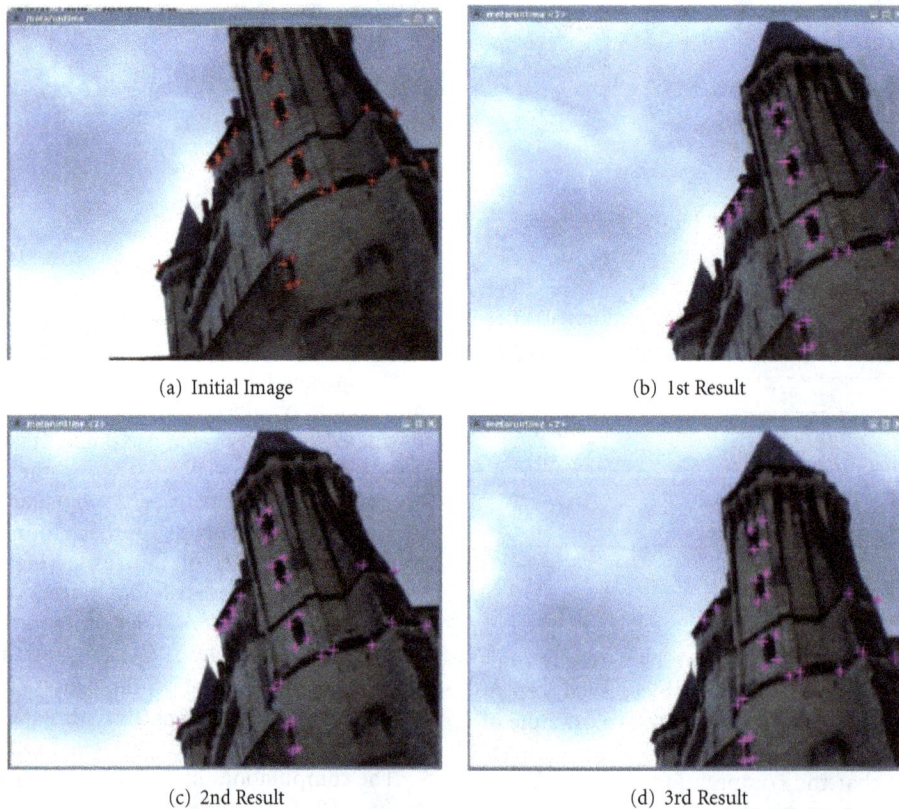

(a) Initial Image

(b) 1st Result

(c) 2nd Result

(d) 3rd Result

FIGURE 10: Matching results for autometed initialization approach-Non coplanr points.

FIGURE 11: Our hybrid sensor mounted on a tripod.

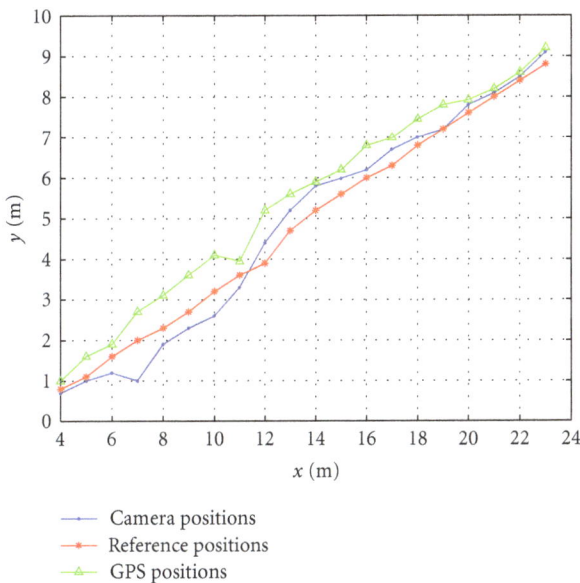

FIGURE 12: The straight line test: reference position versus camera position versus GPS position.

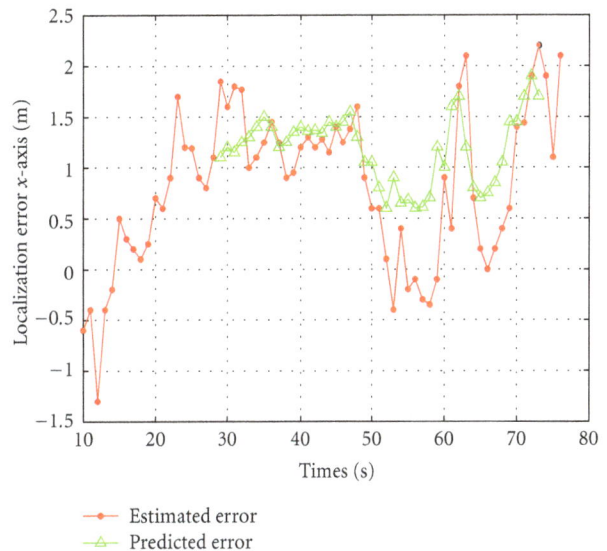

FIGURE 13: Predicted and estimated localization errors (on x-axis).

the total time of the whole algorithm (practically, it is divided by two). Comparing to other similar approaches [15], our proposed automated initialization technique is more flexible and provides best real times performances.

In the next experiments, we have evaluated the performance of our GPS-IMU localization system. The first experiment considers a straight line as a truth data. The origin of this line is defined in front of the origin of the world reference frame. The line is sampled in several positions and for each sample we do some acquisitions, namely images and GPS positions. The sensors are mounted on a tripod to ensure more stability (see Figure 11). The reference positions are measured with a telemeter which accuracy is about 0.15 m. In addition, for each acquired image, we calculated the position and the orientation of the camera.

From the GPS data and the transformation estimated during the calibration step, we deduce the absolute position with respect to the world reference frame associated to the

real scene. By comparing the different estimated positions to the reference positions, we find a mean offset about (1.8374 m; 1.4810 m). The same GPS positions compared to the camera's positions give a mean error equal to (1.7321 m; 1.4702 m) with a standard deviation (1.8314 m; 1.0116 m). Figure 12 shows the hole trajectories obtained from the GPS and camera positions computations compared to the reference trajectory.

The second experiment focused on the relative position between two successive fixed positions. In average the offset between the reference position and that obtained with the GPS is about 0.7817 m with a standard deviation equal to 1.06 m. Similar values are given by the vision subsystem, that is, an offset mean about 0.8743 m with a standard deviation of 0.9524 m. Therefore, these results demonstrate that the movement provided by the two subsystems is consistent. The third experiment performed several continuous recordings of GPS/camera positions. The two sensors are time-stamped in order to synchronize them and to retrieve the set of data acquired at the same time. The positions given by the

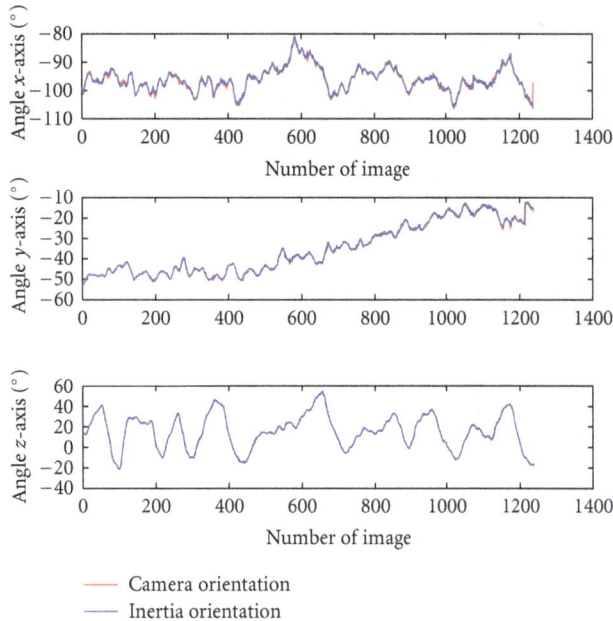

FIGURE 14: The camera's orientation versus inertial sensor's orientation.

vision and the GPS without correction are compared and the obtained errors are about 0.9235 m in the x-axis (with a standard deviation of 0.6669 m) and 0.8170 m in the y-axis (with a standard deviation of 0.6755 m). In addition, in order to study the error prediction approach we first used a set of 76 data acquired in continuous manner to perform the error training. Then, the Gaussian process is used with the last 30 data to predict errors. The mean offset between the predicted error and the real one is about ($\mu_x = 0.2742$ m; $\sigma_x = 0.4799$) and ($\mu_y = 0.5757$ m; $\sigma_y = 0.5097$ m) (see Figure 13). The positions provided by the GPS receiver are then corrected using this predicted error. This allows improving the 3D localization provided by the Auxiliary subsystem.

To assess the accuracy of the inertial sensor, we compared the orientations produced from the gyroscope to those computed by the vision pose estimation algorithm. For that, a video with several orientations in an outdoor environment has performed. Both orientations have the same behavior (Figure 14).

However, in some cases, we found that external factors can affect the inertial measurements, particularly in defining the local reference frame where the x-axis is in the direction of the local magnetic north. This causes errors in the orientation estimation. To solve this problem the rotation between the local reference frame associated to inertial sensor and the world reference frame is re-estimated continuously. The behavior of the whole system is also tested. The initialization process allows having the matching of the 3D visible points from the 3D model with their projections in the first view. From this 2D/3D matching, the set of 2D points are defined and tracked frame to frame. For each frame, the wire frame model is registered using the positions and orientations obtained from the hybrid localization

system. In Figure 15, the green color projection is obtained from the positions and orientations provided by the vision subsystem. The wire frame model is well superimposed on the real view which demonstrates the accuracy of the camera pose estimation. In magenta, the projected model is obtained with the positions and orientations provided by the Auxiliary subsystem. Figure 15 show that when vision fails, the localization system switches to the Auxiliary subsystem to provide 3D localization. The localization is corrected with the predicted error which contributes to improve the estimation.

Figure 16 show that during the occlusion of the tracked points, the GPS-IMU localization system provides always an estimation of the position and orientation of the camera. Therefore, even when a total occlusion occurred, the system can provide a rough estimation of the 3D localization. This would not be the case if we used only the camera.

Furthermore, we evaluated the performance of our 3D localization system in this urban scene. The 3D model of the building is known, it is composed of 120 natural points defined by their 3D coordinates within the world coordinates frame. Several trials in different locations were recorded. The data coming from the three sensors are time-stamped and stored in data file. The camera pose estimated by the visual tracking was used as ground truth for the performance evaluation. We have defined a set of 2D/3D point's correspondences from the first sequence frame. The interest points are then tracked frame to frame using the KLT feature tracking algorithm [17, 18]. The 2D/3D correspondences are then updated and used to estimate the 3D localization. We have compared our approach with a data fusion method using an Extended Kalman Filter. Computational results show the effectiveness of our approach, since it obtains better accuracy. Figure 17 shows the recovered camera trajectory in world coordinate system; we note that the trajectory estimated by our localization approach is closer to the ground truth then the one estimated by fusing data coming from the three sensors.

Finally, the performances of our localization system are compared to the state-of-the-art results reported by DiVerdi and Höllerer [22] with their GroundCam system which also combines a camera, a GPS receiver and an orientation tracker. The authors run their system along a residential street for approximately 90 seconds and compare the estimated trajectory to a hand-labelled ground truth. The reported RMS is about 5.5 m. We run our system in the similar conditions around our institution building. The obtained RMS is about 1.55 m which shows that our localization system generates results significantly better.

7. Conclusion

In this paper, we presented an original solution for 3D camera localization using multi-sensors technology. The system combines a camera, a GPS and an inertial sensor; it is designed to work in outdoor environments. Instead to fusion all data, the proposed system is based on an assistance scheme. It is composed of two parts which work

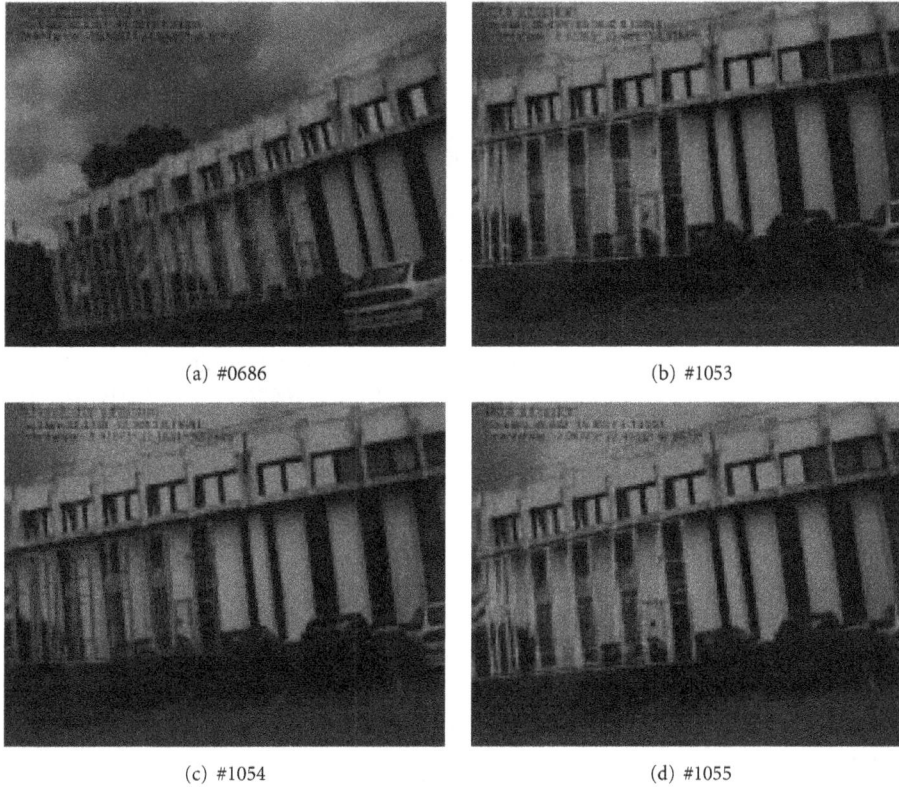

(a) #0686

(b) #1053

(c) #1054

(d) #1055

FIGURE 15: Registration of the 3D model using the poses obtained with our hybrid system.

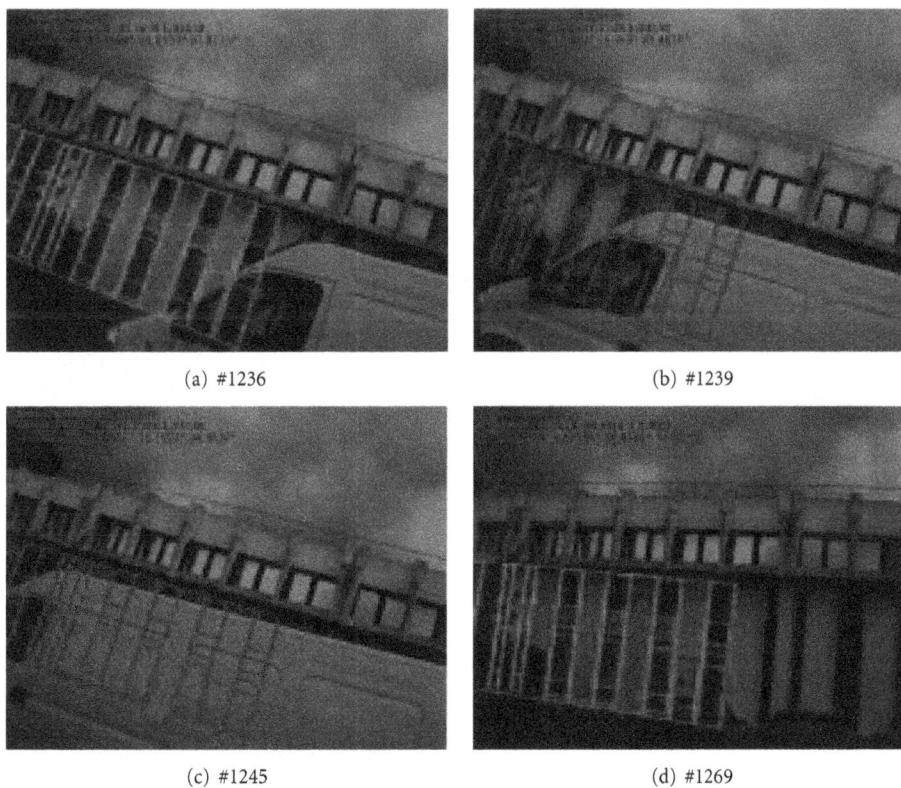

(a) #1236

(b) #1239

(c) #1245

(d) #1269

FIGURE 16: Registration of the 3D model using the auxiliary subsystem: Occlusion case.

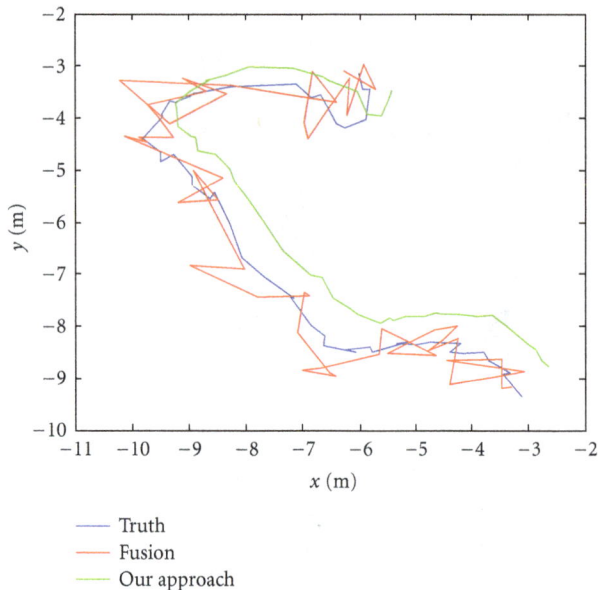

FIGURE 17: Trajectory in world coordinate system.

in a complementary manner and controlled by a finite state machine allowing continuous 3D localization. The vision subsystem, representing the main part, uses a point-based visual tracking. Once the vision fails, the system switches to Auxiliary subsystem which is composed of the GPS/inertial sensors. The Auxiliary subsystem is less accurate then the vision subsystem, especially the GPS positioning. Hence, a prediction stage is performed to improve the accuracy of the Auxiliary subsystem. The 3D localization provided by the two subsystems is used to learn, on-line, the errors made by the Auxiliary subsystem. The two subsystems interact continuously to each other. The obtained results are quite satisfactory with respect to the purpose of Mobile Augmented Reality systems. They have shown that the proposed system has quite good accuracy compared to other approaches.

The system was tested in outdoor environment and has demonstrated its capacity to adapt itself to the several conditions occurred in such environments. For example, when a total occlusion of the scene model is occurred, the Auxiliary system takes over the 3D localization estimation until the vision becomes operational. However to increase the robustness and the efficiency of the whole system, improvements must be made in several parts. Actually, within the implemented vision-based method, the tracked points must be always visible. So, one challenge is to develop a tracking method which can handle visual occlusions and update automatically the set of tracked points by adding, in real time, new visible points. In addition, other markerless tracking approaches can be combined with the point tracker such as edge-based methods [23] or fiducials based approaches [24, 25] to improve the accuracy of the vision-based pose estimation. Also, the fusion process can be optimized if we consider the motion dynamic of the camera given by the IMU sensor. On the other hand, the experiments

have shown that the GPS signal can be obstructed when the user is quite near the buildings. So, when the system switches to the Auxiliary subsystem, the position could not be estimated. This problem can be solved by adding other kinds of positioning sensors which can replace the GPS (RFID, WIFI, etc.). The main idea is to develop a ubiquitous tracking system composed of a network of complementary sensors which can be solicited separately and in real time in terms of the situations occurred in the environments.

References

[1] T. Vieville, F. Romann, B. Hotz et al., "Autonomous navigation of a mobile robot using inertial and visual cues," in *Proceedings of the IEEE/RSJ International Conference on Intelligent Robots and Systems*, pp. 360–367, July 1993.

[2] S. You, U. Neumann, and R. Azuma, "Orientation tracking for outdoor augmented reality registration," *IEEE Computer Graphics and Applications*, vol. 19, no. 6, pp. 36–42, 1999.

[3] M. Ribo, P. Lang, H. Ganster, M. Brandner, C. Stock, and A. Pinz, "Hybrid tracking for outdoor augmented reality applications," *IEEE Computer Graphics and Applications*, vol. 22, no. 6, pp. 54–63, 2002.

[4] J. D. Hol, T. B. Schön, F. Gustafsson, and P. J. Slycke, "Sensor fusion for augmented reality," in *Proceedings of the 9th International Conference on Information Fusion*, pp. 1–6, Florence, Italy, July 2006.

[5] F. Ababsa, "Advanced 3D localization by fusing measurements from GPS, inertial and vision sensors," in *Proceedings of the IEEE International Conference on Systems, Man and Cybernetics (SMC '09)*, pp. 871–875, San Antonio, Tex, USA, October 2009.

[6] G. Bleser and D. Stricker, "Advanced tracking through efficient image processing and visual-inertial sensor fusion," in *Proceedings of IEEE International Conference on Virtual Reality (VR '08)*, pp. 137–144, March 2008.

[7] F. Ababsa, J. Y. Didier, M. Mallem, and D. Roussel, "Head motion prediction in augmented reality systems using Monte Carlo particle filters," in *Proceedings of the 13th International Conference on Artificial Reality and Telexixtance (ICAT '03)*, pp. 83–88, Tokyo, Japan, 2003.

[8] F. E. Ababsa and M. Mallem, "Hybrid three-dimensional camera pose estimation using particle filter sensor fusion," *Advanced Robotics*, vol. 21, no. 1-2, pp. 165–181, 2007.

[9] G. Reitmayr and T. W. Drummond, "Initialisation for visual tracking in urban environments," in *Proceedings of the 6th IEEE and ACM International Symposium on Mixed and Augmented Reality (ISMAR '07)*, Nara, Japan, November 2007.

[10] Z. Hu, U. Keiichi, H. Lu, and F. Lamosa, "Fusion of vision, 3D gyro and GPS for camera dynamic registration," in *Proceedings of the 17th International Conference on Pattern Recognition (ICPR '04)*, pp. 351–354, Washington, DC, USA, August 2004.

[11] M. Aron, G. Simon, and M. O. Berger, "Use of inertial sensors to support video tracking," *Computer Animation and Virtual Worlds*, vol. 18, no. 1, pp. 57–68, 2007.

[12] M. Maidi, F. Ababsa, and M. Mallem, "Vision-inertial tracking system for robust fiducials registration in augmented reality," in *Proceedings of IEEE Symposium Computational Intelligence for Multimedia Signal and Vision Processing (CIMSVP '09)*, pp. 83–90, Nashville, Tenn, USA, April 2009.

[13] D. G. Lowe, "Fitting parameterized three-dimensional models to images," *IEEE Transactions on Pattern Analysis and Machine Intelligence*, vol. 13, no. 5, pp. 441–450, 1991.

[14] R. M. Haralick, H. Joo, C. N. Lee, X. Zhuang, V. G. Vaidya, and M. B. Kim, "Pose estimation from corresponding point data," *IEEE Transactions on Systems, Man and Cybernetics*, vol. 19, no. 6, pp. 1426–1446, 1989.

[15] G. Bleser and D. Stricker, "Advanced tracking through efficient image processing and visual-inertial sensor fusion," *Computers and Graphics*, vol. 33, no. 1, pp. 59–72, 2009.

[16] D. G. Lowe, "Distinctive image features from scale-invariant keypoints," *International Journal of Computer Vision*, vol. 60, no. 2, pp. 91–110, 2004.

[17] C. Harris, *Tracking with Rigid Models, Active Vision*, MIT Press, Cambridge, Mass, USA, 1993.

[18] M. A. Fischler and R. C. Bolles, "Random sample consensus: a paradigm for model fitting with applications to image analysis and automated cartography," *Communications of the ACM*, vol. 24, no. 6, pp. 381–395, 1981.

[19] C. Tomasi and T. Kanade, "Detection and tracking of point features," Carnegie Mellon University Technical report CMU-CS-91-132, 1991.

[20] O. D. Faugeras and G. Toscani, "Camera calibration for 3D computer vision," in *Proceedings of the International Workshop on Industrial Applications of Machine Vision and Machine Intelligence*, pp. 240–247, 1987.

[21] C. Williams, "Prediction with Gaussian processes: from linear regression to linear prediction and beyond," Tech. Rep., Neural Computing Research Group, 1997.

[22] S. DiVerdi and T. Höllerer, "GroundCam: a tracking modality for mobile mixed reality," in *Proceedings of IEEE International Conference on Virtual Reality (VR '07)*, pp. 75–82, March 2007.

[23] F. Ababsa and M. Mallem, "Robust camera pose estimation combining 2D/3D points and lines tracking," in *Proceedings of IEEE International Symposium on Industrial Electronics (ISIE '08)*, pp. 774–779, Cambridge, UK, July 2008.

[24] F. Ababsa and M. Mallem, "A robust circular fiducial detection technique and real-time 3D camera tracking," *Journal of Multimedia*, vol. 3, no. 4, pp. 34–41, 2008.

[25] J. Y. Didier, F. Ababsa, and M. Mallem, "Hybrid camera pose estimation combining square fiducials localisation technique and orthogonal iteration algorithm," *International Journal of Image and Graphics*, vol. 8, no. 1, pp. 169–188, 2008.

15

Reconstruction of Riser Profiles by an Underwater Robot Using Inertial Navigation

Luciano Luporini Menegaldo,[1] Stefano Panzieri,[2] and Cassiano Neves[3]

[1] *Biomedical Engineering Program (PEB/COPPE), The Alberto Luiz Coimbra Institute for Graduate Studies and Research in Engineering, Universidade Federal do Rio de Janeiro, Avenida Horacio Macedo 2030, Bloco H-338, 21941-914 Rio de Janeiro, RJ, Brazil*
[2] *Department of Informatics and Automation, Universita Roma Tre, Via della Vasca Navale, 79, I 00146 Roma, Italy*
[3] *Subsin Engineering, Rua Beneditinos, 16, 12th floor, 20081-050 Rio de Janeiro, RJ, Brazil*

Correspondence should be addressed to Luciano Luporini Menegaldo, lmeneg@peb.ufrj.br

Academic Editor: Jorge Manuel Dias

This paper proposes a kinematic model and an inertial localization system architecture for a riser inspecting robot. The robot scrolls outside the catenary riser, used for underwater petroleum exploration, and is designed to perform several nondestructive tests. It can also be used to reconstruct the riser profile. Here, a realistic simulation model of robot kinematics and its environment is proposed, using different sources of data: oil platform characteristics, riser static configuration, sea currents and waves, vortex-induced vibrations, and instrumentation model. A dynamic finite element model of the riser generates a nominal riser profile. When the robot kinematic model virtually scrolls the simulated riser profile, a robot kinematic pattern is calculated. This pattern feeds error models of a strapdown inertial measurement unit (IMU) and of a depth sensor. A Kalman filter fuses the simulated accelerometers data with simulated external measurements. Along the riser vertical part, the estimated localization error between the simulated nominal and Kalman filter reconstructed robot paths was about 2 m. When the robot model approaches the seabed it assumes a more horizontal trajectory and the localization error increases significantly.

1. Introduction

One of the key elements of deep-water petroleum exploration is the production riser. Risers are the ducts that transport petroleum, water or gases from the exploitation well up to the production platform. Either rigid or flexible types of risers may be used in the oil field. Both types are submitted to a broad spectrum of failure causes [1]: mechanical loads, aging, corrosion, erosion, temperature effects, installation or fabrication nonconformities, and so forth. Therefore, the availability of inspection tools to assess riser integrity status *in situ* is highly desirable. Such procedures are performed mainly by visual inspection with remotely operated vehicles (ROVs) [2] or autonomous underwater vehicles (AUVs) [3]. In some cases, sensors are installed directly on fixed points of the riser surface to measure strain and riser motion [4]. Other types of nondestructive testing (NDT) techniques can

be used, such as magnetic, radiographic, or ultrasound methods [5]. In these cases, however, the operational constraints for using human operators are a major problem. A few papers address robotic devices specifically designed for underwater riser inspection. Psarros and his collaborators [6] proposed a robot that moves along the riser by using a mechanism composed of two parts. One part stays attached to the riser body, and the other part moves towards the riser's side, in a cyclical manner.

A major technical problem in robotic underwater inspection is the navigation and/or localization of the robot in a highly dynamic sea environment. Navigation is especially critical for AUVs and somewhat critical for ROVs. Lee at al. [7] addressed this problem by using several sensors fused by a multirate Extended Kalman Filter (EKF). The sensors set included a strapdown inertial platform, a Doppler velocity log (DVL), magnetic compass, and a depth sensor. However, they had sonar transducers installed in an underwater reference station and in the remote vehicle. Jouffroy and Opderbecke [8] addressed the problem of measuring the horizontal position of a ROV by using a gyro-Doppler together with an ultrashort baseline (USBL) acoustic positioning system. Diffusion-based observers were used to process a trajectory segment, instead of a typical point-by-point localization. He et al. [9] proposed an approach based on an invariance extended Kalman filter (IEKF) to address the problems of using sonar in shallow waters. In the case when the robot is mechanically linked to the inspected structure, the key problem is to localize precisely where it is at every instant of time. Such localization coordinates are associated to NDTs data and the flaws position can be precisely determined.

Recently, our group designed and built a prototype of a robotic device specifically designed to perform nondestructive testing (NDT) in production risers [10]. The robot has neutral floatability and *embraces* the riser by moving along its outside (Figure 1), using a pair of thrusters for propulsion as well as polymeric wheels to guarantee sliding and correct alignment with the riser surface. In Figure 2 it is possible to observe how the robot attaches the riser by opening and closing its motorized arms. This operation is assisted by a human diver. It communicates with the operator's computer by means of an umbilical cable that transmits power, images and control commands. The dimensions of the robot and additional parameters used along the work are shown in Table 2. This robot will be able to perform several NDT procedures, such as ultrasound, imaging, and mechanical vibration measurements.

This paper proposes a kinematic model of the robot performing a riser profile cast mission, in a realistic simulated environment. Initially, a riser dynamic profile is estimated using a finite element model of the riser subjected to sea and ship motions. The nominal robot kinematic path (including position, velocity and acceleration), as it scrolls by the riser, is *contaminated* with experimental errors, simulated by IMU and depth sensor models. The simulated sensor data is used by a Kalman filter to estimate the original robot path. This path is a good estimate of the actual riser profile, if robot mission time is small, compared to platform motion.

The obtained profile can be used as an imposed displacement data for some structural analysis software based on finite elements techniques, that allows stress to be calculated. In addition, the localization algorithm can be used to associate each NDT measurement with its riser geometrical coordinate. These two aspects are intimately connected, and the localization algorithm can be used either to cast the profile, for fast robot runs, or localize the NDTs.

Reproducing the expected environmental conditions, to test the proposed approach, in a laboratory experiment is essentially impracticable. Field tests, by his turn, should require a expensive positioning system such as a 3D sonar, which does not operate at the required frequency resolution, due to the presence of vortex induced vibrations (VIVs). Therefore, a simulation of the riser application, together with simulated sensors, was used to assess the performance of the localization algorithm.

FIGURE 1: Robot prototype.

Actually, a particular environmental and riser configuration scenario is being addressed in this paper. However, the approach is likely to be applicable to similar situations. Other devices that move along subsea pipe systems, such as flowlines, jumpers, and umbilicals, could employ some of the main ideas presented in this paper. No additional localization devices, such as sonar beacons, are needed.

The localization problem formulation used a standard Kalman filter as a sensor fusion algorithm based on a simple kinematic model of a strapdown IMU fused with a depth sensor. More sophisticated sensor fusion algorithms or state-space models for the system (e.g., dynamical models) could also be tested in future implementations, but the problem formulation would be probably quite similar.

2. Riser Simulation Conditions

The particular sea and ship motion conditions selected for running the simulations corresponded to a *severe* condition, relatively similar to that typically found in Campos Basin, in the southeast of South America coast. Actually, they were designed to be worse than the most severe scenario in which a robot is expected to operate (Table 1). Additionally, under milder sea conditions, the localization performance should be expectable to be better than shown here.

Data from a flexible free-hanging riser installed in a PETROBRAS (Brazilian State oil company) Turret Floating Production Storage Offloading (FPSO) oil platform, which is currently in operation in the Campos Basin, was used as the inputs for the riser simulation software FLEXCOM. This is a finite elements software customized for nonlinear static and dynamic analysis of offshore systems, used worldwide by the petroleum industry from the last 20 years, and validated against experimental tests and other finite elements packages [11, 12]. The software allows riser responses to be simulated with several kinds of platform characteristics, sea current profiles, hydrodynamic loads, regular and irregular waves, and so forth. Two situations were studied: static and

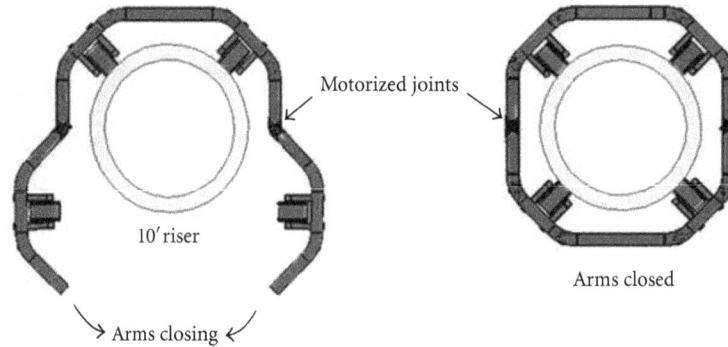

FIGURE 2: Transversal view of the robot frame showing how it attaches the riser, by opening and closing its motorized arms. The polymeric free wheels that effectively touch the riser surface are also shown.

TABLE 1: Parameters used in the simulations.

Parameter	Value
Robot length	1133 mm
Robot max. outside diameter	800 mm
Robot mass	73 kg
Riser outer diameter	295.5 mm
Riser inner diameter	203.5 mm
Riser length	1530 m
Riser internal pressure	60 bar
Outer riser drag coefficient (Cd)	1.2
Water depth	1180 m
Seabed axial friction	0.35
Seabed transversal friction	0.9
Seabed vertical stiffness	104.3 N/m^2
Wave height	5 m
Wave period	10 seconds
FPSO length	330 m

dynamic. In the static analysis, only the equilibrium configuration of the riser is considered, without motions other than from the robot itself. Three positions of the platform with relation to the wheel head were considered: standard and with a ship offset of 150 m in the directions *near* and *far*. For the dynamic analysis, sea conditions with regular waves, reaching the ship 45° obliquely, were used to generate the riser motion profiles.

To estimate numerically the localization system performance and calculate the associated displacement errors, the arrangement shown in Figure 3 was used. By simulating the kinematic model (Section 4), a *physical* profile of Euler angles, inertial accelerations, and depth was calculated. These simulated variables are then contaminated by noise, using the IMU and depth sensor instrumentation error models (Section 5), providing realistic sensor outputs. The trajectory run by the robot is estimated by a sensor fusion algorithm (Kalman filter), using such noisy sensor data (Section 6). Finally, both *physical* and estimated riser shapes

are compared, to estimate the localization error along the run (Sections 7 and 8).

3. Architecture of the Localization System

The proposed localization system, that will be simulated numerically here, is shown in Figure 4. An IMU measures three accelerations, angular velocities, and Euler angles from the robot as it scrolls along the riser. Accelerations are measured in the local reference frame (see definition in Section 4). Using the Euler angles, the accelerations are transformed to the global reference system, using the classic *strapdown* inertial navigation approach [13]. The simulated measurements from the IMU are fused by the Kalman Filter (KF) with the processed simulated external measurements from the depth sensor. Its output is an estimated state vector, that includes robot position, in the global reference frame.

4. Generation of Static and Dynamic Sensor Profiles

A simulated profile of inertial sensor physical excitation was obtained. First, the *nominal* or *static* riser geometry was obtained from the FEM (finite element model) analysis. The global reference system $(\mathbf{X_G}, \mathbf{Y_G}, \mathbf{Z_G})$ was positioned in the turret center, such that X axis points to an arbitrary direction (e.g., north), Z points up, and Y is orthogonal to both (Figure 5). This reference frame is attached to the FPSO, being thus a slowly moving frame. However, due to the very low frequency of the FPSO movement compared to robot mission time span, the global reference frame was considered as being inertial.

The local reference system $(\mathbf{X_l}, \mathbf{Y_l}, \mathbf{Z_l})$ is localized in the geometric center of the riser section that coincides with the position, relative to robot height, where the inertial sensor is expected to be installed, and moves together with the robot. The center of the local reference system was defined, relatively to the global reference system, using the same nodes of the FEM mesh, which coordinates in the global reference system are $P_i^G = (x(i), y(i), z(i)), i = 1, \ldots, N$ (N is number of nodes in the mesh and the superscript G express the reference system). Each $\mathbf{Z_l}$ vector is found from a unit vector

TABLE 2: Identified parameters for the IMU sensor model[1].

Variable	$\tau_r(R_{xx}(s))$	$\tau_r(\text{Allan}(s))$	Low cut (Hz)	High cut (Hz)	σ_{bias}	σ_{bw}
Roll°	629	20*	5	20	0.1635	0.0072
Pitch°	578	20*	6	20	0.1553	0.0072
Yaw°	1403	20*	5	12	0.2144	0.0095
AccX g	0.644*	10	0.3	3	$5.2768e-4$	0.0041
AccY g	0.581*	10	0.25	3	0.0011	0.0037
AccZ g	0.645*	10	0.3	3	$6.4654e-4$	0.0033
AngRateX°/s	0.767*	15	—	—	0.0076	0.0076
AngRateY°/s	0.754*	20	—	—	0.0074	0.0074
AngRateZ°/s	0.745*	10	—	—	0.0057	0.0057

[1] R_{xx} means that used time constant is from autocorrelation, and *Allan* from Allan variance plot. Low cut and high cut correspond to the cutting-off frequencies of the low-pass and high-pass filter used to calculate σ_{bias} and σ_{bw}, respectively. σ_{bw} and σ_{bias} have the same unities of their respective measurement variables.
*Used in the simulations.

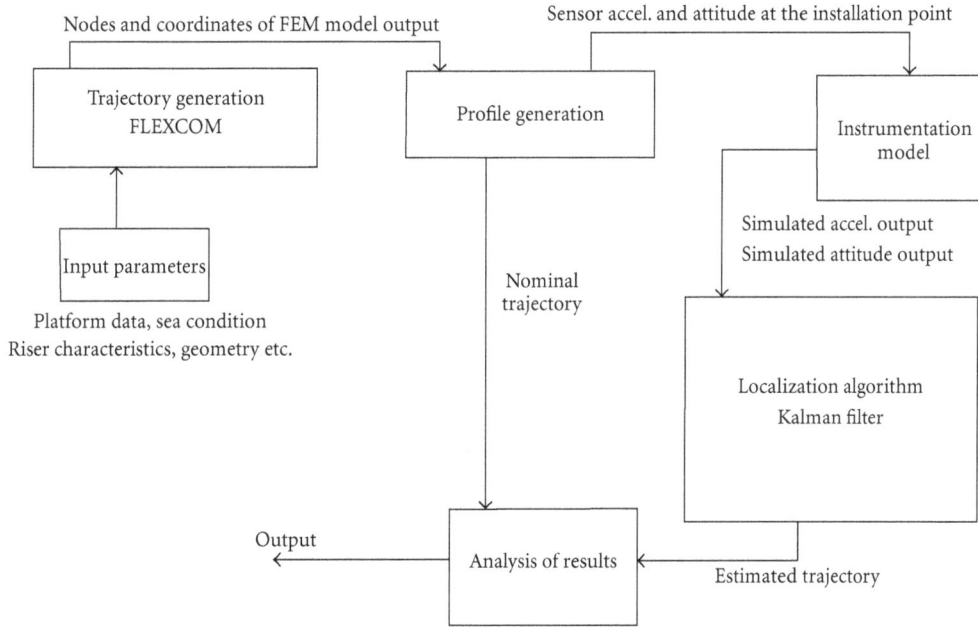

FIGURE 3: Overview of the system simulation analysis. A dynamic riser profile is generated through an FEM analysis. Using robot's kinematical and instrumentation model, the expected sensor readings are used to reconstruct the riser profile by the localization algorithm, which is compared to the original profile from FEM output.

D_u that connects node i to node $i + 1$ of the FEM mesh. The negative sign for defining Z_l in (2) is due to the fact that D_u points down:

$$D = \begin{bmatrix} x(i+1) - x(i) & y(i+1) - y(i) & z(i+1) - z(i) \end{bmatrix},$$
(1)

$$D_u = \frac{D}{\|D\|},$$

$$Z_l = -D_u.$$
(2)

In the sequence, a particular Y_l was chosen (along with its orthogonal counterpart X_l), such that it could represent one plausible mission trajectory. In the present version of the model, the robot communicates with the operator in the platform through an umbilical cable, and only a small amount of spin can be allowed, to prevent the cable from curling along the riser. Therefore, Y_l is simultaneously orthogonal to a point sufficiently far in the X direction and to $-Z_l$. The X_l vector was orthogonal to both Z_l and Y_l. For each local reference system i, a directors cosine matrix (DCM) was defined as follows:

$$\text{DCM} = \begin{bmatrix} X_{1x} & Y_{1x} & Z_{1x} \\ X_{1y} & Y_{1y} & Z1_z \\ X_{1z} & Y_{1z} & Z1_z \end{bmatrix},$$
(3)

where $X_l = (X_{1x}, X_{1y}, X_{1z})^T$ is the unit vector that defines X_l, and similarly Y_l and Z_l.

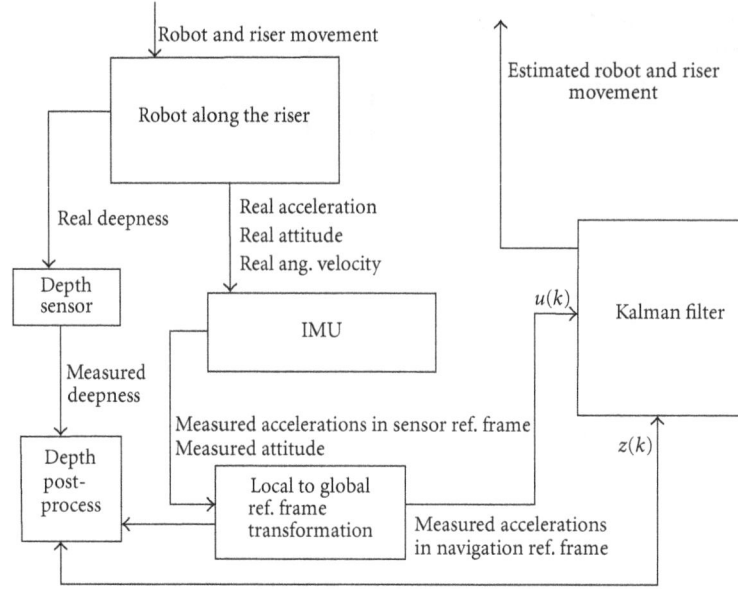

FIGURE 4: Architecture of the proposed localization system architecture. The robot scrolling along the riser generates a set of kinematic variables that is measured by the IMU. The accelerations are transformed to the global reference system, using IMU attitude outputs, and considered as the inputs of the KF. To compensate the drift caused by integrated sensor noise, the KF fuses the IMU with depth sensor data, that is, an absolute measurement. Before entering in the KF, the external measurements are processed by the method described in Section 6, using IMU attitude data.

Euler angles ψ, θ, and ϕ were defined between the $\mathbf{X_G}$ and $\mathbf{X_I}$, $\mathbf{Y_G}$ and $\mathbf{Y_I}$, $\mathbf{Z_G}$ and $\mathbf{Z_I}$ axis, respectively. In every step, these angles were derived from the DCM matrix, by using the formulas presented by [14]. The resulting transformation matrix from local to global reference system $R_I^G = R_x R_y R_z$ was posteriorly verified to be the same as DCM, in order to determine if singularities were present. A fixed point $P_S = (0, 0.5, 0)$ expressed in the local reference frame corresponds to the robot body point where the IMU will be possibly installed. This point was arbitrarily chosen within a reasonable Y distance from the riser section center, where the origin of the local reference frame is positioned. The coordinates of P_S, expressed in the global coordinate frame, were assumed as the sensor displacement profile.

4.1. Sea Current Effect. A linear current profile $\mathbf{V_{cur}}$ from the maximum sea current velocity (V_{curmax}) at sea level to 0 at seabed [15] was considered as being aligned with the riser catenary plane. V_{curmax} was assumed as 1.68 m/s, from Campos Basin data:

$$\mathbf{V_{cur}}(\mathbf{i}) = RZ_{cat} \begin{bmatrix} 0 \\ \dfrac{V_{curmax}(z(N-1) - z(i))}{z(N-1) - z(1)} \\ 0 \end{bmatrix} \qquad (4)$$

$$i = 1, \ldots, N,$$

where RZ_{cat} is the rotation matrix associated with the catenary angle in the $X_G Y_G$ plane, given by the following expression:

$$\phi_{cat} = \arctan\left(\frac{x(N) - x(1)}{y(N) - y(1)}\right),$$

$$RZ_{cat} = \begin{bmatrix} \cos(\phi_{cat}) & \sin(\phi_{cat}) & 0 \\ -\sin(\phi_{cat}) & \cos(\phi_{cat}) & 0 \\ 0 & 0 & 1 \end{bmatrix}. \qquad (5)$$

The robot can move freely along the riser in the $\mathbf{Z_I}$ direction, but it was constrained in the other directions because it *embraced* the riser on the outside. If no water current was present, the nominal robot velocity propelled by a pair of thrusters should be $\|\mathbf{V_r}\|$ in the direction of $\mathbf{Z_I}$. However, due to the presence of the current, the absolute velocity of the robot was found by considering the sea current velocity component that is projected over the robot's trajectory, which can change the robot's progression velocity:

$$\mathbf{V_{r\_abs}} = -\mathbf{V_r} + \mathbf{V_{cur}} \cdot \mathbf{Z_I}, \qquad (6)$$

where $\mathbf{V_{r\_abs}}$ is the absolute velocity and $\cdot$ the dot product. From a preliminary study, $\|\mathbf{V_r}\|$ was estimated as 1 m/s. Figure 6 illustrates robot $\mathbf{Z_I}$ velocity in the particular sea current and thrusters conditions adopted in this paper.

The effect of the longitudinal sheer between the robot and the riser due to the transversal current was not taken into account. This current component was expected essentially to

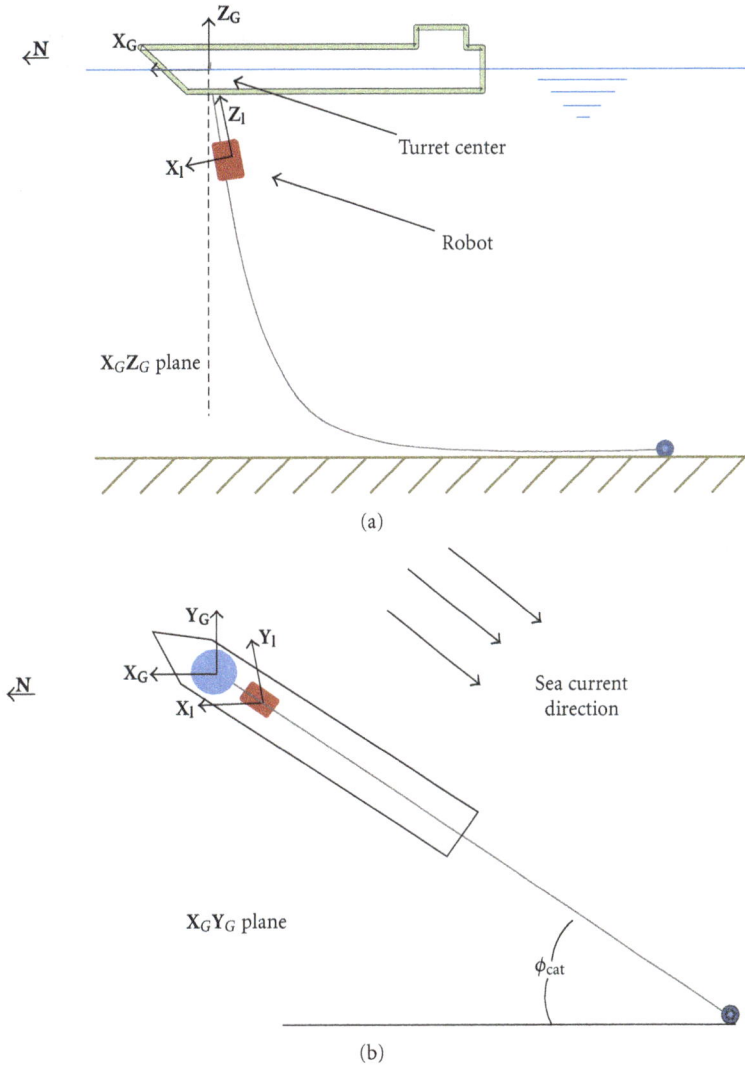

FIGURE 5: Side and upper views of the FPSO and riser configuration. Global $(\mathbf{X_G}, \mathbf{Y_G}, \mathbf{Z_G})$ and local $(\mathbf{X_I}, \mathbf{Y_I}, \mathbf{Z_I})$ reference systems position and orientation are shown. The FPSO and riser catenary plane are considered as being aligned with the sea current, forming an angle ϕ_{cat} with the $\mathbf{X_G}$ direction in $X_G Y_G$ plane.

increase the normal force that the robot applied to the outer surface of the riser. Because the riser was tightly fitted among the robot's rigid free wells to avoid longitudinal and torsional slipping, the increase in the shear force of the wheels that could decelerate the robot was considered to be negligible.

Since all the elements of the FEM mesh have approximately the same length, the time steps are no longer uniformly distributed with such variable velocity profile. The resulting variable time array was calculated by:

$$t_{var}(i) = \frac{\|\mathbf{D(i)}\|}{\|\mathbf{V_r}\|}, \quad i = 1, \dots, N. \tag{7}$$

This nonuniform time array was inconvenient for future calculations of velocity and acceleration profiles, and a new set of $(x(i), y(i), z(i))$ was found by spline interpolation using a uniform time array with the same limits of t_{var}. In the

sequence, the local reference systems $(\mathbf{X_I}, \mathbf{Y_I}, \mathbf{Z_I})$, the DCM matrix and the Euler angles were recalculated using this new set of coordinates, which were sampled by a uniform time array. All profiles were resampled at 5 Hz by spline interpolation, such that the high frequency riser motions could be followed (next subsection). The total mission time was $T_m = 1526$ s, and a total of $M = T_m(\mathrm{s}) \times 5(\mathrm{Hz}) = 7630$ points were generated.

4.2. Riser Motion Effect. The dynamic FEM analysis allows finding, along the time, the altered geometry of the riser, with respect to the nominal static profile. The finite elements model had 170 beam elements with flexion, axial, and torsional deformations. Each element had a nominal length of 10 m. Dynamic FEM analysis provided the $\mathbf{X_G}$, $\mathbf{Y_G}$, and $\mathbf{Z_G}$ coordinates of only ten nodes with respect to time at

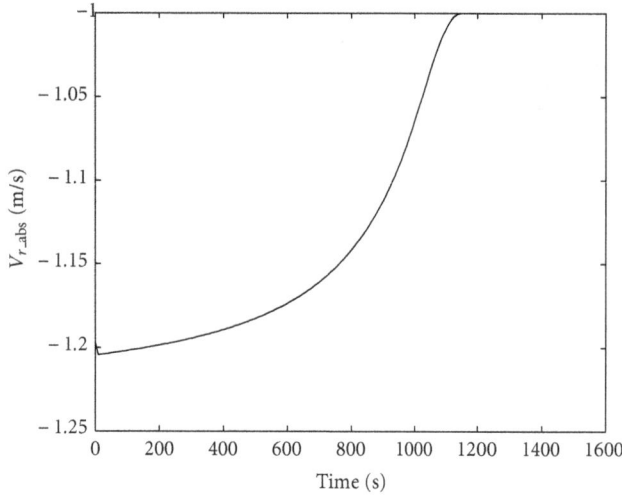

FIGURE 6: Velocity of the robot in $\mathbf{Z_l}$ direction considering a linear sea current profile.

f_s = 20 Hz sampling rate. The analysis was run for a 100 second time span, but the first 50 seconds were disregarded to avoid the transient effect of the FEM solution. The second half of the time window was then replicated to reach the total mission time. Thus, the offset, phase, and amplitude parameters were preserved. This adjustment provided a matrix of ten lines (one for each node) and $T_m \times f_s$ columns, which was resampled by successive spline interpolations. The first interpolation reduced the number of columns to M, the second expanded the lines up to the original FEM mesh (171 nodes) and the third resampled the lines again up to M. Therefore, three $M \times M$ perturbation matrices Per_x, Per_y, and Per_z were obtained, one for each coordinate $\mathbf{X_l}$, $\mathbf{Y_l}$, and $\mathbf{Z_l}$. In these matrices, each row was a particular riser deviation from the nominal profile and each column was a time step. The three components of a displacement vector $\mathbf{d(i)}$ were defined for each node i:

$$\mathbf{d(i)} = \begin{bmatrix} \text{Per}_x(i,i) & \text{Per}_x(i,i) & \text{Per}_x(i,i) \end{bmatrix}, \quad i = 1, \ldots, M. \tag{8}$$

This vector was decomposed into its normal $\mathbf{d_n}$ and tangential $\mathbf{d_t}$ parts. Because the robot was free to move along the riser, only the normal component of $\mathbf{d}$ vector was effectively transmitted to the robot:

$$\begin{aligned} \mathbf{d_t} &= (\mathbf{d} \cdot \mathbf{Du})\mathbf{Du}, \\ \mathbf{d_n} &= \mathbf{d} - \mathbf{d_t}. \end{aligned} \tag{9}$$

Therefore, the new coordinates x, y, and z (in the local reference frame) of the riser path were given by the following sequence of $\overline{P^l_{i\,\text{new}}}$ points, $i = 1$,

$$\overline{P^l_{i\,\text{new}}} = \overline{P^l_{i\,\text{nominal}}} + \mathbf{d_n}. \tag{10}$$

4.3. Vortex-Induced Vibration (VIV) Effect.

The sea current passing through a circular cylinder produces vortex-shedding in the wake, which causes the structure to vibrate. This complex fluid-structure interaction is called Vortex Induced Vibration (VIV), and it occurs predominantly on the cross-flow direction [16]. Simulating this effect is an arduous numerical problem. In this paper, we used experimental data obtained from a scale-model available at the Open-Source VIV Data Repository of the Center for Ocean Engineering at MIT [17]. Cross-flow displacement data that is available for tests performed in a bare cylinder, 20 mm diameter and 10 m length, which was donated by ExxonMobil, was used in our simulations. Several fluid velocities and both linearly sheared and linear flow conditions may be used. We chose a *strong* condition of regular flow, of approximately 1 m/s, which provided greater displacements compared to the sheared flow for the same nominal velocity.

To adapt the experimental data to the riser that was being analyzed, the displacement was scaled by the test riser diameter and multiplied by the actual riser diameter. The frequency of shedding (f_{st}) in cylinders with cross-flow is given by the following:

$$f_{\text{st}} = \frac{\text{St}U}{D}, \tag{11}$$

where St is the Strouhal number, U is the fluid velocity, and D is the cylinder diameter. Keeping St fixed, the ratio of f_{st} between the experiment and the riser was 14.78. This factor was used to scale the time vector, such that the frequency vector of the data spectrum, which was used to simulate riser VIV, was divided by this quantity. The set of ten points along the cylinder where displacements were measured was associated with the closest nodes of the FEM model. A window of data, without transient effects, was replicated ten times until the total mission time was achieved, similar to the previous section, and the same interpolation procedure was applied. Finally, the displacement caused by VIV was rotated to become perpendicular to the catenary plane and added to (10).

Figure 7 shows the RMS profiles of the riser displacements in the $\mathbf{X_G}$, $\mathbf{Y_G}$, and $\mathbf{Z_G}$ directions as a function of the normalized riser length. The figure also shows the RMS of VIV perpendicular to the catenary plane. The number 0 is for the wheel head, and 1 corresponds to the turret.

4.4. Sensor Velocity and Acceleration Profiles.

To simulate IMU output, the expected physical acceleration at the sensor installation point must be found. This acceleration was used to feed the sensor model to find realistic sensor signals. The acceleration at point P_S expressed in the global reference frame was given by the following well-known kinematic equation:

$$\mathbf{a}_{P_S G} = \begin{bmatrix} \ddot{x} \\ \ddot{y} \\ \ddot{z} \end{bmatrix} + \boldsymbol{\alpha} \times P_S + \boldsymbol{\omega} \times \boldsymbol{\omega} \times P_S. \tag{12}$$

The local acceleration $\mathbf{a}_{P_S l}$ and Coriolis terms are zero because the sensor is fixed in the robot body. $\boldsymbol{\alpha}$ is the angular

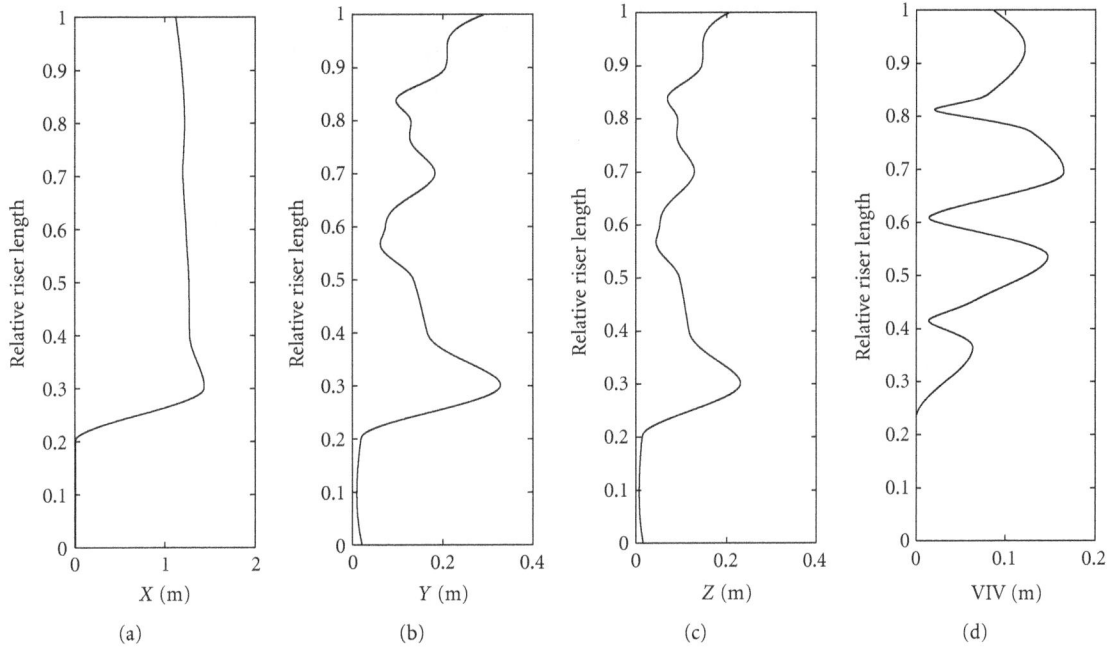

FIGURE 7: RMS of riser motion in $\mathbf{X}_l$, $\mathbf{Y}_l$, and $\mathbf{Z}_l$ directions and VIV perpendicular to the catenary plane.

acceleration and $\boldsymbol{\omega}$ is the angular velocity. Angular velocity expressed in the global reference frame was calculated by the methods used in [14, 18]:

$$\boldsymbol{\omega} = R_z\,R_y \begin{bmatrix} \dot{\psi} \\ 0 \\ 0 \end{bmatrix} + R_z \begin{bmatrix} 0 \\ \dot{\theta} \\ 0 \end{bmatrix} + \begin{bmatrix} 0 \\ 0 \\ \dot{\phi} \end{bmatrix}. \tag{13}$$

To solve (12) and (13), the time derivatives of displacement and rotation $\dot{x}$, $\ddot{x}$, $\dot{y}$, $\dot{\psi}$, and so forth were found by finite differences and considering all variables in the global reference system.

5. Instrumentation Model

A low-cost microstrain 3DMGX-1 IMU was the project choice. This is a compact and integrated device, suitable for a high-depth submarine application, where the electronic case must be as slender as possible, for mechanical structural reasons. It delivers 3D accelerations, angular velocities, and attitude/orientation matrix in a single-serial channel. The error characteristics of each output were modeled as a wide-band noise plus a first order moving bias Markov process [19]:

$$Y = (1 + \mathrm{SFE})U + \mathrm{OFS} + b + w_{\mathrm{bw}}, \tag{14}$$

where

Y: simulated *corrupted* sensor output,

U: simulated *clean* physical signal,

SFE: Scale Factor Error,

OFS: Offset Error,

b: moving bias, 1st order Markov process,

w_{bw}: wide-band sensor noise.

The wide-band sensor noise was defined as $w_{\mathrm{bw}} = \sigma_{\mathrm{bw}}\nu$, $\nu = N[0,1]$ (white noise, zero mean). σ_{bw}^2 is the wide-band noise variance. Scale factor error, 0.5% by sensor specifications, was considered as follows:

$$\mathrm{SFE} = 0.005 \times \mathrm{sgn}(U(-1,1)), \tag{15}$$

such that $U(-1,1)$ is a uniform random variable in $[-1,1]$ and sgn is the *signum* function. This equation introduces a random error limited to 0.5% in the sensor output signal amplitude, when substituted in (14).

The moving bias was found by integrating the following finite difference equation with the Euler method:

$$b_{n+1} = b_n + \Delta t((-1/\tau_r)b_n + \sigma_{\mathrm{sensor\_bias}}\nu),$$
$$\sigma_{\mathrm{sensor\_bias}} = \sqrt{\frac{2f_s\sigma_{\mathrm{bias}}^2}{\tau_r}}, \tag{16}$$

where

Δt: sampling period (s),

f_s: sampling frequency (Hz),

$\sigma_{\mathrm{sensor\_bias}}^2$: bias input noise variance,

σ_{bias}^2: random walk variance,

τ_r: time constant (s).

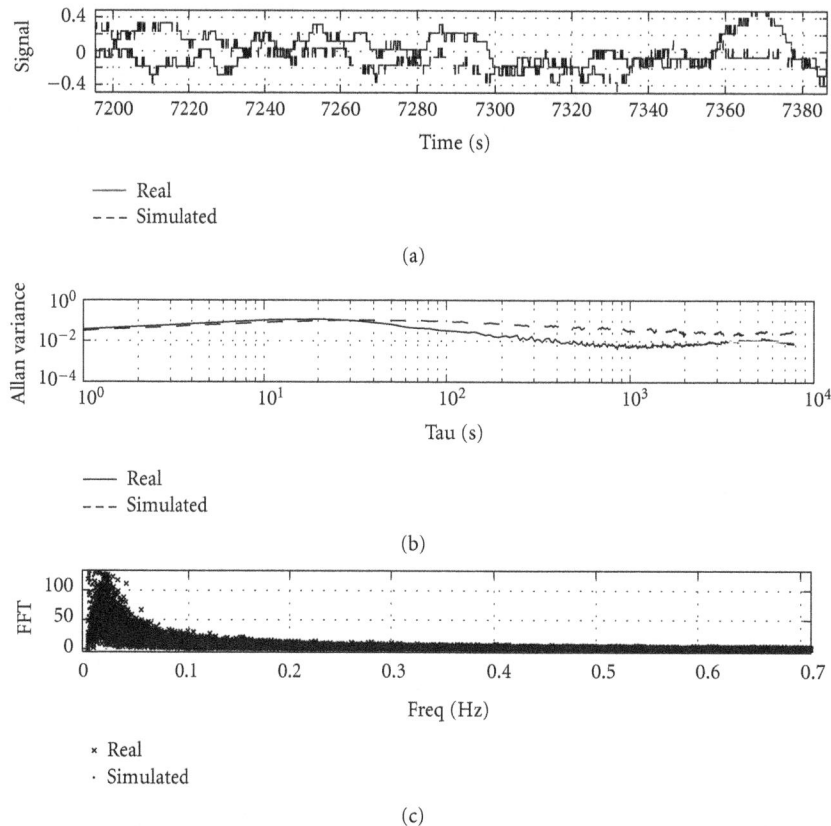

FIGURE 8: Comparative results for real and simulated IMU sensor error for Pitch angle. (a) Window of the time series (degrees). (b) Allan variance plot (degrees). (c) FFT of time series (degrees^2).

5.1. Determination of Sensor Parameters. An experiment was performed to collect the error characteristics by keeping the sensor stationary in laboratory while recording signals. Total acquisition time was 4 h 21 min, with a 75 Hz sampling frequency, after the internal temperature was stabilized. The first determined parameter was the time constant τ_r. Two techniques were employed: the Allan variance plot [20] and the autocorrelation function [19]. Both techniques gave different results (see Table 2). According to the likeness between the real and the simulated signals, one of the two time-constant techniques was selected. For the attitude sensor, τ_r determined by Allan variance was chosen, whereas for acceleration and angular velocity, autocorrelation provided the best results.

Except for angular velocity (AngRate), the experiment time series was low-pass filtered before calculating σ_{bias}, to better characterize the low-frequency component of the error. For σ_{bw}, the high-frequency part was used. The cut-off frequencies were chosen by trial and error, and the agreement between the simulated and original error signals was observed for each case. Time series, Allan variance plot [21] and fast Fourier transform (FFT) were used to test the agreement between original and simulated error signals. In the case of attitude simulation, the Euler angles were discretized in the interval comprising $-180.0328°$ to $180°$, with steps of $0.1°$, corresponding to sensor resolution. For

acceleration, the resolution of the sensor was 0.2 mg, and the angular velocity was $0.01°/s$. However, amplitude discretization was not implemented. Figure 8 shows the results for Pitch angle and Figure 9 shows the results for acceleration in the X_l direction. For other angles, directions and angular velocities, the results were similar.

The water pressure-based depth sensor specified for this robot was the Digiquartz 8CB4000-I Depth Sensor (Paroscientific, Inc., Redmond, WA), provided a 0.01% accuracy and a resolution of 10^{-8} m. In the simulations, we considered that the depth measures were corrupted by the addition of a Gaussian noise with $\sigma = 0.2$ meters, according to the estimates made by Jalving [22].

6. Localization Algorithm

The localization problem consisted of estimating a set of states, which included robot position, by reading accelerations as inputs and considering the depth sensor signals as the external measurements. Since a strapdown navigation scheme is used, the localization algorithm assumes that the accelerations measured in the local frame are transformed into the global frame, by using the rotation matrices calculated from Euler angles measured by the IMU. It is also assumed that if a DC-acceleration sensible accelerometer is being used, the vertical acceleration was compensated

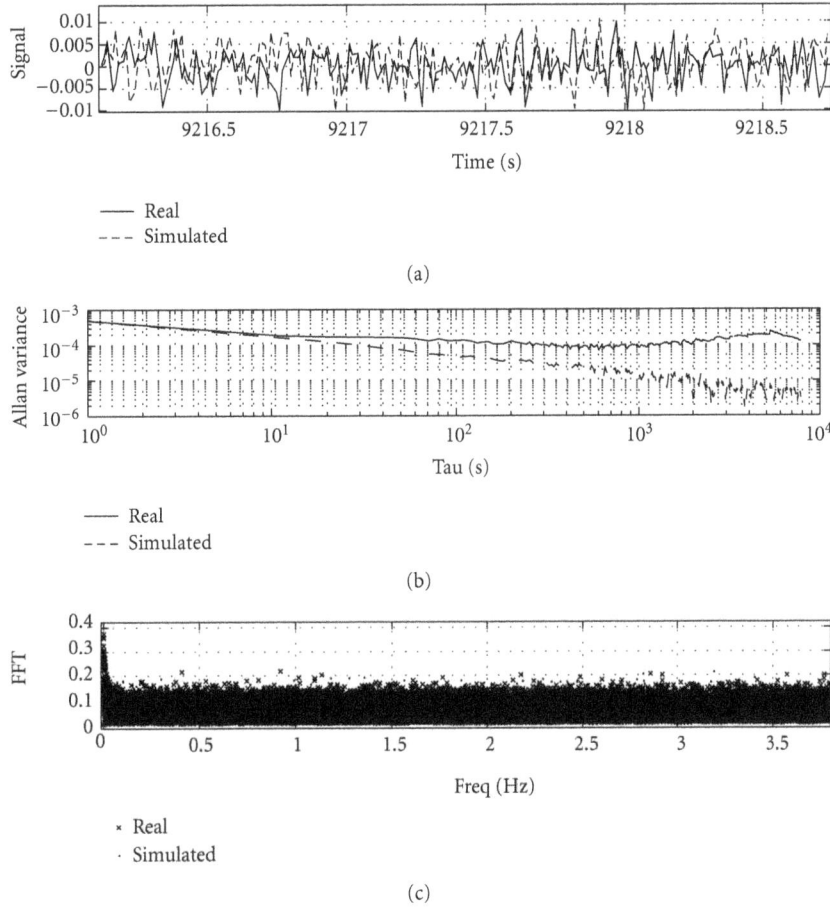

FIGURE 9: Comparative results for real and simulated IMU sensor error for $\mathbf{X}_l$ acceleration, in g (Earth acceleration of gravity at sea level). (a) Window of the time series (g). (b) Allan variance plot (g). (c) FFT of time series (g^2).

for gravity. The problem state equations are formulated as [23, 24]:

$$\mathbf{X}_{k+1} = A\mathbf{X}_k + B\mathbf{U}_k + C\mathbf{w}_k,$$
$$\mathbf{Z}_k = H\mathbf{X}_k + \mathbf{v}_k. \tag{17}$$

The state vectors and matrices are given by the following:

$$\mathbf{X}_k = \begin{bmatrix} P_x & P_y & P_z & V_x & V_y & V_z & \delta a_x & \delta a_y & \delta a_z \end{bmatrix}_k^T,$$
$$\mathbf{U}_k = \begin{bmatrix} a_x & a_y & a_z \end{bmatrix}_k^T,$$
$$\mathbf{w}_k = \begin{bmatrix} w_1 & w_2 & w_3 & w_4 & w_5 & w_6 & w_7 & w_8 & w_9 \end{bmatrix}_k^T, \tag{18}$$
$$\mathbf{v}_k = \begin{bmatrix} v_1 & v_2 & v_3 \end{bmatrix}_k^T.$$

$A_{[9\times9]}$ is an identity matrix, such that the terms (1,4), (2,5), and (3,6) are equal to T; (4,7), (5,8), and (6,9) assumes the value $-T$; (1,7), (2,8), and (3,9) elements are $-1/2T^2$. $B_{[9\times3]}$ is a matrix of zeros except for the terms (1,1), (2,2), and (3,3) that are equal to $1/2T^2$; (4,1), (5,2), and (6,3) elements are T.

$H_{[3\times9]}$ is a matrix of zeros, with the exception of the (i,i) terms equal to 1. In such expressions, k is the sample number,

$\mathbf{X}$ is the state vector, $\mathbf{U}$ the input vector, $\mathbf{w}$ is the process noise, P is the position, V is the velocity, a is acceleration, δa is the acceleration error, T is sampling period, $\mathbf{Z}$ is the external measurements vector, $\mathbf{v}$ is the sensor noise, and σ^2 is the variance of each associated variable. The observability matrix of this system had a full-rank, that is, 9.

An estimate $\hat{\mathbf{X}}_k$ of the state vector is found by a Kalman filter.

Prediction:

$$\hat{\mathbf{X}}_k^- = A\hat{\mathbf{X}}_{k-1}^- + B\mathbf{U}_k,$$
$$P_k^- = AP_{k-1}^- A^T + Q. \tag{19}$$

Update:

$$K_k = P_k^- H^T (HP_k^- H^T + R)^{-1},$$
$$\hat{\mathbf{X}}_k = \hat{\mathbf{X}}_k^- + K_k\left(\mathbf{Z}_k - H\hat{\mathbf{X}}_k^-\right),$$
$$P_k = (I - K_kH)P_k^-, \tag{20}$$

where P is the state estimate error covariance matrix, R is the sensor noise covariance matrix, Q is the process noise matrix, and K is the Kalman gain. This is a standard, simple and generic implementation of a Kalman filter estimator. More sophisticated approaches could be tested as well, although the results obtained with the standard KF were satisfactory for the intended application, as will be addressed in Section 7. Unlike the kinematic models that are usually used for land robots [25] or those built to operate over flat surfaces [26], the state-space model used here to formulate the KF does not use any *a priori* information of the vehicle characteristics. Nor the dynamical characteristics of the sensors are included in this KF formulation. Thus, in principle, the localization algorithm proposed here could be applied for similar applications with different instrumentation, robot, and environmental characteristics. However, some tuning work on the covariance matrices should likely to be necessary.

6.1. Processing of the External Measurements for the KF. Here, we focus the analysis on the key aspect of this particular localization problem, which is determining the external measurements vector $\mathbf{Z}_k = [O_{kx}^G \ O_{ky}^G \ O_{kz}^G]^T$. O_{kx}^G, and so forth, are the coordinates of the robot position in the global reference frame as measured by an independent absolute sensor. The sensor, that is, in practice, available to perform such measurements is the depth sensor, which can only measure the $\mathbf{Z}_G$ coordinate. The attempts to use this measurement exclusively in the Update phase of the KF gave nondrifting estimates in the vertical direction, as expectable, but not in the horizontal one.

We proposed the following method to estimate the complete $\mathbf{Z}_k$ vector using data from $\mathbf{Z}_{k-1}$, O_{kz}^G (depth sensor signal), and the Euler angles measured by the IMU. Considering Figure 10, the vector joining $\mathbf{Z}_k$ to $\mathbf{Z}_{k-1}$, in the local reference frame centered in $\mathbf{Z}_k$, is given by $[0 \ 0 \ L]^T$. This vector lies in $\mathbf{Z}_l$ direction of the local reference frame 1_k, with modulus L, and coinciding with the instantaneous riser path. In the global reference frame, $\mathbf{Z}_{k-1}$ can be found as:

$$R_{1_k}^G = R_z R_y R_x(\psi, \theta, \phi) = \begin{bmatrix} R_{11} & R_{12} & R_{13} \\ R_{21} & R_{22} & R_{23} \\ R_{31} & R_{32} & R_{33} \end{bmatrix}, \quad (21)$$

$$\mathbf{Z}_{k-1} = \mathbf{Z}_k - R_{1_k}^G \mathbf{Z}_{k-1}^{1_k} = \mathbf{Z}_k - \begin{bmatrix} R_{13}L & R_{23}L & R_{33}L \end{bmatrix}^T. \quad (22)$$

Thus,

$$(\mathbf{Z}_k - \mathbf{Z}_{k-1}) = \begin{bmatrix} R_{13}L & R_{23}L & R_{33}L \end{bmatrix}^T. \quad (23)$$

The same difference calculated in (23) can be expressed by a vector of global coordinate differences:

$$(\mathbf{Z}_k - \mathbf{Z}_{k-1}) = \begin{bmatrix} \Delta X & \Delta Y & \Delta Z \end{bmatrix}^T. \quad (24)$$

Since R_{33} and ΔZ are known, by comparing (23) and (24), it is possible to find L as:

$$L = \frac{\Delta Z}{R_{33}}. \quad (25)$$

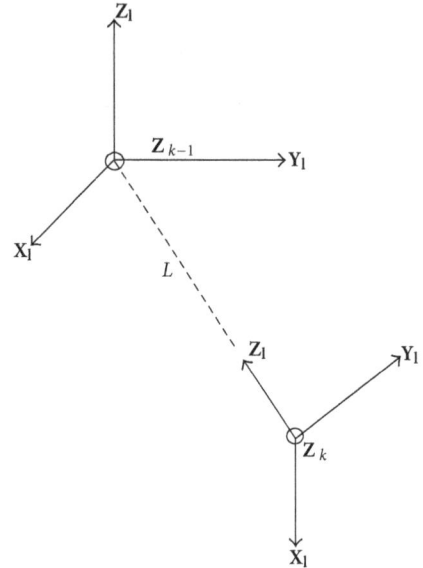

FIGURE 10: Position external measurements $\mathbf{Z}_{k-1}$ and $\mathbf{Z}_k$ and the corresponding local reference frames ($\mathbf{X}_l$, $\mathbf{Y}_l$, $\mathbf{Z}_l$). L is the scalar displacement between the two positions.

Substituting L in (23), the KF external measurements vector $\mathbf{Z}_k$ is found. Alternatively, the displacement L could be measured by an odometer. However, in this case, special constructive care should be taken into account to prevent slippage. In any case, this approach assumes a previous observation in the Correction phase of the KF. However, KF theory assumes conditional independence among the observations. Therefore, only *suboptimal* estimation performance is likely to be expected.

7. Simulation Results

7.1. Static Results. In the static case no riser motion is present. The *standard* position was used as the basis to generate the dynamic analysis and is an intermediary *neutral* configuration between the *near* (ship displaces towards wheel head) and the *far* (the opposite direction) positions. Figure 11 shows the robot trajectory and attitude, represented by the successive local reference frames, for the neutral position. In Figure 12, the associated localization errors are shown, as a function of distance from the seabed (using a grid of 1.680 points, or 1.44 Hz). The absolute errors associated with the neutral, far, and near configurations are shown in Figure 13, in a *semilog* scale. To study the impact in the localization error of the grid refinement grade, an additional error curve was generated, for the neutral configuration, with a 7,630 points (5 Hz) grid, the same used for the dynamic analysis.

Figure 12 shows that the greatest source of error occurs in the $\mathbf{X}_G$ direction, while in the deepness coordinate $\mathbf{Z}_G$, the error is low. In addition, the total error increased as the robot tilted horizontally. This trend was expected due to the relative loss of accuracy of the depth sensor to estimate the L parameter when using (21)–(25). When the depth measure difference ΔZ between the two samples is small,

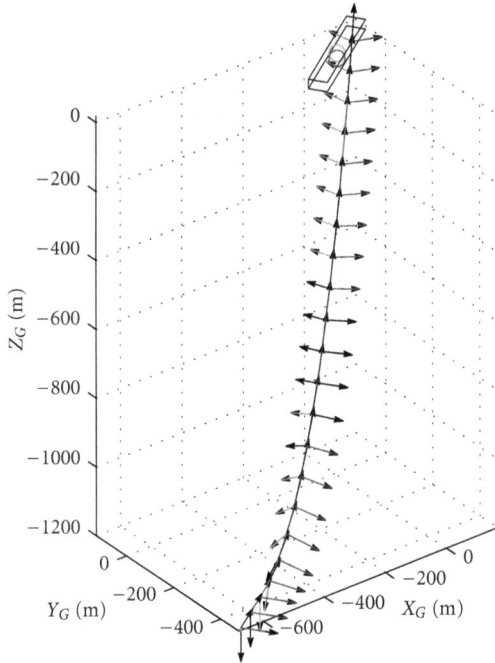

FIGURE 11: Problem neutral configuration, showing the FPSO, the riser profile and the robot attitude. The orthogonal arrows express the successive local reference frames position and orientation along the riser path. Only 1/8 of the 171 FEM mesh nodes used in the static analysis is shown to facilitate visualization.

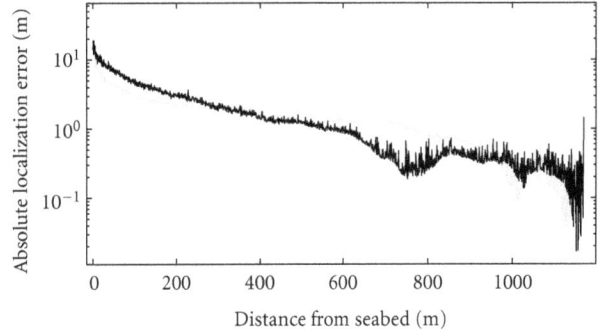

Near
— Far

(a)

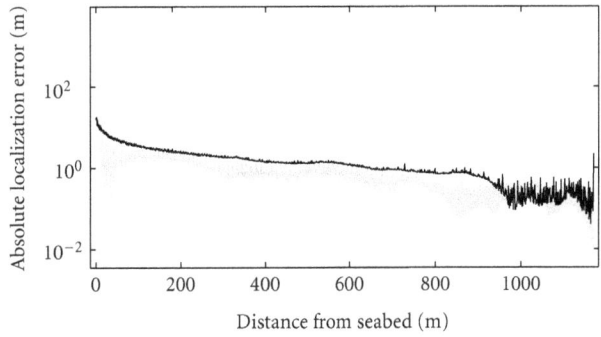

Neutral
— Neutral 5 Hz

(b)

FIGURE 13: Total localization error (in global coordinates) as a function of distance from the seabed, for neutral, near and far configurations with the mesh of 1.680 points (1.44 Hz). The neutral configuration error for the mesh with 7.630 points (5 Hz) is also shown.

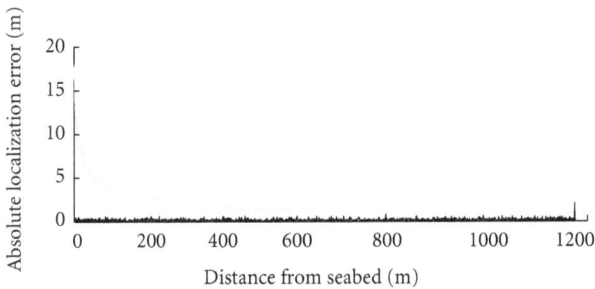

Total
— Error z

(a)

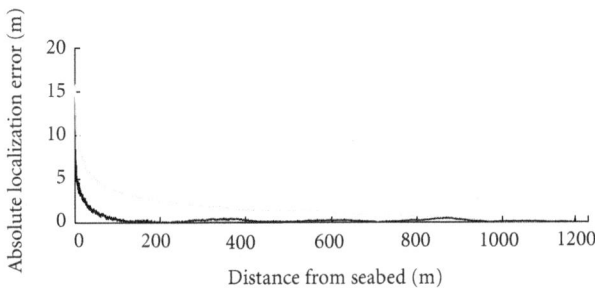

Error x
— Error y

(b)

FIGURE 12: Localization error (in global coordinates) for the static neutral case as a function of the distance from the seabed.

a decrease in signal to Gaussian noise ratio is observed. From an application point of view, the robot should not work too close to the point where the riser periodically touches the seabed, the touch down point (TDP). This safety margin is established as being of 50 m. At this point, the expected positioning error for the static case was approximately 10 m. The error was slightly higher in the far case (Figure 13), where the riser was more horizontal than in the near condition. By increasing the sampling frequency (Figure 12), the error was likely to decrease in most of the path, but when the robot approximates to the TDP, the localization system fails.

7.2. Dynamic Results. By including the wave and VIV effects in the riser nodal displacement profile, the robot trajectory becomes more complex (Figure 14), as expectable. The associated localization errors are shown in Figure 15, using the 5 Hz mesh. Figure 16 shows a zoom in the $X_G Y_G$ plane, close to TDP, of both the robot trajectory and the external measurements, calculated with (21)–(25), as well as the Kalman filter position states estimations. The KF estimated the trajectory accurately, up to the point where the processed

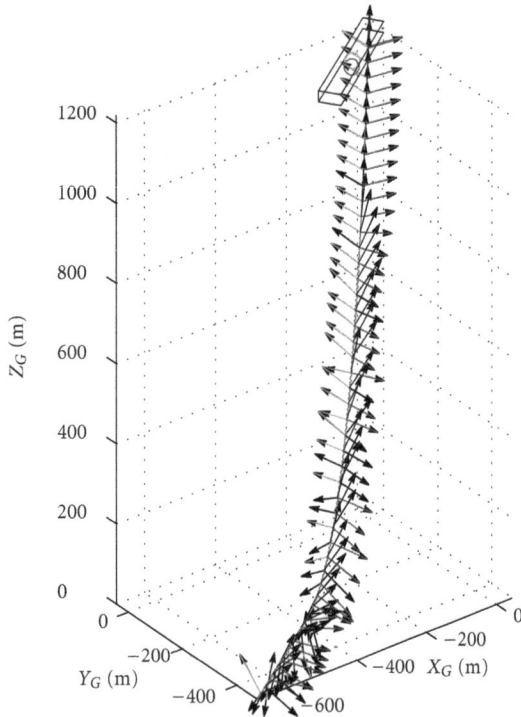

FIGURE 14: Robot attitude profile in the dynamic condition, including wave and VIV effects. Only 1/150 of the 7.630 grid points are shown to keep the figure clearer.

(a)

(b)

FIGURE 15: Localization error (in global coordinates) as a function of the distance from the seabed, for the dynamic condition.

external measurements lost accuracy, below 15 m distance to seabed. The KF estimates essentially followed the external measurements, in part because the KF Update frequency was the same as the Prediction. By comparing the errors in both static and dynamic cases (Figure 17), analogous profiles were found. It means that the Kalman filter was able to closely track the robot's trajectory, even in the dynamic case, and the loss of algorithm localization performance was mainly caused, after all, by depth sensor differential signal to noise ratio.

8. Discussion and Conclusions

The simulations show that the robot trajectory using the localization algorithm, in the dynamic case, presented a wavy pattern. The pattern corresponds to the path that is traveled by the robot, and not the nominal riser profile, as expected. If an estimate of the riser catenary shape is desired, the obtained path could be used to fit a smooth profile curve.

In the more vertical part of the riser (above 15 m from seabed), the average estimated error (standard dev.) was 0.76 (0.47) m (Figure 15). This error can be considered sufficiently small for localizing the riser in the oil field and to feed nodal displacement constraints in a structural Finite Elements tension analysis. This accuracy is also satisfactory for localizing flaws detected by the NDTs.

The localization algorithm depends strongly on the external measurements, which are provided by the depth sensor and the IMU, using (21)–(25). However, close to

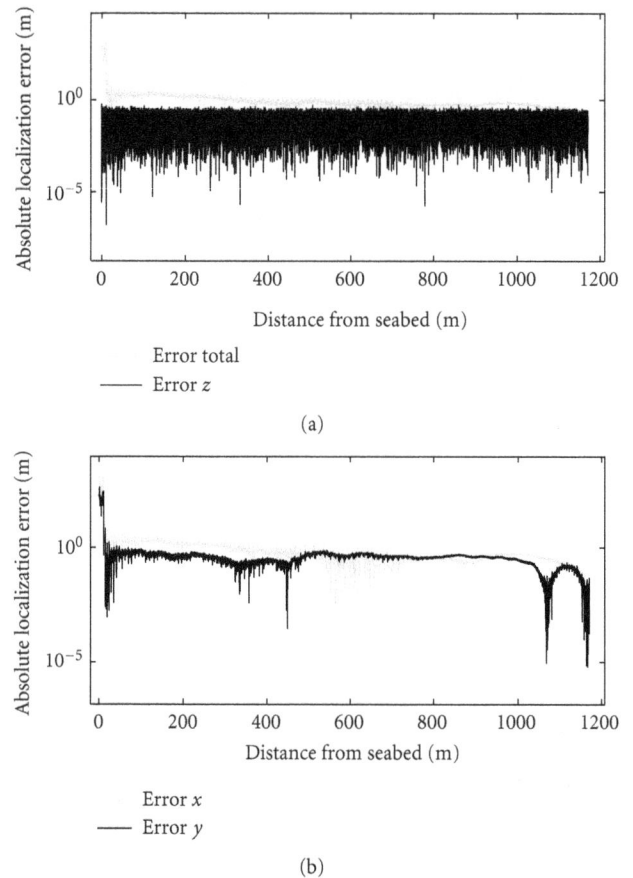

seabed, the simulated robot trajectory tilts towards a more horizontal attitude, and following the riser section which is closer to the TDP. In this part of the trajectory, Gaussian noise from the depth sensor more significantly corrupts the relative depth measurements. This effect increases when a period between two successive measurements becomes shorter. The 5 Hz is the minimum estimated rate necessary to reproduce VIV and the wave motion data. On the other hand, the minimum safety distance between the robot and the seabed coincides with the condition which the localization algorithm loose accuracy. As a possible extreme limit, at 15 m from the seabed, the expected error should be also around 15 m (Figure 15).

In our opinion, the simple and standard KF implementation that was presented in this study is sufficient to provide the localization accuracy required for the proposed application. In the case of rigid risers and other predominantly vertical offshore structures, the observed loss of accuracy is not expected to occur. The kinematic model used to formulate the sensor fusion problem was linear, and therefore the standard KF implementation is appropriate. Other models could be proposed, incorporating robot, riser or sea dynamic features. In such cases, using an extended Kalman filter or a particle filter should be necessary for handling the associated nonlinearities. In any case, the external measurements

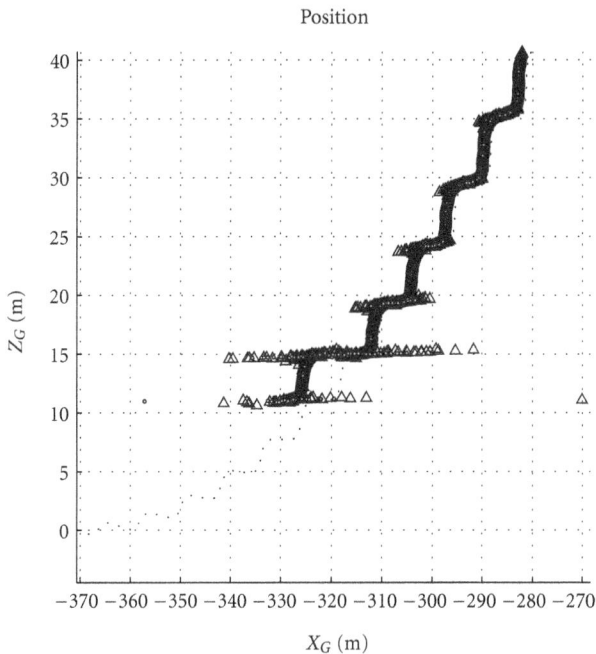

FIGURE 16: $X_G Z_G$ plane close to TDP for the dynamic condition, showing simulated robot trajectory (dashed), simulated and processed external measurements from depth sensor and IMU models (Δ) and Kalman Filter estimates (clearer dots).

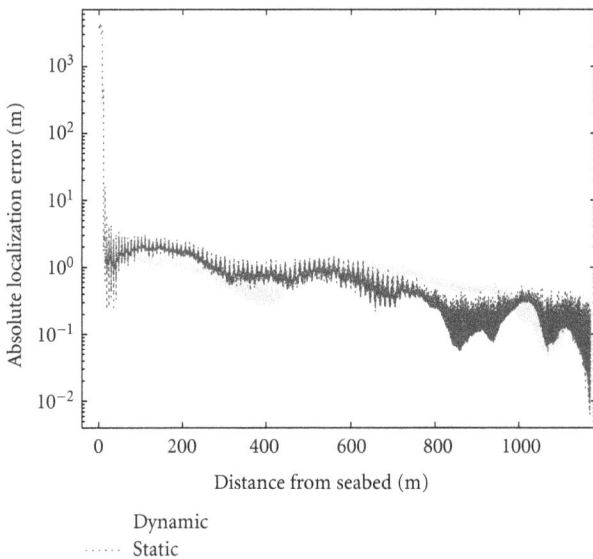

Dynamic
...... Static

FIGURE 17: Mean localization error (in global coordinates) as a function of distance from the seabed, for both static and dynamic conditions.

reliability seems to be key aspect for localization accuracy in this problem.

Using a higher grade low drift IMU instead of a low-cost one could eliminate the need for external measurements in a critical region but at the price of increasing payload, cost and volume. Acoustic positioning systems could be used to

localize the robot in the most horizontal parts of the trajectory. However, these systems work at a very low sampling frequency (0.1 to 1 Hz) [8] and, at best, they could only probably cast the external motion envelope of the robot. More practically, improved accuracy could be achieved in the section close to the TDP by using an odometer to estimate L in (25). As such, the mechanical setup of an odometer should be accomplished very carefully to prevent slippage. Preinstalled RFID (Radio Frequency Identification) [27] tags or visual marks along the riser, signaling the actual length, are possibilities that could be explored. Switching the depth sensor with other kinds of sensors for external measurements, close to TDP, could be performed by using a fuzzy expert system [28]. In future studies, experimental laboratory and field tests should be performed, using an acoustic localization system to assess the bounds of the navigation errors estimated in this work.

Acknowledgments

The authors are gratefully thankful to CAPES (Coordenacao de Aperfeicoamento de Pessoal de Nivel Superior), FINEP (Financiadora de Estudos e Projetos), CNPq (Conselho Nacional de Desenvolvimento Cientifico e Tecnologico), and FAPERJ (Fundacao de Amparo a Pesquisa do Estado do Rio de Janeiro) for financial support.

References

[1] P. O'Brien and and J. Picksley, "State-of-art flexible riser integrity issues," Tech. Rep. 2-1-4-181/SR01, 2001, UKOOA by MSC, Rev. 04.

[2] G. Chapin, "Inspection and monitoring of girassol hybrid riser towers," in *Proceedings of the Offshore Technology Conference (OTC '05)*, Huston, Tex, USA, 2005, OTC paper no. 17696.

[3] K. Hamilton and J. Evans, "Subsea pilotless inspection using an autonomous underwater vehicle (SPINAV): concepts and results," in *Proceedings of the OES/IEEE Oceans (OCEANS '05)*, pp. 775–781, Brest, France, June 2005.

[4] M. Ordonez, M. O. Sonnaillon, D. Murrin, N. Bose, and W. Qiu, "An advanced measurement system for vortex-induced-vibrations characterization in large-scale risers," in *Proceedings of the MTS/IEEE Conference Oceans (OCEANS '07)*, pp. 775–781, Vancouver, BC, Canada, 2007.

[5] E. Veith, C. Bucherie, J. L. Lechien, J. L. Larrouse, and B. Rattoni, "Inspection of offshore flexible riser with electromagnetic and radiographic techniques," in *Proceedings of the 15th World Conference on Nondestructive Testing*, pp. 404–408, Rome, Italy, 2000.

[6] D. Psarros, V. A. Papadimitriou, P. Chatzakos, V. A. Spais, and K. Hrissagis, "A service robot for Subsea flexible risers: analysis and systematic design," *IEEE Robotics and Automation Magazine*, vol. 17, no. 1, pp. 55–63, 2010.

[7] P.-M. Lee, B.-H. Jun, K. Kim, J. Lee, T. Aoki, and T. Hyakudome, "Simulation of an inertial acoustic navigation system with range aiding for an autonomous underwater vehicle," *IEEE Journal of Oceanic Engineering*, vol. 32, pp. 327–345, 2007.

[8] J. Jouffroy and J. Opderbecke, "Underwater vehicle navigation using diffusion-based trajectory observers," *IEEE Journal of Oceanic Engineering*, vol. 32, pp. 313–326, 2007.

[9] C. He, E. Jorge, and L. M. Zurk, "Enhanced kalman filter algorithm using the invariance principle," *IEEE Journal of Oceanic Engineering*, vol. 34, no. 4, pp. 575–585, 2009.

[10] M. F. Santos, L. L. Menegaldo, and M. O. Brito, "An outer device for universal inspectionof risers," US Patent Application based on Brazilian PI 0705113-1, 2008.

[11] R. O'Grady, H.-J. Bakkenes, D. Lang, and A. Connaire, "Advancements in response prediction methods for deep water pipe-in-pipe flowline installation," in *Proceedings of the Offshore Technology Conference (OTC '08)*, Houston, Tex, USA, 2008, OTC paper no. 19400.

[12] P. J. O'Brien and J. F. McNamara, "Significant characteristics of three-dimensional flexible riser analysis," *Engineering Structures*, vol. 54, no. 4, pp. 223–233, 1989.

[13] D. H. Titterton and J. L. Weston, *Strapdown Inertial Navigation Technology*, Edited by Herts, IEEE, London, UK, 2nd edition, 2004.

[14] B. Siciliano, L. Sciavicco, L. Villani, and G. Oriolo, *Robotics: Modelling, Planning and Control*, Springer, London, UK, 3rd edition, 2009.

[15] A. L. S. Pinho, *Tension reduction in TLP platforms rigid risers*, M.S. thesis, Civil Engineering Department, Federal University of Rio de Janeiro, Rio de Janeiro, Brazil, 2001.

[16] C. L. Cunff, F. Biolley, E. Fontaine, S. Etienne, and M. L. Fancchinetti, "Vortex-inducedvibrations of raisers: theoretical, numerical and experimental investigation," *Oil and Gas Science and Technology (IFP)*, vol. 5, pp. 59–69, 2002.

[17] M. I. of Technology, "Vortex induced vibration data repository," 2011, http://oe.mit.edu/VIV/.

[18] H. I. Weber, "Inertial Systems: basic foundations," in *Proceedings of V SBEIN 5th Brazilian Symposium on Inertial Engineering, Tutorial A, (in Portuguese)*, Rio de Janeiro,Brazil, 2007.

[19] W. S. Flenniken IV, *Modeling inertial measurement units and analyzing the effect of their errors in navigation applications*, M.S. thesis, Auburn University, Auburn, Ala, USA, 2005.

[20] D. W. Allan, "Statistics of atomic frequency standards," *IEEE Proceedings*, vol. 54, pp. 221–230, 1966.

[21] IEEE, "IEEE Recommended Practice for Inertial Sensor Test Equipment, Instrumentation, Data Acquisition, and Analysis," IEEE Std 1554-2005, IEEE Aerospace and Electronic Systems Society, 2005.

[22] B. Jalving, "Depth accuracy in seabed mapping with underwater vehicles," in *Proceedings of the OES/IEEE Oceans (OCEANS '99)*, pp. 973–978, Seattle, Wash, USA, September 1999.

[23] D. D. S. Santana, *Terrestrial trajectory estimation using a low cost measurement unity and sensor fusion*, M.S. thesis, Mechatronic Engineering Department, Polytechnic School, University of Sao Paulo, Sao Paulo, Brazil, 2005.

[24] H. Qi and J. B. Moore, "Direct Kalman filtering approach for GPS/INS integration," *IEEE Transactions on Aerospace and Electronic Systems*, vol. 38, no. 2, pp. 687–693, 2002.

[25] T. Sasaki, D. Brščić, and H. Hashimoto, "Human-observation-based extraction of path patterns for mobile robot navigation," *IEEE Transactions on Industrial Electronics*, vol. 57, no. 4, pp. 1401–1410, 2010.

[26] L. L. Menegaldo, G. A. N. Ferreira, M. F. Santos, and R. S. Guerato, "Development and navigation of a mobile robot for floating production storage and offloading ship hull inspection," *IEEE Transactions on Industrial Electronics*, vol. 56, no. 9, pp. 3717–3722, 2009.

[27] S. S. Saad and Z. S. Nakad, "A standalone rfid indoor positioning system using passive tags," *IEEE Transactions on Industrial Electronics*, vol. 58, no. 5, pp. 1961–1970, 2011.

[28] S. H. P. Won, F. Golnaraghi, and W. W. Melek, "A fastening tool tracking system using an IMU and a position sensor with Kalman filters and a fuzzy expert system," *IEEE Transactions on Industrial Electronics*, vol. 56, no. 5, pp. 1782–1792, 2009.

16

Computationally Efficient Iterative Pose Estimation for Space Robot Based on Vision

Xiang Wu and Ning Wu

Shenzhen Graduate School, Harbin Institute of Technology, Harbin, Heilongjiang 150001, China

Correspondence should be addressed to Ning Wu; aning.wu@gmail.com

Academic Editor: Farhad Aghili

In postestimation problem for space robot, photogrammetry has been used to determine the relative pose between an object and a camera. The calculation of the projection from two-dimensional measured data to three-dimensional models is of utmost importance in this vision-based estimation however, this process is usually time consuming, especially in the outer space environment with limited performance of hardware. This paper proposes a computationally efficient iterative algorithm for pose estimation based on vision technology. In this method, an error function is designed to estimate the object-space collinearity error, and the error is minimized iteratively for rotation matrix based on the absolute orientation information. Experimental result shows that this approach achieves comparable accuracy with the SVD-based methods; however, the computational time has been greatly reduced due to the use of the absolute orientation method.

1. Introduction

Vision based methods have been applied to estimate the pose of space robot since 1990s. In these methods, the relative position and orientation between a camera and a robot target are determined with a set of n feature points expressed in the three dimensional (3D) object coordinates and their two dimensional (2D) projection in the camera coordinate. The error in position and orientation is usually optimized using the noniterative or iterative algorithms. The noniterative algorithms give an analytical solution for the optimization [1–3], and a typical example of these algorithms includes the method to represent feature points as a linear combination of four virtual control points based on their coordinates [4]. The noniterative methods are generally less time consuming than the iterative methods with acceptable accuracy; however, they are sensitive to observation noise such as image noise, different lighting conditions, and even occlusion by outliers. The iterative approaches, however, achieve better accuracy than the noniterative methods by solving the rotation matrix with a nonlinear least-square method iteratively.

A typical iterative method is the Levenberg-Marquardt (L-M) algorithm [5–7], and it has been widely used and accepted as a standard algorithm for least-square problem in photogrammetry. The L-M method is in essentially the combination of the steepest descent method and the Gauss-Newton method in different optimization stages. The steepest descent method is used at the early stage of optimization when the current value of error is still far from the minimum, while the Gauss-Newton method is used at the later stage of optimization when the solution is relatively close to the target. The combination of the two methods at different stages is in fact a coarse search from a globalwise followed by a fine search within a local area. The use of the steepest descent method in the early stage of optimization helps to find a guaranteed convergence direction and locate the solution within a small area, while the Gauss-Newton approach finds the optimized solution with fast speed. The L-M algorithm offers a way to find an optimized solution for the iterative approach; however, since the L-M method is a general-purposed optimization method, it can be improved significantly to suit the specific requirement of the pose estimation for a faster converging speed and a better noise-tolerant solution.

The iterative optimization method specially designed for pose estimation purpose has been considered based on the target pose and the depths of the feature points [8]. This method calculates the depth information and absolute orientation of target, respectively, and the perspective nonlinearity can be reduced by introducing the depth variable. However, this method requires hundreds of iterations before it can reach a convergence point [8].

Another type of iterative algorithm, orthogonal iterative (OI), was proposed by Lu et al. [9] to estimate the object-space collinearity error with a different objective function. Instead of using the depth of the object, this algorithm uses scene points to improve the calculation of the translation vector, and it achieves a higher accuracy and more computation efficiency. However, the corruption in the input data can cause a considerable error in the rotation matrix, and thus the accuracy of the OI method is affected [10]. The OI algorithm was further developed by Zhang et al. by introducing depth update in the computation of the translation vector in a two-stage iterative process [11, 12]. Higher accuracy than the OI algorithm can be achieved from this method by refining the error of absolute orientation [10].

The essential process of the above optimization algorithms specially designed for pose estimation is to solve the absolute orientation problem, which can be applied with quaternions [13–15] and Singular Value Decomposition (SVD) [15, 16] methods, respectively. The SVD method has achieved a great performance and has been used extensively because of its closed form solution and enhanced orthogonality; however, the computational load makes it difficult to implement in real-time system.

In order to overcome this problem, this paper introduces the FOAM method [17] to calculate the absolute orientation for pose estimation. The experimental result shows that the performance of accuracy and noise resistance will be shown to be comparable with the SVD method; however, the computational efficiency is considerably better. The structure of this paper has been organized as follows. Section 2 of this paper further introduces the problem, and Section 3 presents our solution for the problem. Section 4 shows the experimental result, and finally a conclusion is drawn.

2. Theory

In pose estimation of space robot, we usually have a target coordinate frame $OXYZ$ and a camera coordinate frame $O'X'Y'Z'$, and they are defined as illustrated in Figure 1, respectively.

It can be seen that the center of the projection from the object is at the origin O' and the optical axis points to the positive Z' axis. Supposed that a lens with the focal length of f is located at the origin, the plane $Z' = f$ is then considered as the image plane of the camera on which the feature points are projected. If the coordinates of feature points, P_i ($i = 1, \ldots, n$) on the target are denoted as $\mathbf{l}_i = (x_i, y_i, z_i)^T$ in the target coordinate frame, and its corresponding projection on the camera axis is expressed as $\mathbf{m}_i = (x_i', y_i', z_i')^T$, then the rotation matrix and the translation vector from the target

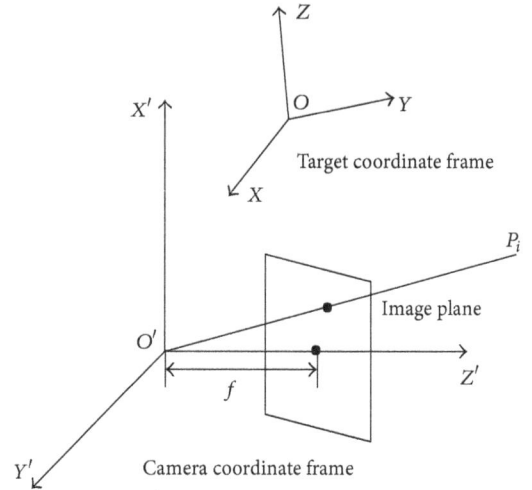

FIGURE 1: Coordinate frames in the pose estimation problem.

coordinate frame to the camera coordinate frame can be written in the relationship such as

$$\mathbf{m}_i = A\mathbf{l}_i + \mathbf{t}, \tag{1}$$

where $A = (\mathbf{a}_1, \mathbf{a}_2, \mathbf{a}_3)^T$ denotes the rotation matrix and $\mathbf{t} = (t_x, t_y, t_z)^T$ is the translation vector.

If the image point P_i' in camera axis frame represents the projection from the feature point P_i, and its coordinate is written as $\mathbf{w}_i = (u_i, v_i, f)^T$, according to the idealized pinhole camera model, the relationship between the coordinates of P_i^T and P_i can be expressed as

$$\mathbf{w}_i = \frac{f}{\mathbf{a}_3^T \mathbf{l}_i + t_z} \mathbf{m}_i = \frac{f}{\mathbf{a}_3^T \mathbf{l}_i + t_z} (A\mathbf{l}_i + \mathbf{t}), \tag{2}$$

and $\mathbf{w}_i$ in (2) is regarded as a collinearity equation in image space, and the orthogonal projection of $\mathbf{m}_i$ to $\mathbf{w}_i$ can be written as

$$A\mathbf{l}_i + \mathbf{t} = W_i (A\mathbf{l}_i + \mathbf{t}), \tag{3}$$

where

$$W_i = \frac{\mathbf{w}_i \mathbf{w}_i^T}{\mathbf{w}_i^T \mathbf{w}_i} \tag{4}$$

denotes the projection matrix to the vector $\mathbf{w}_i$.

The object-space collinearity error for each feature point is then formulated for optimization, such as

$$e_i = \mathbf{m}_i - \widehat{W}_i \mathbf{m}_i = (A\mathbf{l}_i + \mathbf{t}) - \widehat{W}_i (A\mathbf{l}_i + \mathbf{t}), \tag{5}$$

where

$$\widehat{W}_i = \frac{\widehat{\mathbf{w}}_i \widehat{\mathbf{w}}_i^T}{\widehat{\mathbf{w}}_i^T \widehat{\mathbf{w}}_i} \tag{6}$$

represents the observed projection matrix to vector $\widehat{\mathbf{w}}_i$, and $\widehat{\mathbf{w}}_i$ denotes the observation of $\mathbf{w}_i$. Considering each of the feature points, the following objective function is defined:

$$E(A, \mathbf{t}) = \sum_{i=1}^{n} \|e_i\|^2$$
$$= \sum_{i=1}^{n} \left\| (A\mathbf{l}_i + \mathbf{t}) - \widehat{W}_i (A\mathbf{l}_i + \mathbf{t}) \right\|^2. \tag{7}$$

The objective function in (7) can be minimized by finding a suitable A and $\mathbf{t}$.

Since the objective function in (7) is the second norm of both the translation vector $\mathbf{t}$ and the rotation matrix A, the error of the absolute orientation can be found by taking the derivative of (7) with respect to $\mathbf{t}$ and making it equal to zero, such as

$$\frac{\partial E(A, \mathbf{t})}{\partial \mathbf{t}} = 0 \implies n \left(I - \frac{1}{n} \sum_{i=1}^{n} \widehat{W}_i \right) \mathbf{t} = \sum_{i=1}^{n} \left(\widehat{W}_i - I \right) A\mathbf{l}_i. \tag{8}$$

The left side of (7) can be rewritten as,

$$\mathbf{x}^T \left(I - \frac{1}{n} \sum_{i=1}^{n} \widehat{W}_i \right) \mathbf{x} = \frac{1}{n} \sum_{i=1}^{n} \left(\|\mathbf{x}\|^2 - \mathbf{x}^T \widehat{W}_i \mathbf{x} \right)$$
$$= \frac{1}{n} \sum_{i=1}^{n} \left(\|\mathbf{x}\|^2 - \mathbf{x}^T \widehat{W}_i^T \widehat{W}_i \mathbf{x} \right) \tag{9}$$
$$= \frac{1}{n} \sum_{i=1}^{n} \left(\|\mathbf{x}\|^2 - \left\| \widehat{W}_i \mathbf{x} \right\|^2 \right).$$

Since $\widehat{W}_i \mathbf{x}$ is the projection of $\mathbf{t}$ on $\widehat{\mathbf{w}}_i$, and in this case $\mathbf{t}$ cannot project on itself, we have $\|\mathbf{x}\| \ge \| \widehat{W}_i \mathbf{x} \|$, or $\|\mathbf{x}\| - \| \widehat{W}_i \mathbf{x} \| \ge 0$; therefore, the term $n(I - (1/n) \sum_{i=1}^{n} \widehat{W}_i)$ in (8) is nonsingular, and the optimal position of $\mathbf{t}$ can be written as a function of the rotation matrix A such as

$$\mathbf{t}(A) = \frac{1}{n} \left(I - \frac{1}{n} \sum_{i=1}^{n} \widehat{W}_i \right)^{-1} \sum_{i=1}^{n} \left(\widehat{W}_i - I \right) A\mathbf{l}_i. \tag{10}$$

If $\widehat{\mathbf{m}}_i(A) = \widehat{W}_i(A\mathbf{l}_i + \mathbf{t}(A))$, then (7) can be rewritten as a function of A, such that

$$E(A) = \sum_{i=1}^{n} \|A\mathbf{l}_i + \mathbf{t}(A) - \widehat{\mathbf{m}}_i(A)\|^2. \tag{11}$$

It can be seen from (11) that the object-space collinearity error function can be minimized with respect to the rotation matrix A only.

In the same nonlinear least-square format as shown in (11), another objective function can be formulated in order to find the optimal orientation matrix A and translation vector $\mathbf{t}$, such that

$$L(A, \mathbf{t}) = \sum_{i=1}^{n} \|\mathbf{m}_i - (A\mathbf{l}_i + \mathbf{t})\|^2. \tag{12}$$

If the mean values of the feature points in object coordinate and camera coordinate are calculated as $\bar{\mathbf{l}} = (1/n) \sum_{i=1}^{n} \mathbf{l}_i$ and $\bar{\mathbf{m}} = (1/n) \sum_{i=1}^{n} \mathbf{m}_i$ respectively, (12) can be rewritten in the form of,

$$L(A, \mathbf{t}) = \sum_{i=1}^{n} \|\mathbf{m}_i - (A\mathbf{m}_i + \mathbf{t})\|^2$$
$$= \sum_{i=1}^{n} \left\| \mathbf{m}_i - \bar{\mathbf{m}} - A\left(\mathbf{l}_i - \bar{\mathbf{l}} \right) + \left(\bar{\mathbf{m}} - A\bar{\mathbf{l}} - \mathbf{t} \right) \right\|^2 \tag{13}$$
$$= \sum_{i=1}^{n} \left\| \mathbf{m}_i' - A\mathbf{l}_i' \right\|^2 + n \left\| \bar{\mathbf{m}} - A\bar{\mathbf{l}} - \mathbf{t} \right\|^2,$$

where $\mathbf{l}_i' = \mathbf{l}_i - \bar{\mathbf{l}}$ and $\mathbf{m}_i' = \mathbf{m}_i - \bar{\mathbf{m}}$. Since the second term of (13) can be set to zero by assigning $\mathbf{t} = \bar{\mathbf{m}} - A\bar{\mathbf{l}}$, the objective function $L(A, \mathbf{t})$ can be optimized by minimizing the first term, such that

$$G(A) = \sum_{i=1}^{n} \left\| \mathbf{m}_i' - A\mathbf{l}_i' \right\|^2$$
$$= \sum_{i=1}^{n} \left\| \mathbf{m}_i' \right\|^2 + \sum_{i=1}^{n} \left\| A\mathbf{l}_i' \right\|^2 - 2 \operatorname{tr} \left(A^T \sum_{i=1}^{n} \mathbf{m}_i' \mathbf{l}_i'^T \right) \tag{14}$$
$$= \sum_{i=1}^{n} \left\| \mathbf{m}_i' \right\|^2 + \sum_{i=1}^{n} \left\| A\mathbf{l}_i' \right\|^2 - 2 \operatorname{tr} \left(A^T B \right),$$

where $B = \sum_{i=1}^{n} \mathbf{m}_i' \mathbf{l}_i'^t$. It is obvious that $G(A)$ in (14) can be minimized with a maximum value of $\operatorname{tr}(A^T B)$, and it can be found by using the SVD method. In traditional SVD method, B can be decomposed as (U, Σ, V), where U and V are orthogonal to each other, and $\Sigma = \operatorname{diag}[\sigma_1, \sigma_2, \sigma_3]$ $(\sigma_1 \ge \sigma_2 \ge \sigma_3 \ge 0$ are the singular values of B), such that

$$B = U' S V'^T, \tag{15}$$

where $S = \operatorname{diag}[s_1, s_2, s_3]$, $(s_1 = \sigma_1; s_2 = \sigma_2; s_3 = (\det U)(\det V)\sigma_3)$, $U' = U \operatorname{diag}[1, 1, \det U]$, and $V' = V \operatorname{diag}[1, 1, \det V]$. Therefore, we have

$$\operatorname{tr}\left(A^T B \right) = \operatorname{tr}\left(A' U' S V'^T \right)$$
$$= \operatorname{tr}\left(S U'^T A V' \right). \tag{16}$$

It can be seen that the maximum of $\operatorname{tr}(A^T B)$ in (16) is $\operatorname{tr}(S)$ with $U'^T A V' = E$, where E is an identity matrix. Therefore, the optimal A can be obtained by using $A_{\text{opt}} = U' V'^T$. However, the computation of the matrices U' and V' is too time consuming and this method cannot be implemented for real-time applications.

It is noted that the calculation of the singular values of B can be replaced by the combination of $\det(B)$, $\|B\|$ and $\operatorname{adj}(B')$,

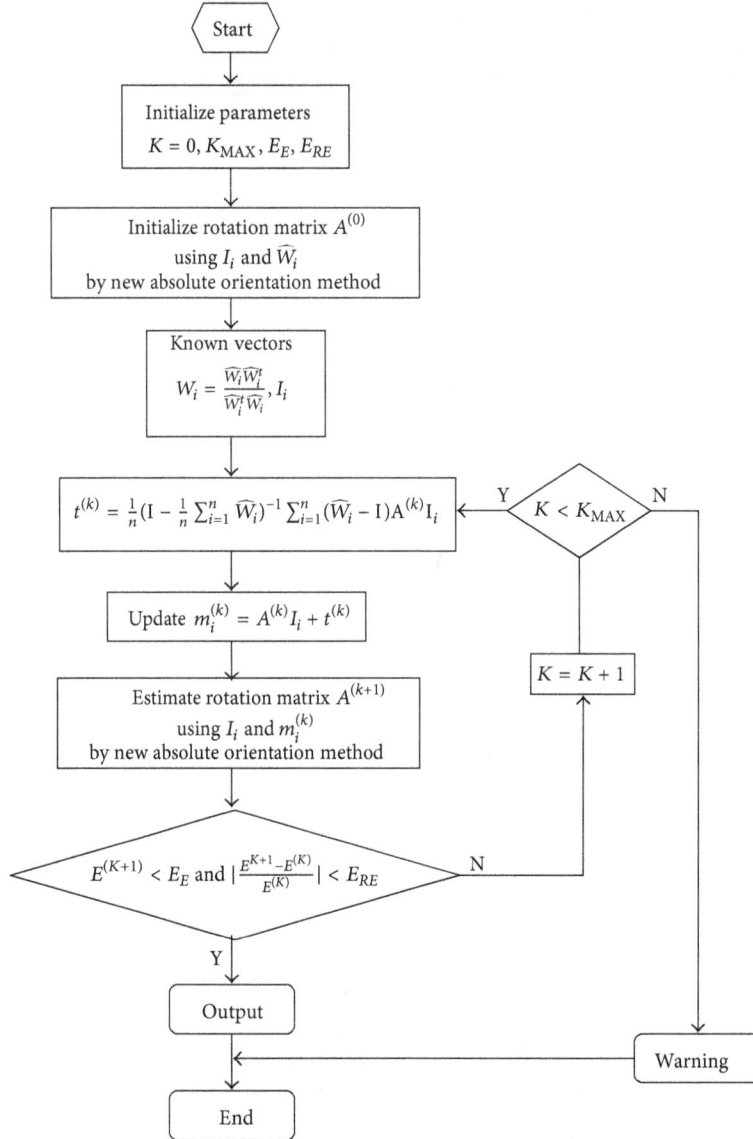

FIGURE 2: Flow chart of the algorithm.

and therefore A_{opt} can be calculated without the need for the SVD operation, such that

$$\det(B) = s_1 s_2 s_3,$$

$$\|B\|^2 = s_1^2 + s_2^2 + s_3^2,$$

$$\operatorname{adj}\left(B'\right) = U'\operatorname{diag}\left[s_2 s_3, s_1 s_3, s_1 s_2\right]V'^T, \qquad (17)$$

$$BB^T B = U'\operatorname{diag}\left[s_1^2, s_2^2, s_3^2\right]V'^T.$$

If we let

$$\lambda = \operatorname{tr}\left(A_{\text{opt}}^T B\right) = s_1 + s_2 + s_3, \qquad (18)$$

then the following equation can be formed in order to obtain A_{opt}, such that

$$\left(\lambda^2 - \|B\|^2\right)^2 - 8\lambda\det(B) - 4\|\operatorname{adj}(B)\|^2 = 0, \qquad (19)$$

It is noted that (18) is one of the four solutions and the largest root of (19). Therefore, A_{opt} can be derived from (19) such that

$$A_{\text{opt}} = \frac{\left(\alpha + \|B\|^2\right)B + \lambda\operatorname{adj}\left(B'\right) - BB^T B}{\alpha\lambda - \det(B)}, \qquad (20)$$

where $\alpha = (1/2)(\lambda^2 - \|B\|^2)$. With the optimal A being found, the minimum $L(A, \mathbf{t})$ in (13) can be calculated as,

$$L_{\min} = \sum_{i=1}^n \|\mathbf{m}_i\|^2 + \sum_{i=1}^n \|\mathbf{l}_i\|^2 - 2\sum_{i=1}^3 s_i. \qquad (21)$$

The procedures of the computationally efficient post estimation method can be summarized as follows.

First of all, calculate the kth optimal translation vector $\mathbf{t}^{(k)}(A^{(k)})$ using the kth rotation matrix $A^{(k)}$; secondly, update

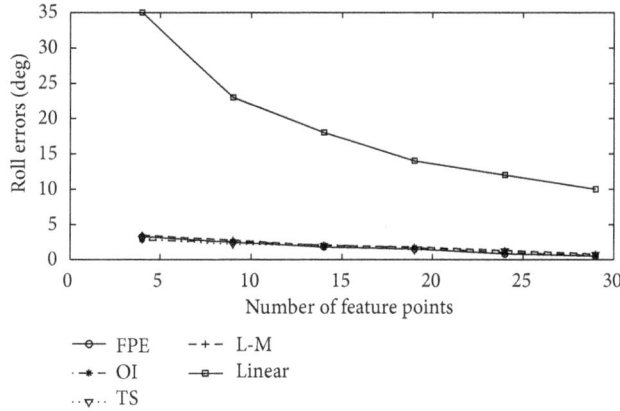

FIGURE 3: Rotation errors against a number of feature points. FPE denotes the proposed Fast Pose Estimation algorithm, OI means Lu's orthogonal iterative algorithm, TS is Zhang's two stage algorithm, L-M represents the Levenberg-Marquardt method, and the classic linear method is denoted as Linear. The same abbreviations are applied to the figures below too.

the camera-frame coordinates $\mathbf{m}_i^{(k)}(A^{(k)})$ with $\mathbf{m}_i^{(k)}(A^{(k)}) = A^{(k)}\mathbf{p}_i + \mathbf{t}^{(k)}(A^{(k)})$; thirdly, calculate the $(k + 1)$th optimal rotation matrix $A^{(k+1)}$ with $\mathbf{l}_i$ and $\mathbf{m}_i^{(k)}$ based on the proposed absolute orientation method. Repeat the process above until the absolute and relative object-space collinearity errors are less than the predefined thresholds. The initial value $A^{(0)}$ is computed with $\mathbf{w}_i$ and $\mathbf{l}_i$ based on the weak perspective approximation [9]. The flowchart of the iterative algorithm is shown in Figure 2.

3. Experiment

In this section, the performance-like accuracy, noise resistance, and computation efficiency for the proposed method are tested and compared with the methods such as Lu's OI algorithm, Zhang's two-stage algorithm, the classic linear transform method [18], and the L-M method, respectively.

The above methods are tested in programming languages such as Matlab 7.0 and VC++ 2008 with LAPACK 3.3.1 (linear algebra package), respectively, and the tests are run on a PC with a CPU of Intel Pentium D E5500 (with clock frequency 2.8 GHz and with 2 GB of random access memory). The operating system is Microsoft Windows XP Professional with service pack 3.

The feature points for the testing are generated randomly within a space of $[-4, 4] \times [-4, 4] \times [-4, 4]$ in the object coordinate frame. The rotation matrix A is chosen randomly by the Euler angles: yaw, pitch, and roll within $(-90, 90) \times (-90, 90] \times [0, 360)$ degree. The translation vector $\mathbf{t}$ is distributed randomly within a space of $[5, 15] \times [5, 15] \times [10, 200]$ in the camera coordinate frame. The focal length on Figure 1 is set to $f = 1$. The coordinates of the image points $\mathbf{w}_i$ are calculated based on $\mathbf{l}_i$, A, and $\mathbf{t}$, and Gaussian white noise is added to $\mathbf{w}_i$ to generate observed points $\widehat{\mathbf{w}}_i$. The standard deviation σ of the noise is a function of the SNR (Signal-to-Noise Ratio) defined by $SNR = -20\lg(\sigma(t_z/10))$. In this way,

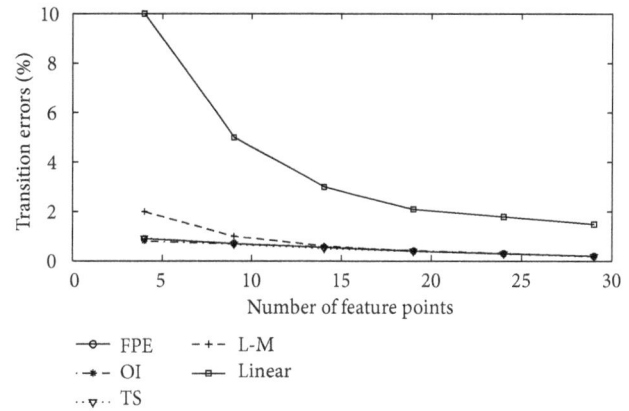

FIGURE 4: Translation errors versus the number of feature points.

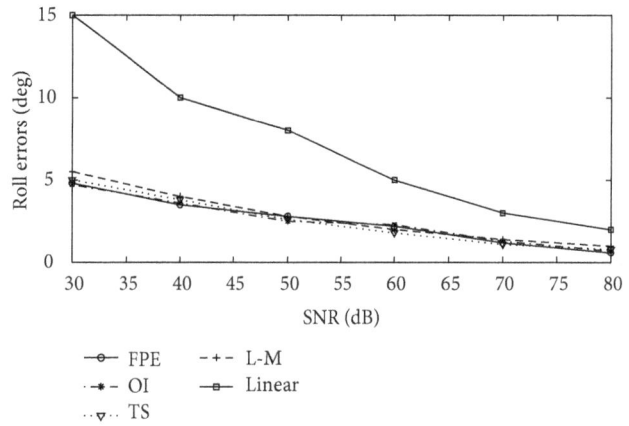

FIGURE 5: Rotation errors versus SNR.

1000 sets of $\widehat{\mathbf{w}}_i$ are generated for each set of $\mathbf{l}_i$, A, and $\mathbf{t}$, and the final result is the average value of the 1000 set of $\widehat{\mathbf{w}}_i$.

In the first test, we select $n = 10$ feature points for the optimization, and the SNR is varied from 30 dB to 80 dB with the interval of 10 dB. The errors of Euler angles and translation vector are recorded on the camera, and the result of each of the steps is averaged from 500 sets of $\mathbf{l}_i$, A, and $\mathbf{t}$.

In the second test, the SNR is set to be 60 dB, and the number of feature pints selected varies from 4 to 29 with the interval of 5. The time of computation is recorded as well as the errors of Euler angles and translation vector. The result of each step is taken from the average value of 500 sets of $\mathbf{l}_i$, A and $\mathbf{t}$.

Figure 3 shows the rotation error against the number of feature points, and Figure 4 gives the translation error versus the number of feature points. Figure 5 describes the rotation error against the SNR, and Figure 6 shows the translation error versus the SNR. In these figures, the rotation errors are represented by the roll angle errors because both yaw and pitch have the same effect. It can be seen from these figures that four of the approaches have comparable accuracy and noise-resistance capability, but the performance of the Linear method is relatively poor.

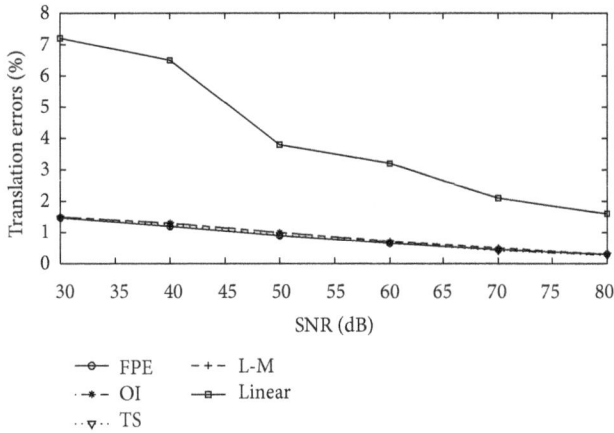

FIGURE 6: Translation errors against SNR.

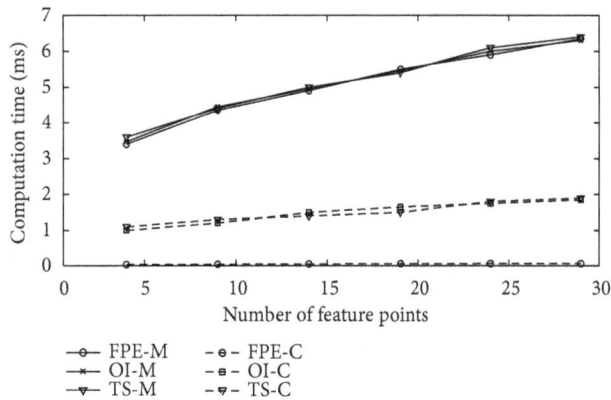

FIGURE 7: Computation time versus a number of feature points. "FPE-M", "OI-M", and "TS-M" mean that the algorithms are tested in MATLAB language, while "FPE-C", "OI-C", and "TS-C" mean that the algorithms are tested in C++ language.

Figure 7 shows the computation time of three of the algorithms required to complete the calculation in both Matlab and C++ environments. Since the accuracy and noise resistance of the Linear method are not satisfactory, this method is not considered in the comparison of computation time. Also, because the computation time of the L-M method is too time consuming, the time data shown in Figure 7 is far beyond the range of the time coordinate, and therefore the L-M method is not included in the comparison as well. It can be seen from Figure 7 that the programmes run on C++ platform are generally faster than in Matlab environment. The proposed Fast Pose Estimation algorithm does not show its advantage in Matlab because the SVD calculation has been optimized in the built-in function, while the matrix operations in the Fast Pose Estimation algorithm have not been improved. However, when the tests are carried on based on the LAPACK in C++, the SVD calculation and other matrix operations are optimized equally, and it can be seen from the result that the proposed Fast Pose Estimation algorithm performs much faster than the other two approaches. Since, in real world application, our software for embedded system

is generally designed using C++ language, the performance of the algorithms in C++ environment would be concerned the most.

4. Conclusions

In this paper, a computationally efficient pose estimation algorithm is proposed based on vision data. In this approach, a new absolute orientation method is designed to replace the time-consuming SVD calculation with the operations such as Frobenius norm, determinant, and adjoint of matrix. Experimental results show that the computation time required for pose estimation with the proposed method is much less than the original SVD approach in C++ programming language, while the performance such as accuracy and noise resistance is maintained as similar with the original method. Therefore, the proposed fast post estimation method is more suitable for real world applications such as the embedded systems used in satellites or other space missions.

References

[1] X. S. Gao, X. R. Hou, J. Tang, and H. F. Cheng, "Complete solution classification for the perspective-three-point problem," *IEEE Transactions on Pattern Analysis and Machine Intelligence*, vol. 25, no. 8, pp. 930–943, 2003.

[2] Z. Y. Hu and F. C. Wu, "A note on the number of solutions of the noncoplanar P4P problem," *IEEE Transactions on Pattern Analysis and Machine Intelligence*, vol. 24, no. 4, pp. 550–555, 2002.

[3] D. Nistér, "An efficient solution to the five-point relative pose problem," *IEEE Transactions on Pattern Analysis and Machine Intelligence*, vol. 26, no. 6, pp. 756–770, 2004.

[4] V. Lepetit, F. Moreno-Noguer, and P. Fua, "EPnP: an accurate O(n) solution to the PnP problem," *International Journal of Computer Vision*, vol. 81, no. 2, pp. 155–166, 2009.

[5] D. G. Lowe, "Fitting parameterized three-dimensional models to images," *IEEE Transactions on Pattern Analysis and Machine Intelligence*, vol. 13, no. 5, pp. 441–450, 1991.

[6] J. Weng, N. Ahuja, and T. S. Huang, "Optimal motion and structure estimation," *IEEE Transactions on Pattern Analysis and Machine Intelligence*, vol. 15, no. 9, pp. 864–884, 1993.

[7] X. B. Cao and S. J. Zhang, "An iterative method for vision-based relative pose parameters of RVD spacecrafts," *Journal of Harbin Institute of Technology*, vol. 37, no. 8, pp. 1123–1126, 2005.

[8] R. M. Haralick, H. Joo, C. N. Lee, X. Zhuang, V. G. Vaidya, and M. B. Kim, "Pose estimation from corresponding point data," *IEEE Transactions on Systems, Man and Cybernetics*, vol. 19, no. 6, pp. 1426–1446, 1989.

[9] C. P. Lu, G. D. Hager, and E. Mjolsness, "Fast and globally convergent pose estimation from video images," *IEEE Transactions on Pattern Analysis and Machine Intelligence*, vol. 22, no. 6, pp. 610–622, 2000.

[10] S. Umeyama, "Least-squares estimation of transformation parameters between two point patterns," *IEEE Transactions on Pattern Analysis and Machine Intelligence*, vol. 13, no. 4, pp. 376–380, 1991.

[11] Z. Shijie, L. Fenghua, C. Xibin, and H. Liang, "Monocular vision-based two-stage iterative algorithm for relative position and attitude estimation of docking spacecraft," *Chinese Journal of Aeronautics*, vol. 23, no. 2, pp. 204–210, 2010.

[12] S. J. Zhang, X. B. Cao, F. Zhang, and L. He, "Monocular vision-based iterative pose estimation algorithm from corresponding feature points," *Science in China, Series F*, vol. 53, no. 8, pp. 1682–1696, 2010.

[13] B. K. P. Horn, H. M. Hilden, and S. Negahdaripour, "A closed-form solution of absolute orientation using orthonomal matrices," *Journal of the Optical Society of America A*, vol. 5, pp. 1127–1135, 1988.

[14] B. K. P. Horn, "Closed-form solution of absolute orientation using unit quaternion," *Journal of the Optical Society of America A*, vol. 4, pp. 629–642, 1987.

[15] M. W. Walker, L. Shao, and R. A. Volz, "Estimating 3-D location parameters using dual number quaternions," *CVGIP: Image Understanding*, vol. 54, no. 3, pp. 358–367, 1991.

[16] K. S. Arun, T. S. Huang, and S. D. Blostein, "A least-squares fitting of two 3-D point sets," *IEEE Transactions on Pattern Analysis and Machine Intelligence*, vol. 9, pp. 698–700, 1987.

[17] F. L. Markley, "Attitude determination using vector observations: a fast optimal matrix algorithm," *Journal of the Astronautical Sciences*, vol. 41, no. 2, pp. 261–280, 1993.

[18] Y. I. Abdel-Aziz and H. M. Karara, "Direct linear transformation into object space coordinates in close-range photogrammetry," in *Proceedings of the Symposium Close-Range Photogrammetry*, pp. 1–18, Urbana, Ill, USA, 1971.

Permissions

The contributors of this book come from diverse backgrounds, making this book a truly international effort. This book will bring forth new frontiers with its revolutionizing research information and detailed analysis of the nascent developments around the world.

We would like to thank all the contributing authors for lending their expertise to make the book truly unique. They have played a crucial role in the development of this book. Without their invaluable contributions this book wouldn't have been possible. They have made vital efforts to compile up to date information on the varied aspects of this subject to make this book a valuable addition to the collection of many professionals and students.

This book was conceptualized with the vision of imparting up-to-date information and advanced data in this field. To ensure the same, a matchless editorial board was set up. Every individual on the board went through rigorous rounds of assessment to prove their worth. After which they invested a large part of their time researching and compiling the most relevant data for our readers. Conferences and sessions were held from time to time between the editorial board and the contributing authors to present the data in the most comprehensible form. The editorial team has worked tirelessly to provide valuable and valid information to help people across the globe.

Every chapter published in this book has been scrutinized by our experts. Their significance has been extensively debated. The topics covered herein carry significant findings which will fuel the growth of the discipline. They may even be implemented as practical applications or may be referred to as a beginning point for another development. Chapters in this book were first published by Hindawi Publishing Corporation; hereby published with permission under the Creative Commons Attribution License or equivalent.

The editorial board has been involved in producing this book since its inception. They have spent rigorous hours researching and exploring the diverse topics which have resulted in the successful publishing of this book. They have passed on their knowledge of decades through this book. To expedite this challenging task, the publisher supported the team at every step. A small team of assistant editors was also appointed to further simplify the editing procedure and attain best results for the readers.

Our editorial team has been hand-picked from every corner of the world. Their multi-ethnicity adds dynamic inputs to the discussions which result in innovative outcomes. These outcomes are then further discussed with the researchers and contributors who give their valuable feedback and opinion regarding the same. The feedback is then collaborated with the researches and they are edited in a comprehensive manner to aid the understanding of the subject.

Apart from the editorial board, the designing team has also invested a significant amount of their time in understanding the subject and creating the most relevant covers. They scrutinized every image to scout for the most suitable representation of the subject and create an appropriate cover for the book.

The publishing team has been involved in this book since its early stages. They were actively engaged in every process, be it collecting the data, connecting with the contributors or procuring relevant information. The team has been an ardent support to the editorial, designing and production team. Their endless efforts to recruit the best for this project, has resulted in the accomplishment of this book. They are a veteran in the field of academics and their pool of knowledge is as vast as their experience in printing. Their expertise and guidance has proved useful at every step. Their uncompromising quality standards have made this book an exceptional effort. Their encouragement from time to time has been an inspiration for everyone.

The publisher and the editorial board hope that this book will prove to be a valuable piece of knowledge for researchers, students, practitioners and scholars across the globe.

List of Contributors

G. Palmieri
Facolta di Ingegneria, Universita degli Studi e-Campus, 22060 Novedrate, Italy

M. Palpacelli, M. Battistelli and M. Callegari
Dipartimento di Ingegneria Industriale e Scienze Matematiche, Universita Politecnica delle Marche, 60121 Ancona, Italy

Jacopo Zenzeri, Dalia De Santis and Vishwanathan Mohan
RBCS Department, Istituto Italiano di Tecnologia, Via Morego 30, 16163 Genoa, Italy

Pietro Morasso
RBCS Department, Istituto Italiano di Tecnologia, Via Morego 30, 16163 Genoa, Italy
DIBRIS Department, University of Genoa, Viale Causa, 13 16145 Genoa, Italy

Maura Casadio
DIBRIS Department, University of Genoa, Viale Causa, 13 16145 Genoa, Italy

Giovanni Gerardo Muscolo
R&D Department, Creative Design Laboratory, Humanot s.r.l., via Modigliani 7-59100 Prato, Italy

Kenji Hashimoto
Department of Modern Mechanical Engineering, Waseda University, 17 Kikui-cho, Shinjuku-ku, Tokyo 162-0044, Japan

Atsuo Takanishi
Department of Modern Mechanical Engineering, Waseda University, 17 Kikui-cho, Shinjuku-ku, Tokyo 162-0044, Japan
Humanoid Robotics Institute, Waseda University, 2-2Wakamatsu-cho, Shinjuku-ku, Tokyo 162-8480, Japan

Paolo Dario
The Bio Robotics Institute, Scuola Superiore Sant'Anna, Viale Rinaldo Piaggio 34, 56025 Pontedera, Italy

Christian Brecher, Thomas Breitbach, Simon Muller and Werner Herfs
Laboratory for Machine Tools and Production Engineering (WZL), RWTH Aachen University, 52074 Aachen, Germany

Marcel Ph. Mayer, Barbara Odenthal and Christopher M. Schlick
Institute for Industrial Engineering and Ergonomics, RWTH Aachen University, 52062 Aachen, Germany

Ying Bai
Johnson C. Smith University, Charlotte, NC 28216, USA

Nailong Guo
Benedict College, Columbia, SC 29204, USA

Gerald Agbegha
Georgia Gwinnett College, Atlanta, GA 30043, USA

Ryan A. Beasley
Department of Engineering Technology and Industrial Distribution, Texas A&M University, 3367 TAMU, College Station, TX 77843, USA

Abbi Hamed
Department of Advanced Robotics, Chiba Institute of Technology, 2-17-1 Tsudanuma, Narashino, Chiba 285-0016, Japan

Sai Chun Tang
Department of Radiology, Brigham and Women's Hospital, Harvard Medical School, 221 Longwood Avenue, Boston, MA 02115, USA

Hongliang Ren
Department of Bioengineering, National University of Singapore, Singapore 117575

Alex Squires, Javad Mohammadpour and Zion Tsz Ho Tse
Driftmier Engineering Center, College of Engineering, The University of Georgia, Athens, GA 30602, USA

Chris Payne
Department of Mechanical Engineering, Imperial College London, London SW7 2AZ, UK

Ken Masamune
Advanced Therapeutic and Rehabilitation Engineering Laboratory, Faculty of Engineering, The University of Tokyo, Suite No. 83A3, Building No. 2, Hongo, Bunkyo-ku, Tokyo 113-8656, Japan

Guoyi Tang
Advanced Material Institute, Graduate School at Shenzhen, Tsinghua University, Shenzhen, Guangdong 518055, China

Yueyue Deng, Pierre-Philippe J. Beaujean, Edgar An and Edward Carlson
Department of Ocean and Mechanical Engineering, Florida Atlantic University, 777 Glades Road, Boca Raton, FL 33431, USA

Huan Tan
Department of Electrical Engineering and Computer Science, Vanderbilt University, Nashville, TN 37240, USA

Qian Du
Institute of Robotics and Automatic Information System, Nankai University, Tianjin 300071, China

Na Wu
Graduate School of Decision and Technology, Tokyo Institute of Technology, Tokyo 152-8552, Japan

Xiaona Wang and Max Q. H. Meng
Department of Electronic Engineering, The Chinese University of Hong Kong, Shatin, NT, Hong Kong

M. L. Seto
Defence R&D Canada, Dartmouth, Nova Scotia, Canada

Yu Liu, Zainan Jiang and Hong Liu
State Key Laboratory of Robotics and System, Harbin Institute of Technology, Harbin 150001, China

Wenfu Xu
Mechanical Engineering and Automation, Harbin Institute of Technology Shenzhen Graduate School, Shenzhen 518057, China

Jared Wood
Vehicle Dynamics Lab and Center for Collaborative Control of Unmanned Vehicles, Department of Mechanical Engineering, University of California, Berkeley, 6141 Etcheverry Hall, Berkeley, CA 94720-1740, USA

J. Karl Hedrick
Vehicle Dynamics Lab, Department of Mechanical Engineering, University of California, Berkeley, 6141 Etcheverry Hall, Berkeley, CA 94720-1740, USA

F. Ababsa, I. Zendjebil, J.-Y. Didier and M. Mallem
Laboratoire IBISC, EA 4526, Universite d Evry-Val-d Essonne, 40 rue du Pelvoux, 91020 Evry, France

Luciano Luporini Menegaldo
Biomedical Engineering Program (PEB/COPPE), The Alberto Luiz Coimbra Institute for Graduate Studies and Research in Engineering, Universidade Federal do Rio de Janeiro, Avenida Horacio Macedo 2030, Bloco H-338, 21941-914 Rio de Janeiro, RJ, Brazil

Stefano Panzieri
Department of Informatics and Automation, Universita Roma Tre, Via della Vasca Navale, 79, I 00146 Roma, Italy

Cassiano Neves
Subsin Engineering, Rua Beneditinos, 16, 12th floor, 20081-050 Rio de Janeiro, RJ, Brazil

Xiang Wu and Ning Wu
Shenzhen Graduate School, Harbin Institute of Technology, Harbin, Heilongjiang 150001, China